MANUFACTURING IN THE NINETIES

VNR COMPETITIVE MANUFACTURING SERIES

Quality Control/Reliability

PRACTICAL EXPERIMENT DESIGN, 2E by William J. Diamond
VALUE ANALYSIS IN DESIGN by Theodore C. Fowler
A PRIMER ON THE TAGUCHI METHOD by Ranjit Roy
RELIABILITY AND MAINTAINABILITY MANAGEMENT by Balbir S. Dhillon and Hans Reiche
APPLIED RELIABILITY by Paul A. Tobias and David C. Trindad
GUIDE TO RELIABILITY ENGINEERING: Data, Analysis, Applications, Implementation and Management by R. Sundararajan
ENGINEERING APPLICATIONS OF RELIABILITY ANALYSIS by Shu-Ho Dai and Ming-O Wang

Product and Process Design

INDUSTRIAL ROBOT HANDBOOK: Case Histories of Robot Use in 70 Industries by Richard K. Miller
ROBOTIC TECHNOLOGY: Principles and Practice by Werner G. Holzbock
MACHINE VISION by Nello Zeuch and Richard K. Miller
DESIGN OF AUTOMATIC MACHINERY by Kendrick W. Lentz, Jr.
TRANSDUCERS FOR AUTOMATION by Michael Hordeski
MICROPROCESSORS IN INDUSTRY by Michael Hordeski
DISTRIBUTED CONTROL SYSTEMS by Michael P. Lucas
BULK MATERIALS HANDLING HANDBOOK by Jacob Fruchtbaum
MICROCOMPUTER SOFTWARE FOR MECHANICAL ENGINEERING by Howard Falk

Management

DEVELOPING PRODUCTS IN HALF THE TIME by Preston G. Smith and Donald G. Reinertsen
MANUFACTURING IN THE NINETIES: How to Become a Mean, Lean, World-Class Competitor by Harold J. Steudel and Paul Desruelle
THE MANUFACTURING CHALLENGE: From Concept to Production by S. E. Stephanou and Fred Spiegl
WORKING TOWARDS JUST-IN-TIME by Anthony Dear
GROUP TECHNOLOGY: Foundation for Competitive Manufacturing by Charles S. Snead
COMPETITIVE MANUFACTURING by Stanley Miller
STRATEGIC PLANNING FOR THE INDUSTRIAL ENGINEERING FUNCTION by Jack Byrd and L. Ted Moore
PRODUCT LIABILITY HANDBOOK: Prevention, Risk, Consequence and Forensics of Product Failure edited by Sam Brown
TOTAL MATERIALS MANAGEMENT: The Frontier for Cost-Cutting in the 1990s by Eugene L. Magad and John Amos

MANUFACTURING IN THE NINETIES

How to Become a Mean, Lean, World-Class Competitor

Harold J. Steudel, Ph.D., P.E.

Paul Desruelle

COMPETITIVE Manufacturing SERIES

VNR VAN NOSTRAND REINHOLD
New York

Library of Congress Catalog Card Number 91-18726
ISBN 0-442-00182-7

I(T)P Van Nostrand Reinhold is an International Thomson Publishing company.
ITP logo is a trademark under license.

Manufactured in the United States of America

Van Nostrand Reinhold
115 Fifth Avenue
New York, NY 10003

International Thomson Publishing GmbH
Königswinterer Str. 418
53227 Bonn
Germany

International Thomson Publishing
Berkshire House,168-173
High Holborn, London WC1V 7AA
England

International Thomson Publishing Asia
221 Henderson Bldg. #05-10
Singapore 0315

Thomas Nelson Australia
102 Dodds Street
South Melbourne 3205
Victoria, Australia

International Thomson Publishing Japan
Kyowa Building, 3F
2-2-1 Hirakawacho
Chiyoda-Ku, Tokyo 102
Japan

Nelson Canada
1120 Birchmount Road
Scarborough, Ontario
M1K 5G4, Canada

BBR 16 15 14 13 12 11 10 9 8 7 6

Library of Congress Cataloging-in-Publication Data

Steudel, Harold J.
Manufacturing in the nineties : how to become a mean, lean, and world-class competitor / Harold J. Steudel, Paul Desruelle.
p. cm.—(VNR competitive manufacturing series)
Includes index.
ISBN 0-442-00182-7
1. United States—Manufactures—Management. 2. United States—Manufactures—Quality control. I. Desruelle, Paul. II. Title.
HD9725.S68 1991
658.5—dc20

91-18726
CIP

Dedicated
with love to
Carol
and
Marie José

Contents

Preface

The 1980s was an era of renaissance in manufacturing, born largely from the pressing need of many companies to compete in a growing global marketplace. The term "World-Class Manufacturing" (WCM) became the trademark of this movement as many companies began to implement simplified manufacturing methods and new organization structures to increase their competitive position. While competitive and cost forces were significant for many companies, in retrospect, it was generally a period in which changes in manufacturing were implemented in an environment of world-wide economic growth and prosperity. In the 1990s, global competition will continue to increase, but the challenge now is how to become a mean (tough, resilient, assertive), lean (agile, efficient), world-class competitor in an environment of tighter financial constraints.

This is a "how to" book that addresses the implementation of simple, affordable, yet effective solutions for transforming traditional manufacturing organizations into world-class competitors. While the concepts for WCM are not new, many of the ideas and methods explained in this book are new and useful for applying these concepts. Through a framework of the ten integrated components of WCM, the book presents a straightforward approach for implementing rapid and continuous improvements in manufacturing in order to achieve the larger strategic goal of sustainable competitive advantage and profitability. The book addresses how to work together—management and employees, a company and its vendors, and sales, marketing, design engineering, production, and human resource—in order to utilize resources effectively for producing products that are more attractive to the customer than the competition's. It addresses the corporate environment and the cultural changes required so that resources and energies can be redirected toward bringing products to the marketplace that provide relatively better quality, performance, and other characteristics recognized as valuable. The book also discusses how to establish "win-win" arrangements—an absolute necessity for achieving the commitment and dedication required of all parties for implementing WCM.

The book is not a theory-based treatment of world-class manufacturing. No attempt is made to link these practical methods and effective practices that can provide a competitive advantage to supporting theories in psychology, sociology, organizational behavior, or marketing. Rather, this book presents the practical ideas and knowledge that they never told you enough about in school. Likewise, this book is not intended to be inspirational per se, since if one needs a book to be inspired to change, instead of by the necessities of competing in the global marketplace, there is little hope for long-term survival. However, it is intended to be used as a permanent reference and as a source of practical ideas for companies on the course of change implementation.

The book is based upon the premise that the success and profitability of a manufacturing enterprise depends not on macroeconomic factors but on the ability to achieve a sustainable competitive advantage by providing products to the marketplace that offer better value to the customer. Creating products with better value is achieved through rapid and continuous improvements in product quality and manufacturing productivity. Such improvements are possible only through management's initiatives and direction, working through the cooperative efforts of a knowledgeable and motivated workforce. The purpose of this book is thus to provide the knowledge needed to formulate vision and direction, knowledge on how to begin and proceed efficiently and effectively, and knowledge for maintaining momentum in rapid and continuous improvement. Toward this end, it discusses such topics as senior managements' needs for leading their company in the competitive environment of the 1990s, department managers' needs for defining the changes necessary for achieving the strategic goals of the company, and team members' needs for understanding and implementing the functional changes in manufacturing technology, methods, and practices. In addition, this book will help

- Vendors to be better prepared for establishing solid relationships with their industrial customers;
- Local, state, and national elected representatives to understand the infrastructure and workforce needs of the companies that can create wealth in their region, state, or the nation; and
- Engineering and business students to be well prepared for the new environment that they will have to implement.

The ideas and methods to meet these needs transcend national boundaries and focus on the universal characteristics that make any manufacturing company competitive. In short, the book presents practical knowledge on "what to do" and "how to do it" in order to become a mean, lean, and world-class competitor in the nineties.

This book could not have been written without the assistance and contributions of many people, particularly the hundreds of managers, engineers, and shop floor people who have worked with us to identify and implement ideas and methods that truly yield a competitive advantage in manufacturing. Likewise, we have learned much about group technology and cellular manufacturing from Dr. Urban Wemmerlöv—Wisconsin Distinguished Professor of Business—whose sharp thinking and depth of knowledge has been a fountain of intellectual refreshment and nourishment during our eight years of professional interactions. Many ideas in this book are also due to enthusiastic efforts of Mr. L. Gene Berg—colleague, partner, and friend. Similarly, the cell formation algorithm presented in Chapter 4 is adapted from an IJPR article co-authored with Dr. Arvind Ballakur of Bell Labs. The pioneering research and ongoing development efforts of Dr. Taeho Park—directed at the simulation-based design and evaluation of manufacturing workcells—is likewise evident in Chapter 5. Many of the examples on the successful implementation of Total Maintenance, in Chapter 9, have been provided by Mr. Yves Duretz, Jeumont-Schneider Industrie, France. To these people, and the many unnamed others whose ideas and efforts are embodied in this book, we are grateful.

Writing a book is not only a significant intellectual undertaking, but also one that requires much logistical support. The editorial prowess of Mrs. Carolyn Berg is evident throughout the book, contributing greatly to making the ideas flow in a readable fashion. The assistance of both Mrs. Sue Bader and Ms. Myrna Kasdorf, wizards of word-processing technologies, was invaluable. Their patience and productivity in the face of constant revisions was remarkable, and is greatly appreciated.

Finally, thanks to our families and closest friends. Without their encouragement, understanding, and desire to go shopping, the writing of this book would have been much less fun.

MANUFACTURING IN THE NINETIES

1

Establishing a Strategic Plan for Becoming a World-Class Manufacturer

Meeting the challenges of worldwide competition in manufacturing now requires operating policies, practices, and systems that eliminate waste and create value for the customer. Offering more value for a lot less money—often referred to as "value pricing"—is a crucial tactic and requirement for manufacturing companies in the nineties. Customers are now demanding that manufacturers provide high-quality products at substantially lower prices. Cost pressures caused by energy, materials, labor costs, and many other factors continue to increase for the manufacturer; yet, competitive forces no longer tolerate steadily climbing price increases. Competitiveness and profitability can only be maintained if manufacturers cut costs by becoming better at what they do and how they do it. This involves redesigning, remanufacturing, repackaging, and repricing to provide low-cost quality products and service without going broke. In the nineties, the term "world-class manufacturing" will be more than a popular buzzword; it is and will be the basis of a strategy for both survival and success.

Chapter 1 addresses top management, discussing what it means to be and takes to become world-class in manufacturing. Using the structure of the twenty characteristics of world-class manufacturing companies, the chapter shows that becoming world-class is not simply the responsibility of the manufacturing function, but rather requires the involvement and commitment of sales, marketing, engineering, customer service, accounting, finance, and all other functional groups. The competitive necessities and financial benefits are likewise made apparent, as is the fact that being world-class is more about top management's approach to competition than about company size or resources.

The chapter then presents the components and steps required to transform and use the manufacturing function as a competitive weapon. This presentation also provides an overview of the organization and contents of the book. Based on the characteristics and components present in world-class manufacturing, the chapter then addresses how to define a corporate strategic plan based on eliminating waste and creating products that offer better value than the competition. In particular, a seven-step approach is presented for defining a strategic plan that assesses a company's strengths and weaknesses, and establishes a systematic sequence of directed tactical efforts for achieving the company's future vision of excellence. Detailed here is the first and foremost step to the successful organization and operation of a world-class manufacturing company in the nineties: Establishing a strategic plan.

1.1 WORLD-CLASS MANUFACTURING: CHARACTERISTICS AND COMPONENTS

The term *world-class* has been widely used in many different contexts in recent years. Not only does one encounter media about world-class manufacturing, but also about world-class universities, automobiles, sports equipment, clothing, and many other industrial products and consumer goods. Before addressing how to build and sustain a world-class manufacturing organization, a working definition of the term "world-class manufacturing" is called for, along with the need to address the operational characteristics embodied in world-class manufacturing companies.

1.1.1 Definition of World-Class Manufacturing

The purpose of any manufacturing enterprise is to consistently make a profit, pure and simple. Being world-class in manufacturing means that the company can compete successfully and make a profit in an environment of international competition, not only now, but also in the future. To make a profit, a company must sell its products at a price above its costs and still provide a product to the marketplace that offers better value than the competition's product. This value is typically perceived by customers as a combination of price, quality, availability of the product and service, and the product's performance and capability. In essence, being world-class means being capable of bringing products to the marketplace that offer better value than the competition without going broke. Being world-class is also not simply "playing with the big boys," but rather being "top class"—as good as, or better than, the toughest competition worldwide in at least several strategically important areas.

While the definition of world-class manufacturing is easy to understand, it provides limited insight concerning the differentiating characteristics between the policies, practices, and operating systems of world-class versus non-world-class manufacturing companies. The following list contains twenty characteristics of world-class manufacturing organizations, grouped by management/employee involvement, quality, and production operations. These characteristics illustrate management's requirements and the types of competitive gains and financial benefits that result.

Management/Employee Involvement

1. **Visionary leadership and champions:**
   World-class manufacturing companies have a critical mass of managers who are dedicated to continuous improvement and, through leadership and "coaching," articulate a vision of excellence, motivating people to work together to identify and eliminate waste, and thereby create competitive value.

   In this corporate culture, company personnel commonly evolve into new roles. For example, top management's role becomes that of a visionary leader and part-time cheerleader, while middle management plays the roles of both coach and teacher. Supervisors assume the role of facilitator and supporter for the empowered employees who are team members involved in identifying and eliminating waste. To effect the cultural changes required for continuous improvement, management's leadership, visibility, and involvement are crucial. Excellence begins with leadership!
2. **"New culture" goals and thinking:**
   World-class manufacturing companies use benchmarking methods to seek and evaluate the best policies and practices in their industries as a basis for setting agendas that ascend the old traditions and ways of thinking, and strive for previously unreachable goals.

   Competitive comparisons and world-class benchmarks are becoming more important to support product development, quality planning, product and process improvement, and corporate goal setting. To be as good as, or better than, the toughest competition worldwide, it is essential to know the current "standards of excellence."

   A common question of management is "What are the benchmarks of performance and expected benefits of world-class manufacturing?" While benchmarks by nature are a moving target, the following performance measures are appropriate for most discrete part manufacturing companies.
   - Manufacturing lead times—days not weeks.
   - Work-in-process inventory—hours or days not weeks.
   - Inventory turns—fifteen to twenty-five times per year.
   - External customer rejects—fifty parts per million.
   - Internal customer rejects—200–250 parts per million.

- On-time deliveries—95–98 percent.
- Cost of quality—less than 5 percent of sales.
- Setup times—minutes not hours.
- Percentage of workforce on teams—40–60 percent.

The remainder of this book addresses "how to" implement the changes necessary to bring a manufacturing organization up to or beyond these levels of performance. These levels are not unreachable!

3. **Long-term strategic plan and direction:**
World-class manufacturing companies have a rolling, long-term (three to five years) strategic plan, which not only defines corporate objectives, goals, and operational plans for implementing policies and practices that consistently add value to the company's products and services but also identifies the knowledge, tools, and skills required for effective implementation (strategic planning is addressed in section 1.2).

4. **Employee involvement and human resource development:**
World-class manufacturing companies get employees involved at all levels of the organization and have extensive training programs for providing their employees with the knowledge and skills necessary to improve themselves and to understand and implement the many changes and technologies that accompany a philosophy of continuous experimentation and improvement.

The fundamental idea of employee involvement is both simple and proven: that is, if people are treated with equity and respect, provided with meaningful jobs and the responsibility of problem solving and decision making regarding their work, and are given opportunities to learn, then they will satisfy both personal goals and organizational goals. World-class companies know that to achieve beneficial results, people need to be involved and developed. As Edward J. Hay once stated so succinctly, "If you change the system, but people aren't developed to take advantage of it, then nothing happens."

5. **Integrative and holistic objectives:**
World-class manufacturing companies have management policies and practices that promote holistic approaches that dissolve boundaries between and integrate the objectives and activities of different functional areas and divisions; thereby emphasizing that quality, cost, lead time, and customer service are not only compatible and possible, but also necessary.

The development of common corporate goals—accomplished in the strategic planning process—involving all functional areas and divisions is essential for competing successfully in the nineties. The internal costs and inefficiencies associated with "civil wars" destroy value, the key to competitiveness. Only top management can provide the leadership to prevent such internal losses.

6. **Goal-consistent measurement/reward systems:**
World-class manufacturing companies recognize that "what gets measured and rewarded gets done," and use simple performance measurement systems that consider quality, human resource improvement, team efforts, and other selected key variables that reflect adding value to the product; thereby avoiding running the company by strictly short-term dictates that are generated from financially dominated control systems or dictated standards.
The importance of measurement systems that drive an organization toward management's objectives of becoming a world-class manufacturer cannot be overstated. Many good manufacturing improvement projects have failed because the measurement systems in place failed to capture and recognize the efforts of people and/or the true benefits of the change. Accounting thus has a crucial role in developing new measures of performance and in eliminating existing measures that are counterproductive, such as those that only emphasize production and ignore the other important elements of "value."
7. **Product- or customer-focused organization:**
World-class manufacturing companies solicit relationships and establish linkages with university systems, promoting research and educational activities that provide for a long-term competitive advantage.
Complexity in an organization yields a lessening of control and agility, two important attributes for maintaining high levels of customer focus and service. To improve accountability and focus, many companies are decentralizing their operations and organizational structure, making things smaller and more self-contained. For example, instead of having six vice presidents each responsible for a specific job covering all the operations, companies are assigning one general manager (GM) who is responsible for all tasks for a given product or customer line. The GM of each standard business unit (SBU) then reports to the chief operating officer of the company.
8. **Good communication systems and practices:**
World-class manufacturing companies recognize the importance of good communication and strive to establish and maintain simple systems and procedures that provide timely and accurate information flow throughout the manufacturing enterprise.
Information is the lifeblood of an organization. In order to work as a team for "winning" the organizational goals, people want and need good information—both directional and feedback. Imagine going to a basketball game where only the two coaches knew the score. Without a scoreboard for the fans to monitor, the level of personal involvement in watching the players run up and down the floor, shooting baskets, would be greatly diminished.

Providing company personnel with appropriate information in an understandable format is management's responsibility.

9. **Promotion/support of research and education:**
World-class manufacturing companies solicit relationships and establish linkages with university systems, promoting research and educational activities that provide for a long-term competitive advantage.

Quality

10. **Customer-driven product development and marketing:**
World-class manufacturing companies focus on customer-driven strategies for product development and marketing, emphasizing ongoing customer contact and intellectual commitment for defining product concepts and performance and quality specifications.

    Determining current and future customer requirements and expectations is a key issue for the nineties. The challenge for management is to ensure that the "voice of the customer" is heard, communicated throughout the organization, and acted on in a timely and responsive manner. The manner in which customer needs and expectation are determined and then used both for new product development and for improving existing products and services is a major determinant for competing successfully. It's the customer's definition of value that counts!

11. **Cross-functional teams for product design/manufacturing:**
World-class manufacturing companies use cross-functional (for example, design–manufacturing–marketing) teams for responding to and communicating the needs of the customer throughout the organization, coupling decisions in product design, materials, and manufacturing process selection to bring better products to the market more quickly.

    "Design is a strategic activity, by intention or by default. It influences flexibility of sales strategies, speed of field repair, and efficiency of manufacturing. It may well be responsible for the company's future viability."[1] Design is too important an activity to be done solely by engineering. World-class manufacturing companies now recognize that multifunctional teams are currently the best way known to cut through the barriers to good design. Given that 65 percent to 75 percent of a product's manufacturing cost is typically established during the "vision stage" of product development, it is essential to get early team involvement. A team approach to product development and product improvement has allowed many companies to achieve four to sixfold improvements in product

---

[1]Daniel Whitney, *Harvard Business Review*, July-Aug. 1988.

reliability. 70–90 percent reductions in warranty costs, 40–50 percent reductions in workmanship and assembly errors, and 20–40 percent reductions in product cost. It is clearly a key strategy for becoming a world-class manufacturer and a strategy that only top management can drive through an organization since it is much more of a cultural change than anything technical.

12. **Individual responsibility and continuous quality improvement:**[2]
World-class manufacturing companies make quality *everyone's job responsibility*, which allows the quality assurance department to serve as a support and coordination function for fostering continuous quality improvement efforts throughout the organization.
13. **Statistical process control of key product characteristics:**[3]
World-class manufacturing companies focus on controlling the process based on STATISTICAL measures, and encourage decision making at the operating level using local data sources on key variables for comparisons against customer needs.
14. **Emphasis on innovation and experimentation:**
World-class manufacturing companies are innovators, constantly experimenting to improve existing products and processes, and to develop new ones, striving for less variability and greater capability.

    Manufacturing engineers are essential for making process improvements—often in a team environment—on an everyday basis. Dennis Butt—former GM at Kawasaki in Lincoln Nebraska—states it more emphatically as follows: "You cannot do world-class manufacturing without a lot of manufacturing/industrial engineers."
15. **Partnership-like relationships with quality-certified vendors:**
World-class manufacturing companies seek out and establish win-win partnership-like relationships (based heavily on nonprice criteria such as quality and delivery) with one supplier (or a very few) per commodity and seek early vendor involvement in quality improvement and new product development efforts.

    Suppliers are not "important" to success, they are *critical* to success. The consolidation and certification of selected suppliers not only improves the quality and delivery performance of purchased materials, but it also allows for the reduction of internal costs related to inspection, inventories, and procurement personnel. These issues and how to achieve these gains are addressed in chapter 10.

---

[2]Addressed fully in chapter 3.
[3]Addressed fully in chapter 3.

Production Operations

16. **Continuous-flow processing/cellular manufacturing:**
World-class manufacturing companies focus on standardizing, simplifying, and focusing their manufacturing operations and related instructions; thereby reducing complexity and facilitating the effective use of continuous-flow processing concepts for reductions in lead times, work-in-process inventories, and materials handling.

    While continuous process improvement is a key and operative concept for becoming world-class in manufacturing, working solely to improve existing systems will only provide limited incremental gains. In most cases, the entire system needs to be changed to achieve 100 percent, 200 percent, 300 percent improvements. Continuous-flow processing, often implemented through cellular manufacturing, provides such quantum leaps. Improvements in manufacturing lead times from ten to twelve weeks to one to three days are common along with corresponding reductions in work-in-process levels from "weeks" to "days." Furthermore, companies consistently find that lowering inventory levels expose previously hidden problems that can and must be solved.

17. **Demand-based, not capacity-based, processing**
World-class manufacturing companies recognize that adopting an enlarged view of manufacturing operations even at the cost of allowing machines to sit idle some of the time can provide gains in plantwide efficiency and quality; whereas single-minded efforts to push machine capacities to the limit typically produce process-yield imbalances, high levels of in-process inventories, and longer manufacturing lead times.

18. **Quick changeover procedures/small lot sizes:**
World-class manufacturing companies use multidisciplined, multilevel work teams to standardize and simplify changeover procedures, thus reducing equipment downtime during job changeovers and allowing production in smaller lot sizes—a key requirement for flexible production.

19. **Emphasis on standardizing/simplifying before automating:**
World-class manufacturing companies tend to view high technology and automation more as complementary tools than as a manufacturing strategy, focusing on standardizing, simplifying, and proving the integrity of a manufacturing process before automating.

    Automating before standardizing and simplifying adds complexity, thus creating, not solving, problems. In general, world-class manufacturing companies focus on making flexible changes and decisions, and avoid making expensive changes and inflexible decisions.

20. **Preventive/predictive maintenance programs:**
World-class manufacturing companies have a preventive and predictive

maintenance program, typically based on worker involvement efforts, to minimize the occurrence of unplanned machine downtime, which disrupts the continuous flow of processing.

When considering each of these twenty characteristics of world-class manufacturing companies (summarized in Table 1.1) it is evident that the embodied policies, practices, and systems do not merely occur naturally in a manufacturing organization. Rather, such environments only happen through management's leadership and the application of the underlying philosophies of continuously striving to become better, eliminate waste, and create value for the customer. It is also evident that world-class manufacturing environments do not simply happen abruptly. Rather, such environments need to be cultivated and nurtured through a systematic sequence of activities involving focused projects that are well-orchestrated through a clearly defined strategic plan.

1.1.2 Components and Relationships

World-class manufacturing environments evolve most efficiently when a well-defined strategic plan is implemented by management through employee involvement in sustainable policies, practices, and systems. The magnitude of change in most companies is so large that implementation is most effectively accomplished in stages, using focused projects that have clearly defined objectives. Each project is basically a vital component that changes a company's culture to gain increased manufacturing capability and competitive advantage. The problem with projects, however, is that it is too easy to lose an understanding of the total picture and scope of the world-class manufacturing goal. It is natural for people to become lost in the inevitable details and difficulties of project implementation during the required time span, meanwhile losing perspective of how this project fits as an integral component in the ongoing process of continuous improvement. One of management's responsibilities is to communicate not only the objectives of specific projects, but also to articulate a vision of excellence in manufacturing, and show how each of the components interface to comprise the whole.

Figure 1.1 shows the ten components of world-class manufacturing and the fundamental relationships among them. A brief discussion will now be presented addressing the basic concepts and interrelations of each component. The issue of how to successfully implement these components is covered in detail in other chapters of this book.

The first and foundational component is to *establish a strategic plan* by addressing the following elements for implementing world-class manufacturing:

1. How to envision world-class manufacturing in terms of what it is and how it can provide a competitive advantage to the company.

TABLE 1.1. Summary of the Twenty Characteristics of World-Class Manufacturing Companies.

Management/Employee Involvement

1. Visionary Leadership and Champions
2. "New Culture" Goals and Thinking
3. Long-Term Strategic Plan and Direction
4. Employee Involvement and Human Resource Development
5. Integrative and Holistic Objectives
6. Goal-Consistent Measurement/Reward Systems
7. Product or Customer Focused Organization
8. Good Communication Systems and Practices
9. Promotion/Support of Research and Education

Quality

10. Customer-Driven Product Development and Marketing
11. Cross-Functional Teams for Product Design/Manufacturing
12. Individual Responsibility and Continuous Quality Improvement
13. Statistical Process Control of Key Product Characteristics
14. Emphasis on Innovation and Experimentation
15. Partnership-Like Relationships with Quality-Certified Vendors

Production Operations

16. Continuous-Flow Processing/Cellular Manufacturing
17. Demand-Based, Not Capacity-Based, Processing
18. Quick Changeover Procedures/Small Lot Sizes
19. Emphasis on Standardizing/Simplifying Before Automating
20. Preventive/Predictive Maintenance Programs

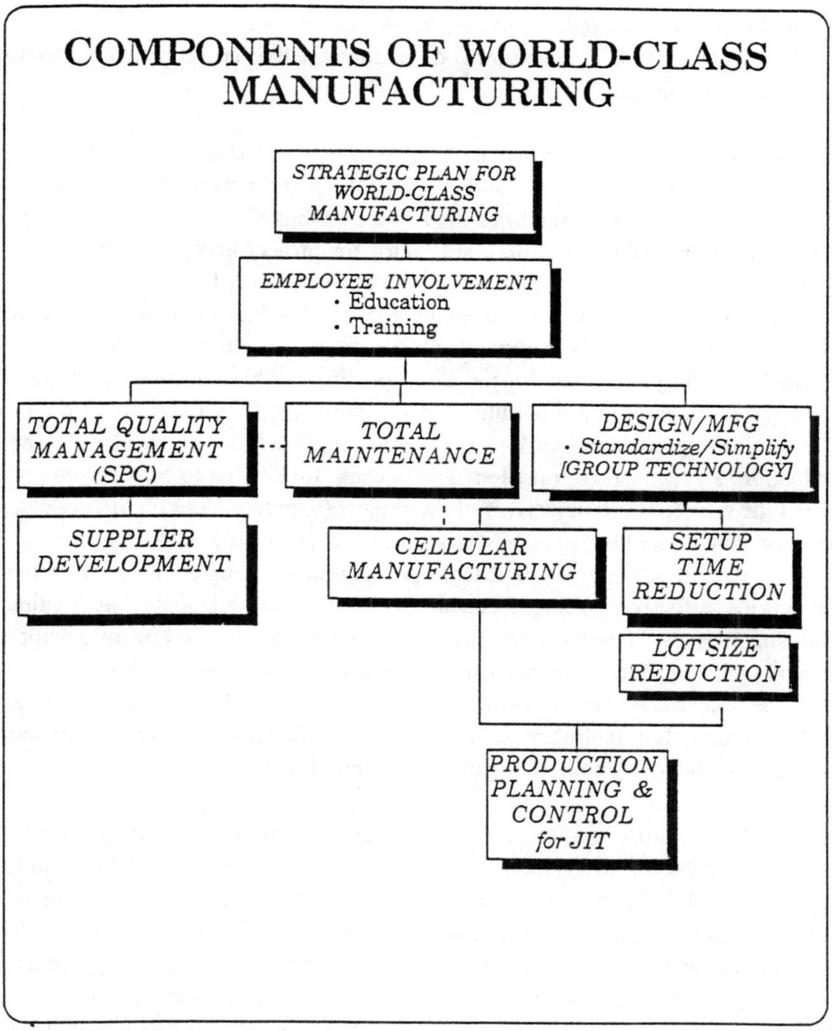

Figure 1.1. Components of World-Class Manufacturing.

2. How to become more competitive, given the company's current strengths and weaknesses.
3. How and where to start implementation.
4. How to articulate and "sell" the vision, implementation plan, and benefits to all employees.
5. How to control focused projects.

6. How to foster implementation and affect change.
7. How to sustain momentum and continue the evolutionary process to manufacturing excellence.

It is essential that a strategic plan be developed by management since the plan will not only provide a strategy for undertaking a long and typically difficult process, but also will demonstrate the foresight and leadership of management to everyone who will be involved—an important factor for project acceptance and success. Developing a strategic plan is discussed further in section 1.2.

The *employee-involvement* component of world-class manufacturing is likewise critical because building excellence in manufacturing is accomplished through building excellence in people—specifically through treating the human resource as an asset, not a liability. Many companies have failed to gain the full benefits of implementing various projects (such as cellular manufacturing, setup reduction, and preventive maintenance) because they failed to educate the people about the concepts and objectives of the project (which reduces the fear of change and resistance), and because they failed to provide training on the new skills and tools needed to do the new job. Educating and training people only after problems arise is less effective. Failing to educate and train people adequately at any time is foolish, since the benefits of the project are greatly diminished or lost. Although investing in the human resource is an ongoing project and requires effort and expense, experience has shown that it is an essential building block for successfully implementing the other components of world-class manufacturing. Chapter 2 addresses how to achieve manufacturing competitiveness using employee involvement.

The *total-quality-management* (TQM) component is a set of policies and practices espousing the principles of "do it right the first time" and "quality is everyone's job." Because TQM addresses both quality and productivity improvement through employee involvement, many companies have started with TQM as the first major thrust on the way to becoming world-class competitors. For example, a small company (less than $100 million in sales) that aggressively implemented TQM realized a 367 percent productivity improvement from 1986 to 1988 and documented savings of $10.3 million *annually*. The implementation of TQM not only is highly dependent on employee involvement, but it also serves as a catalyst for implementing the *supplier-development* and *total-maintenance* components. It is interesting to note that implementing TQM typically requires that different functional areas of the company work together with common goals—a fundamental characteristic of a world-class manufacturing company. The TQM component is addressed in chapter 3.

The *supplier-development* component is ideally intended to extend the quality and productivity activities and related benefits from a company to its vendors, establishing long-term partnership-like relationships with a selected number of vendors on the joint basis of quality, cost, and delivery. Supplier development ideally should follow the development and implementation of major TQM activities in one's own facilities. In the past, some large companies have acted like bullies by insisting that vendors implement improved methods and practices that the large companies were unable or unwilling to implement in their own facilities. Experience has shown that unless a company has made significant progress in implementing TQM into its own operations, and unless the company is willing to invest substantial time and effort in sharing with its vendors what it has learned through practice and experience, then the chance of establishing successful win-win supplier relationships is only mediocre at best. However, when both parties, customer and vendor, continue to contribute their manufacturing energies and strengths to the relationship, the resulting synergy provides each a significant increase in competitive advantage. The underlying principle is that "the best customer–vendor team will win." Chapter 10 discusses designing and implementing supplier development and partnership programs.

The systematic implementation of systems and procedures that foster *standardization and simplification in both design and manufacturing* is another component of world-class manufacturing. These implementations are based in part on group technology (GT), a philosophy that exploits the underlying sameness that exists in the design and manufacture of high varieties of discrete parts. Reducing complexity in manufacturing is not only a desirable goal, but is also a catalyst for achieving many other objectives that are related to adding value to the product by reducing waste and improving quality. Group-technology concepts and procedures provide a means for standardizing the parts that are designed and the methods of manufacture used to make those parts. For example, GT provides the underlying basis to group thousands of part numbers into part families (production families, not product or marketing families) where each family contains similar parts that are processed along the same flow or routing of operations. Likewise, dissimilar machines (for example, a saw, lathe, mill, or grinder) can be grouped together into a unique production unit or workcell to produce each family of parts. Although each workcell produces many different part numbers, all parts are in the same part family and are processed in a repetitive manner. This standardized processing, possible through the grouping of similar parts, provides for simplified control and allows the workcell to be operated like a mini-production or flow line.

In addition to cell formation, many companies use the GT concepts and techniques discussed in chapter 4 to simplify and reduce the number of active part numbers, to simplify and reduce the size of the inventory's master listing and bills of material, to generate highly consistent process plans more quickly, and to standardize tooling and part fixturing. Through standardization and simplification, GT provides the basis for cellular manufacturing and setup reduction.

The *cellular-manufacturing* component of world-class manufacturing is much more than cell formation and plant layout because it encompasses employee involvement activities that make improvements in the quality, cost, and productivity of parts produced in the workcell. Cellular manufacturing creates concentrated miniprocessing and responsibility units, which provide "point-of-manufacturing control." In essence, cellular manufacturing supports the objectives of both just-in-time (JIT) manufacturing and TQM. Thus, while some companies' facilities may already be organized to provide continuous-flow processing and, hence, not require the physical regrouping of their equipment into workcells, the key issues related to both operator cross training and the operator's responsibilities and control of processing for adding value to the product are still fundamental. The methods and procedures required to achieve focused and continuous-flow processing are addressed in chapter 5.

The *setup-reduction* component, which is closely tied to the GT strategies of standardizing and simplifying, is aimed at increasing the flexibility of a manufacturing facility to the point where it can produce exactly what is needed, when it is needed. The objective of setup reduction is not to make fewer setups or even reduce the total annual cost associated with doing setups, but rather to reduce the amount of machine downtime incurred with each job changeover. Since making more parts than can be sold is both costly and wasteful, it is essential to be able to change quickly from one job to the next and thereby have the ability to manufacture parts as needed. Chapter 6 presents the methodology, tools, and team approaches necessary to implement effective setup time reduction programs.

Setup reduction, is just a means to an end, and that end is *lot size reduction*, another component of world-class manufacturing. "The smaller the lot size, the better," is by now a well-known principle of JIT manufacturing. The meaning of small lot size, however, is quite relative. A lot size of 5,000 pieces, for example, might be quite small for a manufacturer of threaded fasteners, but huge for a manufacturer of precision instruments. The key point is that lot sizes should be as small as possible and still meet the customer's demand and delivery requirements. Hence, the component of lot size, as discussed in chapter 7, is highly related to and dependent on both setup reduction and production planning control.

Production planning and control (PP&C) for JIT production encompasses many activities controlling the availability and flow of materials through the organization. As new policies, practices, and operating systems are implemented in the evolution toward world-class manufacturing, PP&C must be revised to reflect the changes in these practices and systems. For example, processing parts in workcells typically reduces the manufacturing lead time of these components by 70–90 percent or more. These reductions need to be incorporated into the material planning system. In addition, significant reductions in manufacturing lot sizes will cause corresponding significant increases in order frequencies for these parts. Furthermore, as companies increase their ability to use point-of-use manufacturing and point-of-manufacturing control, it is desirable to simplify and/or eliminate many of the inventory tracking, shop-floor control, and performance-reporting systems often in place—especially in large companies. In general, many companies' PP&C systems are too complex, too cumbersome, and too costly to operate given the benefits obtained. As companies standardize and simplify their manufacturing facilities and related operations, the PP&C systems require revisions and, more importantly, provide opportunities for simplification and a means to eliminate non-value-adding activities. The many changes in PP&C required to support flexible and continuous-flow manufacturing are presented in chapter 8.

The *total-maintenance* component of world-class manufacturing, as shown in Figure 1.1, is linked to both TQM and the concepts of continuous-flow manufacturing, which are embodied in cellular manufacturing. Total maintenance incorporates the strategies of operator ownership and systematic planned maintenance activities to keep machines from breaking down or malfunctioning during production. Unreliable equipment is very unproductive because it breaks the momentum of machines and people in continuous-flow manufacturing. Once disrupted, it is difficult to regain the momentum and smooth out the resulting imbalances in inventory and material flow. Furthermore, poorly maintained equipment leads to accelerated deterioration and increases in process variability, which results in lower process capabilities for producing quality parts. Total maintenance is addressed in chapter 9.

Overall, it is clear that the components of world-class manufacturing, while not sequential in implementation, are certainly highly interrelated. It is not possible, for example, to be a flexible, world-class, JIT producer with poor equipment and second-rate quality systems. Likewise, good quality alone will not be sufficient if a company is unable to introduce new products and satisfy the needs of the marketplace in a timely manner. In short, being world-class means being a mean, lean, and world-class competitor by bringing products to the marketplace that offer better value than is offered by the competition. Each of the components of world-class manufacturing are essential to accomplish this task, both now and in the future. Striving to be world-class—like learning—needs to go on forever, otherwise stagnation and waste will eventually occur.

1.2 PLANNING FOR IMPLEMENTING EXCELLENCE

One of the characteristics of world-class manufacturing companies is the presence of a critical mass of managers, dedicated to continuous improvement and growth, who through leadership and coaching articulate a vision of excellence and motivate people to work together to eliminate waste and create competitive value. Having already defined the characteristics and components associated with world-class manufacturing companies, the strategies that work in planning for and implementing these characteristics will be discussed. Companies with good strategic plans consistently outperform those companies without them. Success is more likely with a strategic plan because it provides a framework for decision making throughout the entire company. This section presents how to establish a strategic plan that addresses the organizational and implementation issues in light of the resistance and pitfalls that are typically encountered in changing the way people think and behave while achieving the goals of world-class manufacturing.

1.2.1 Getting Organized for Change

The question has been asked numerous times: What is the most essential element for becoming world-class? The answer given—an echo from manufacturers who have achieved recognition for their world-class accomplishments—is quite basic: "You need to have the desire to be the best at what you do and be willing to pay the price to achieve it." The first installment on that price is to invest the time and effort to establish a strategic plan—a systematic sequence of actions to move a company from its current position to where it wants and needs to be in order to establish a competitive advantage. The strategic plan is not a "copy cat" plan, attempting to follow what the competition is doing or already has done; rather, it is a unique and consistent set of plans and policies, which when implemented, provide advantages over the competition. A strategic plan provides a proactive approach to assist a company in making decisions on what to do now about future opportunities and threats, instead of waiting and then reacting to events as they occur. Furthermore, a well-defined and articulated strategic plan provides an excellent means for communicating corporate direction to ensure that all employees are "marching" in the same direction. People need and want to know where the company is going. In essence, it's the game plan for winning and should address the seven elements shown in Table 1.2.

The first element the strategic plan must address is how to envision world-class manufacturing in terms of what it is and how to apply it to achieve a competitive advantage. This involves defining the fundamental purpose and boundaries of the company and establishing a mission statement that reflects the corporate vision of excellence. Experience has shown that manufacturing excellence cannot be implemented by manufacturing managers alone. There is a need to establish a world-class-manufacturing strategy that is based on information, issues, concerns, and perspectives provided by different functional areas within the organization. The corporate strategy for achieving world-class competitiveness must be supported by complementary financial, operations, and marketing plans that provide detailed steps covering how the corporate objectives and goals will be accom-

TABLE 1.2. Seven Elements of a Strategic Plan for World-Class Competitiveness.

Elements	Options for Implementation
1 Define Objectives for WCM and corporate vision of excellence	-Management Training on WCM
2 Assess Existing Enterprise Strengths & Weaknesses for competitive advantage	-With Diagnostic Tools -By Outside Professionals -Benchmarking, Customer Surveys
3 Define Approach (Choose Projects & Areas) -How to Get Started -What to do Next	-Pilot Project -Most Beneficial Steps -Two-Front Attack
4 Define Strategies for Communicating to Employees: -Vision articulation -Implementation Plan -Benefits to be Expected	-Counsel from Outside Professionals -Define Win-Win Plans -Top Management Involvement
5 Define How to Control Focused Projects (Progress Measurement)	-Project Leader (Champion) -Project Management Techniques
6 Define How to Foster Implementation/How to Affect Change GOAL: Entire Work Force Support	-Management Involvement -Employee Education
7 Define How to Sustain Momentum and continue the evolutionary process	-Strong Initial Momentum -Employee Involvement -Continuous Improvement

plished as shown in Figure 1.2. It is thus essential to develop the corporate strategic plan in conjunction with managers from functional areas outside of manufacturing. Involving managers from functional areas that interact with manufacturing also provides a useful outside perspective for the strategic plan. Furthermore, this approach provides a better understanding of the roles and difficulties associated with manufacturing. All too often, people outside of manufacturing look for quick solutions that initially appear to provide relief but actually do not.

In order to effectively envision world-class manufacturing in developing a strategic plan, many companies find benefits in initially implementing a management training program. While most managers today have some exposure to the basic ideas and publicized success stories associated with various components of

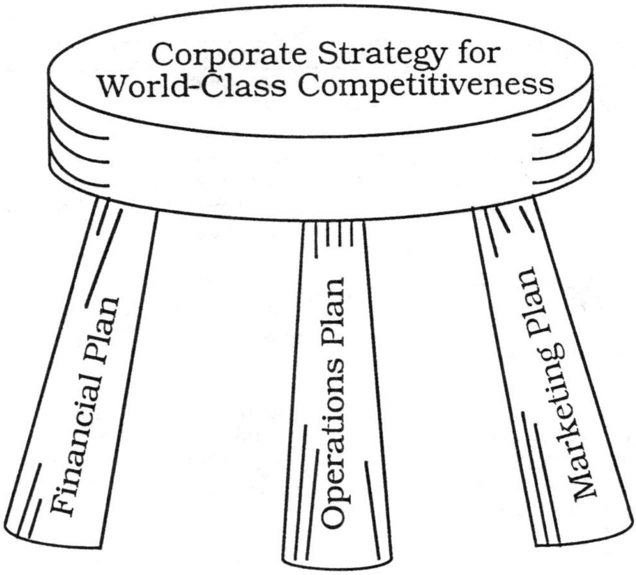

Figure 1.2. The Basis of a Sound Strategic Plan for World-Class Manufacturing Competitiveness.

world-class manufacturing, most managers lack a clear and consistent understanding of the underlying concepts and principles required to build manufacturing excellence in their organization. Furthermore, managers from functional areas outside of manufacturing typically have less knowledge of world-class manufacturing concepts and principles. Since one cannot effectively envision what one does not fully understand and comprehend, a management development program is often a necessary first step. Such programs typically consist of a combination of seminars and/or workshops on the various components of world-class manufacturing, presented either in-house or in conjunction with professional societies and other professional programs.

Beginning with education and training for management development is a bold step for a company to undertake. Time is a precious commodity to managers. There is always more to do, or so it seems, than time available in which to do it. When senior management takes the initiative and makes management education and development a top priority and preface to developing a strategic plan, the impact of this action is immediately apparent, placing it at a level of utmost importance. Investing time and money in management development drives home the point that learning is an underlying basis and foundation for continuous improvement for competitive advantage. Figure 1.3 shows the framework for an approach to

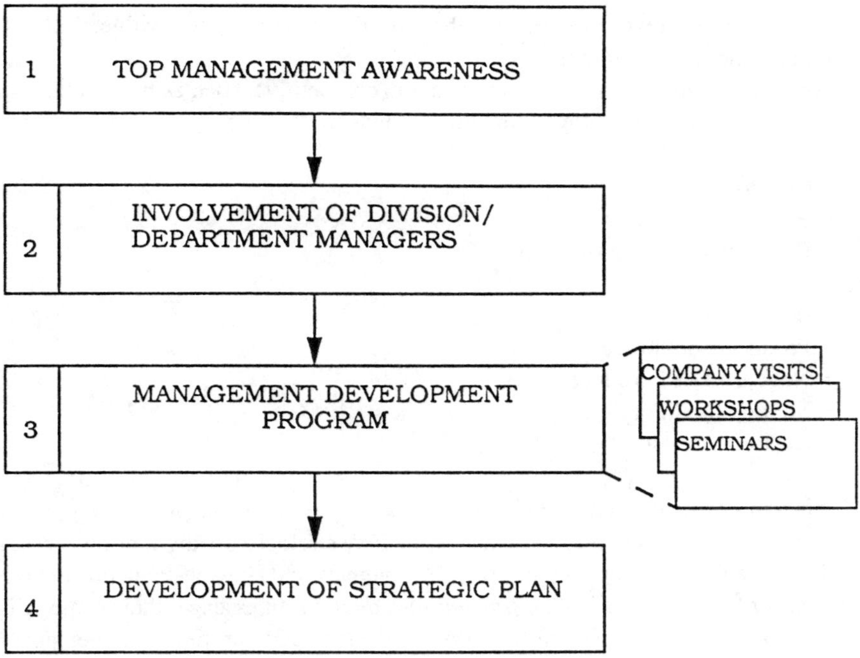

Figure 1.3. An Approach to Management Development.

management's development for strategic planning. It is top management's responsibility to recognize and respond to this fact with leadership and commitment. Quite simply, knowledge, whether obtained through education, personal experience, or the experience of others, is the first key ingredient for transforming an organization into a world-class enterprise.

Once the principles, concepts, and ultimate objectives of world-class manufacturing are understood by the management team, the next issue in developing the strategic plan is to determine how to become more competitive, given the company's current strengths and weaknesses. Using situation analysis, the management team addresses the question of "where is the company competitively positioned now, and what does the future look like?" The internal analysis assesses the strengths and weaknesses of the organization's structure, operating systems and procedures, and effectiveness in creating value and satisfying customer requirements. The external analysis assesses how the company is positioned and viewed relative to the competition, typically based on data obtained from benchmarking, market surveys, and direct feedback from customers. Situation analysis allows the team to define a common view of what strengths the company has and needs to leverage, and what weaknesses exist that provide for competitive vulnerability—both now and in the future.

There are many different ways to become competitive, such as by excelling in one or more of the following competitive demands.

- Cost to produce
- Product performance
- Reliability and workmanship
- Speed of delivery
- Delivery reliability
- Product customization
- New product introduction
- Volume flexibility
- Customer service

Since no company can excel in every measure, the strategic-plan-development team needs to focus and prioritize the competitive strategies that will bring success to their organization. For example, the 3M Company and Hewlett-Packard are both known for their abilities in *new product introduction.* In essence, their strategy is to develop and market innovative products, take the "early profits," and then phase out when the "catch up" companies enter the market, thereby focusing their energy and resources on identifying and developing the next new product. Regardless of the emphasis and focus a manufacturing company may choose, the final objective is the same; namely, how to bring products to the marketplace that offer better value than the competition. There is no generic answer to this question—the strategic-plan-development team needs to address this issue to develop a winning game plan with specific objectives and goals for their unique organization.

The third element of the strategic plan is to address *where and how to start implementation.* There are a number of different approaches that companies use to get started, which are shown in figure 1.4. Depending to a great degree on particular circumstances, one approach may be more attractive than the others.

One approach is essentially to implement all ten components of world-class manufacturing (in stages, of course) at a single facility or location as a pilot project. This approach involves people from all functional areas of the company and is often used to demonstrate that "yes, world-class manufacturing can in fact be implemented successfully here, too." The pilot program thus serves as the role model for continued implementation efforts elsewhere.

A second approach is to focus on implementing only those world-class manufacturing components that will yield the greatest benefits to the company's competitive position. For example, many companies select employee involvement and TQM as the starting point for implementation.

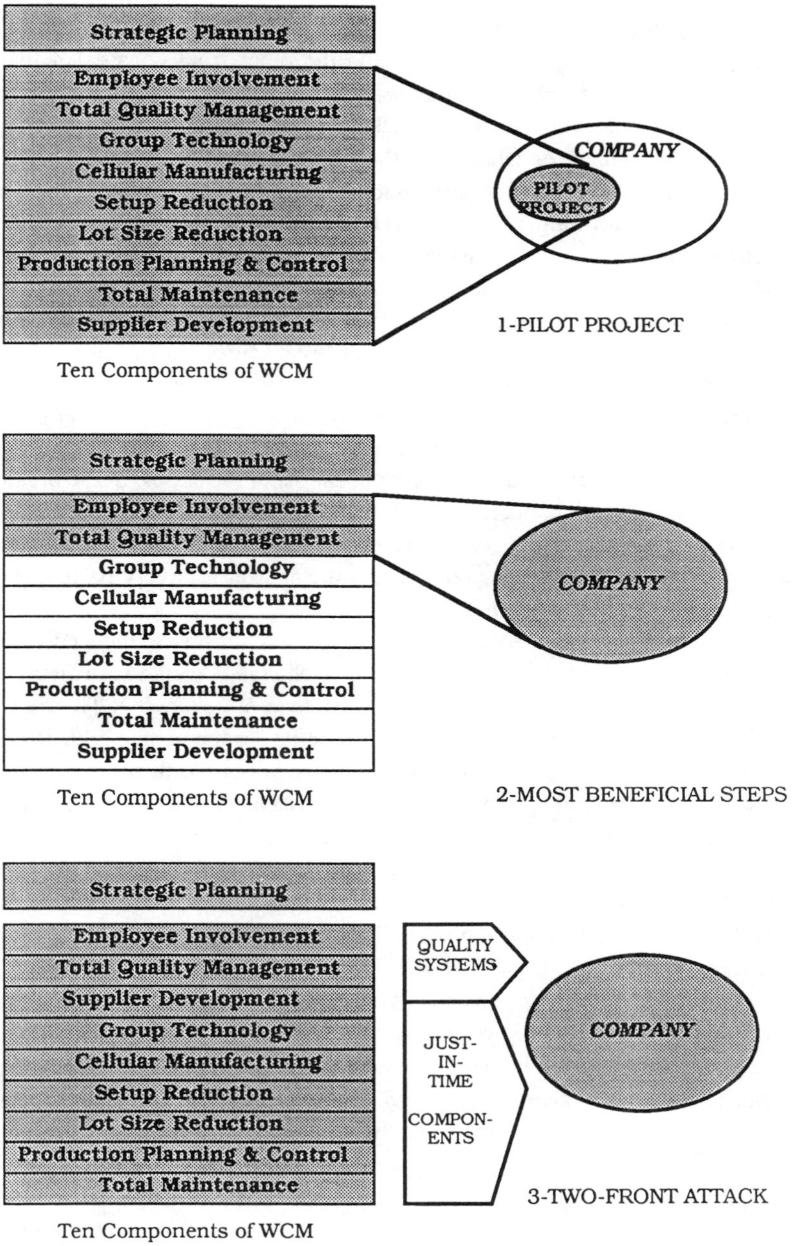

Figure 1.4. Three Possible Ways of Getting Started.

A third approach is to make a two-front attack. One front consists of the components related to quality systems, and the other front consists of the JIT components for implementing continuous-flow processing with quick setup times and small batch sizes. The two-front attack, however, requires more team orientation and training support up front given the size of the task.

The approach that is best for a particular company depends largely on the results obtained from performing an assessment of the manufacturing organization's competitive strengths and weaknesses. The decision can also be based on the physical locations and organizational structures already in place for the operational units of the company. Companies often find that an assessment performed by a professional (who is not affiliated with the organization) provides useful and unbiased input for the strategic planning team in determining the most beneficial starting point for implementing the components of world-class manufacturing.

The fourth issue the strategic plan needs to address is *how to articulate and "sell" the vision, implementation plan, and benefits* to the people in the organization in order to "make it happen."

This is an issue that many companies do a poor job in addressing. The major factor, which really makes things happen, is the answer to a very basic question about human nature: "What's in it for me?"

A major strategy for achieving world-class manufacturing excellence is continuous improvement, which involves continuous change. Change is difficult. It is often threatening, and it is certainly more work than not changing. There simply has to be the possibility of gaining tangible benefits for the people both affecting and affected by the change, or the cost of change is not worth paying. Thus, an important part of the strategic plan needs to be a clear definition of the win–win payoff associated with any implementation plan affecting change. The answer to the question, "What's in it for me?" needs to be defined and established in a clear, concise, and concrete way for everyone in the organization and for each component and related implementation project. The management attitude of "as long as you want a job here, you'll do what I say" does not solicit the type of dedication and commitment needed to become world-class. Similarly, the management attitude of "if you want to keep your job (for whatever reason), then you'll do this" does not generate much excitement or commitment. There are sufficient benefits, both financial and personal, from achieving a level of competitive advantage so that everyone in the company really can "win." It is management's responsibility to articulate these benefits in "selling" their vision for excellence throughout the organization. Companies that ignore this question, or fail to answer it adequately, will eventually lose the battle of implementing the types of changes required to achieve world-class manufacturing excellence.

Selling the change to world-class manufacturing excellence typically requires education to ensure that people understand and appreciate the meaning, importance, and win–win benefits of achieving manufacturing excellence and why it yields the most advantages for everyone involved. This process of education is ongoing and needs to be packaged and presented in stages, often linked to the implementation plans, to ensure that not only does the knowledge spread and permeate to people at all levels of the organization, but also that the knowledge is packaged and presented in a format that is clear and geared to the background of the audience for easy learning. Like any other substantial subject, the principles, concepts, techniques, and benefits of world-class manufacturing cannot be fully absorbed all at once.

Often it is beneficial to engage professional educators and/or consultants to assist in designing and presenting an educational program and presentation format that is appropriate for articulating the vision of world-class manufacturing and the logistics for achieving it. Professional educators and/or consultants are particularly effective when they participate in developing the entire strategic plan along with the management team.

This segment of the strategic plan includes not only defining the content of the educational presentation to "sell the vision," but also designing the format of the presentations. Examples are very important—especially examples taken directly from the experiences of the group in attendance—in order to ensure that people understand and relate to a new way of looking at and ultimately doing their jobs. Experience has also shown that introductory presentations given by top management, often in large assemblies, have greater impact than simple announcements by immediate supervisors. As stated previously, implementing change in manufacturing organizations to achieve new levels of excellence and a competitive advantage requires the leadership, commitment, and courage of top management. Continuous feedback and personal involvement by top management are very important. Creating and presenting the reality and benefits of change is crucial for success; therefore, it must be addressed by a strategic plan.

How to control focused projects for implementing the components of world-class manufacturing is the fifth issue that the strategic plan needs to address. Specifically, the plan should identify ways to detail and use measures of progress and performance that relate directly to the objectives of the project. The well-known techniques of project management can also be appropriate when thus applied. The most important factor is to identify, select, and appoint a project leader or "champion of the vision" who can effectively manage both the technical and the human relations aspects of the job. Since employee involvement is a major factor in implementing changes in an organization, selecting a person who is able to solicit the best in people and develop the work force into a knowledge-based asset is of key importance. The strategic plan should therefore establish criteria that define these job requirements appropriately and avoid the appointment of people based on current position, seniority, or other forms of selective favoritism.

Defining *how to foster implementation and affect change* in the strategic plan is one of the most difficult tasks, requiring considerable courage and commitment from management in overcoming the status quo. The biggest roadblocks to implementing the components of world-class manufacturing and the changes that must accompany it, are the policies, practices, and systems already in place in an organization. For example, many of the cost-accounting, material-tracking, and performance-measurement systems in place are clearly designed to track material and item costs as they move in and out of inventory at each stage of manufacturing, and to report operator performance on an individual basis that is related to individual machine part-production standards. Such systems not only are extremely cumbersome to use once part family workcells are set up and operated by a team of cross-trained workers, but they also fail to measure the performance and benefits of the workcell, namely, less work-in-process inventory, faster manufacturing throughput times, and better-quality parts. In one company, for example, when the cell-implementation team proposed modifying the tracking and cost systems to make them more simple and easier to use with more accurate performance reporting, an outcry arose from accounting claiming that the proposed modifications would ruin the way the company does its budgeting and cost control. Given that there was no recognition for affecting change in the strategic plan (which was itself almost nonexistent) and little involvement and commitment from top management, the current accounting system was kept intact and the issue was effectively ignored. Although this company is not world-class—and not moving in that direction very quickly, in spite of an involvement in cellular manufacturing—the example illustrates the types of obstacles that current policies, practices, and systems typically present. What is needed in the strategic plan is a recognition that policies, practices, and systems will need to be, and will be, changed when manufacturing systems are simplified and streamlined. Clearly, making such changes requires effort and commitment, again emphasizing the value of and need for having managers from various functional areas on the strategic-plan-development team.

How to sustain momentum and continue the evolutionary process to manufacturing excellence once the glamour of doing something new and exciting has worn off is another issue that the strategic plan needs to address. Changing a manufacturing organization from mediocre to world-class requires years, not months, of ongoing incremental efforts. The fact that it is really a never-ending, continuous process can make sustaining momentum a difficult task. Only the burning desire to be the best, not only now but also in the future, can provide the fuel to keep the torch burning. Continuous improvement can only be sustained, and the benefits forthcoming, if never-ending experimentation, improvement, and the search for knowledge becomes part of the culture of the company—a way of life where change is natural and not a source of fear. So, in effect, the best way for management to sustain momentum is to initially devote sufficient energies and provide the enthusiasm and leadership required to really change the culture of the organization, and not simply implement another project.

1.2.2 Summary and Resolution

Developing a strategic plan for achieving manufacturing excellence and a competitive advantage is not an easy process. It requires both knowledge and forethought. Although the issues that need to be incorporated into the strategic plan have been discussed here, there are still many questions that can only be answered by actually doing it. Each company has its own current set of competitive strengths and weaknesses on which a strategic plan must be developed. A good strategic plan, however, makes the difference in the amount of gains achieved in quality, productivity, manufacturing flexibility, and product cost reduction—the key components of value, which determine competitive advantage and profitability. Based on experiences in different types of manufacturing companies, the differences in improvement can be significant—often between 30-60 percent. Thus, the first component of world-class manufacturing—establishing the strategic plan, which is shown in Figure 1.1, is truly the first step toward achieving excellence in one's manufacturing organization.

2

Employee Involvement and Development

To become a world-class competitor, a company must develop excellence in every area, including product development, design, manufacturing, and marketing. Excellence cannot be achieved by management alone; that is, a company cannot compete and survive in global markets without the help of its people and without developing the value of its people. According to the executive director of Matsushita Electric Industrial Company Ltd., Mr. Konosuke Matsushita, addressing western company managers: "...The survival of firms...their continuous existence depend[s] on the day-to-day mobilization of every ounce of intelligence....Only by drawing on the combined brain power of all its employees can a firm face up to the turbulence and constraints of today's environment...." People are the key to the efficient functioning of any organization. Excellence can be achieved only as a result of a companywide team effort that includes management, union leadership, and the rest of the work force.

For many companies, introducing teamwork implies that a new company culture has to be developed, one that promotes the concept of "doing things together" and one that shows how each employee's personal interest and development can be linked to the company's. Bidirectional-vertical communication channels have to be created or reinforced so that objectives are communicated and input on the best way to meet these objectives is received. A passive (or even reactive), often frustrated work force must become involved, active, and even enthusiastic. The work force's ability to adapt to and be involved in technological and organizational changes should be enhanced through continuous education.

Employee involvement (EI), if not done correctly, however, can become expensive and ineffective, lowering productivity, and increasing levels of frustration and costs. Unless employees embrace the concepts of manufacturing excellence through continuous improvement and have the guidance, skills, and support required to identify and implement these improvements, the expected and appealing benefits of EI will not be realized. It must also be clear, first to the company for management, and then to the rest of the employees that EI is a never ending process. It is not a "quick fix" to communication problems, but rather it has to become a *way of life* in the company, leading to an environment where change is natural and not feared.

This chapter addresses how to implement EI and development programs that provide a company with cost-effective team approaches for achieving continuous improvement in all phases of the manufacturing enterprise. The emphasis is on how to implement low-cost, and effective approaches to transform the work force into an asset for achieving competitive advantages. Figure 2.1 illustrates how work

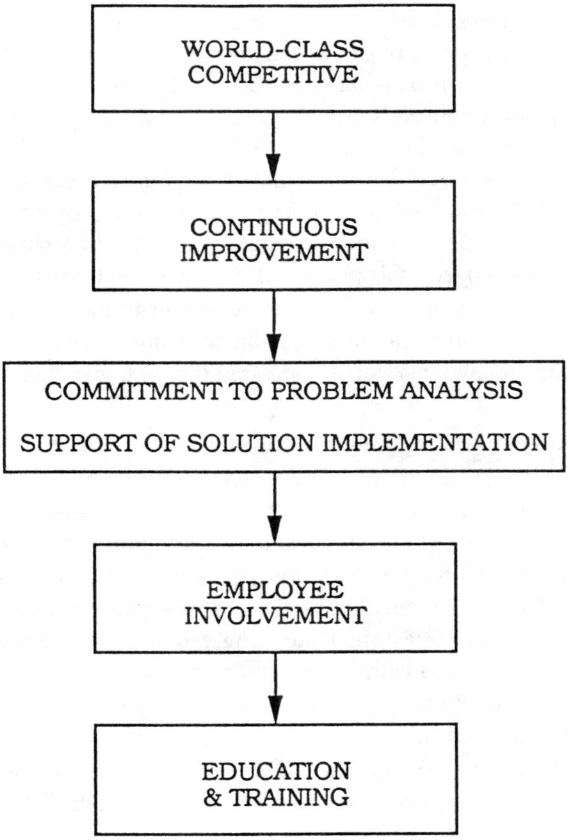

Figure 2.1. The Role of EI in the Strategy to becoming World-Class Competitive.

force education and EI are building blocks of the strategy to becoming world-class competitive. This diagram illustrates that education and training are prerequisites to involving the employees in work teams whose mission is to (1) analyze existing problems, (2) propose alternative solutions to management, (3) be actively involved in the implementation of the solution adopted by management, and (4) as a result, participate in continuous improvement, which is one of the conditions for being mean, lean, and world-class competitive in the nineties. Specifically, chapter 2 discusses how to (1) establish and communicate common goals, (2) develop EI teams and programs, and (3) provide the required education and skills.

2.1 COMMUNICATION AND EMPLOYEE INVOLVEMENT

Communication and trust between management and workers are prerequisites to making any change because fear of the unknown is greater than the dissatisfaction of continuing with currently bad—but known—conditions. When top management decides to change the company's organization and technology in order to become more competitive and profitable, the immediate step, after having defined the strategic plan for excellence, is to articulate this plan to the work force and to consider the work force's feedback for implementing the proposed changes. For those companies without a tradition of open communication channels between management and employees, this phase calls for the definition of a new company culture. Modifying the company culture can be a delicate and risky step because it mandates a profound modification of human behavior to alter long-established relationships. How to take this step is addressed first in this section.

2.1.1 The Vision of a New Company Culture

The company's culture is unique and is highly influenced by the vision and managing methods of its top managers. It is difficult for many manufacturing company managers to accept the idea that they must entirely reconsider and alter their way of working in order to successfully implement world-class manufacturing techniques. The implementation of the new company culture, which is the breeding ground for world-class manufacturing, requires (1) an assessment of the present culture, (2) a clear identification of the prerequisites for change, (3) an awareness of the necessity to acquire a new attitude by every company member, and (4) a list of actions for implementation.

Assessing the company's present culture. A simple test for assessing a company's culture and the way it is influenced by its managers follows:

- Do the company's people have the feeling of belonging to a unique group that shares common work objectives, and do they understand and trust each other?

- Does the union leadership (if any) or work force support productivity improvements?
- Is employee education a continuous process?
- Is delegation well developed at all levels of the organizational structure?
- Do managers spend a large proportion of their time informing their subordinates about the company's objectives and do they listen to their subordinates' comments?

If the answer to all five questions is yes, then it is likely that this company is already a world-class company or at least well on the way to world-class excellence, indicating that the concepts presented here have already been applied. However, if one or more of the answers to the questions above is no, then a new culture for the company should be envisioned.

The prerequisites for change. The main idea that shows the direction in which the company culture should evolve is in "respect for the people." In many companies, hourly workers are considered a resource measured in number of labor hours. This resource is defined as "productive" when it accomplishes such tasks as producing parts, assembling products, setting up, or repairing equipment. This view denies the workers the right to be human because it does not recognize the worker's ability to think about and suggest improvements toward a better and safer working environment[1], better ways to produce, better product design, etc. A debilitating practice is to not call on hourly workers to think, suggesting that there is also no reason to inform them on the modification of their working environment. As a result, people become frustrated and do not understand the objectives of management when changes are introduced in the work place. Many company managers have realized the danger of solely considering people as mere extensions of production machines and have realized that respecting the people, (that is, considering their capability to think and not just perform physical work) is the foundation for any durable improvement of the company's performance.

Respecting the people involves the following prerequisites:

- Recognize that the employees are willing to interest themselves in their work.
- Recognize that the employees are capable of and willing to improve.
- Accept that the employees are represented in continuous improvement and that their interests are clearly supported and negotiated.

---

[1]In this chapter, working environment refers to the entire set of conditions that define the environment of the workers, including working conditions such as lighting and level of noise; job characteristics such as production rate and production operations to perform; as well as the "human environment," including relationships with colleagues, supervisors and subordinates.

The acquisition of a new attitude. This new culture can be implemented only through a redistribution across the company of different roles. Companies have to move away from the Taylor system, which is characterized by hyper-specialized tasks. Workers must be given more responsibility than only executing clearly defined tasks: Their responsibility must be increased horizontally (for example, by training them to use several different machines), and vertically (for example, by assigning them maintenance tasks and by implementing a structure that promotes employee involvement). Given a new vision of their role in the company, workers will enlarge their horizons from simple and clearly defined tasks to being part of an organization that is focused on customer satisfaction. Thus, workers will feel less task-dedicated and more committed to companywide involvement, which benefits both the employees—who become more inclined to learn new skills—and the company management—who then has a much more flexible work force that has the additional potential of becoming an army of internal consultants making suggestions on how to reduce costs and improve the company's profitability. As a part of their role as company or department leaders, managers will grow closer to their employees, inform them, listen to them. These new roles can only be assumed if a new attitude is acquired by everyone in the company. Table 2.1 lists these new attitudes. The modification of the employees' attitude is initially the responsibility of the company's top management. Communication and education are the two major avenues leading to this modification.

Actions for implementation. Table 2.2 lists a series of management actions that foster the development of a new company culture. Two conditions to the implementation of any of these actions are (1) that top management firmly believes in the necessity for a new culture and in its benefits, and (2) that it continuously supports and motivates the rest of the company's personnel for adopting a new attitude.

The management actions listed in Table 2.2 are all oriented toward the diffusion and collection of information (communication) and the diffusion of new ideas (education). These actions are not only necessary to initiate the evolution toward a new company culture, but *they are part of the new culture* and are a normal way of doing business in world-class companies. Only when company management has defined the new culture and has started to take the necessary steps for its implementation can common goals and efforts be shared within the company.

2.1.2 Sharing Common Goals and Efforts

Traditionally, encouraging production workers and workers' representatives, middle managers,top managers, and a company's professionals (such as engineers and accountants) to share common goals and efforts has seldom been attempted except when a superior motive, like the war effort experienced in the United States

TABLE 2.1. New Attitudes to be Acquired by Company Personnel.

Top Management	Define Organization Evolution Inform and Listen to the Work Force Negotiate with the Worker Representatives Motivate Work Force for the Changes
Middle Management & Professional Staff	Acquire Knowledge Outside of Initial Domain Be Able to Evolve Learn to Inform, Teach, and Listen
Supervisors	Motivate workers Adopt a Coaching Attitude
Workers	Be Ready to Change to Become More Autonomous and More Flexible
Worker Representatives	Negotiate with Top Management to Preserve Jobs as well as Company Competitiveness

(between 1941 and 1945), helps to focus everyone's energy on a single objective. Conventional scientific work-organization practices, which clearly define the scope of the various jobs and the corresponding responsibility limits, hinder the development of common goals. This hindrance is aggravated when incentive systems promote local objectives, such as the maximization of the production rate on specific equipment. A vital goal, such as customer satisfaction, is of little concern to a production worker or to a foreman whose objectives are to meet production quotas, and both of whose salaries grow proportionally to the numbers of parts produced. However, it is only when all energies in a group are geared toward the accomplishing clearly defined common goals that there exists the maximum chances to reach these goals.

The ultimate goal for a company is to consistently make a profit and to pay dividends to its investors. All organizational and technological changes should be aimed toward continuous customer satisfaction, which guarantees steady profits. Customer satisfaction is an objective that can be easily understood and can thus be shared by the entire company personnel. As suggested in Figure 2.2, the interme-

TABLE 2.2. Management Actions for Implementing a New Culture.

- Educate Personnel, Starting with the Managers
- Sell the "win-win" benefits of change
- Open Communication Channels Within the Organization

  (Small-Group Activities, Quality Circles, Suggestion Systems, etc.)
- Inform, Inform, and Inform
- Inform as Early as Possible:
  - To Avoid Rejection or Frustration
  - To Have Sufficient Time to Adjust
  - To Have Sufficient Time for Training
  - To Guarantee Involvement and Support
- Listen to the Work Force (Two-Way Communication Channels)
- Provide Ongoing Personal Top Management Feedback
- Negotiate with Employee Representatives
- Delegate Responsibility

diate goals set to attain these final goals are a combination of competitive product quality and cost, and of the company's flexibility in responding to customer demand. Each organizational or technological change should be aimed at meeting the intermediate goals, thereby attempting to meet final goals. It is important to keep in mind that these are long-term goals and that management should be patient in attempting to meet them. Short-term profits, for example, should not be preferred over long-term customer satisfaction.

When changes are to be introduced in the work place, one management task is to clearly explain what long-term goal(s) the changes will benefit and then to work with the employees to find the best way to implement the changes. Involving the

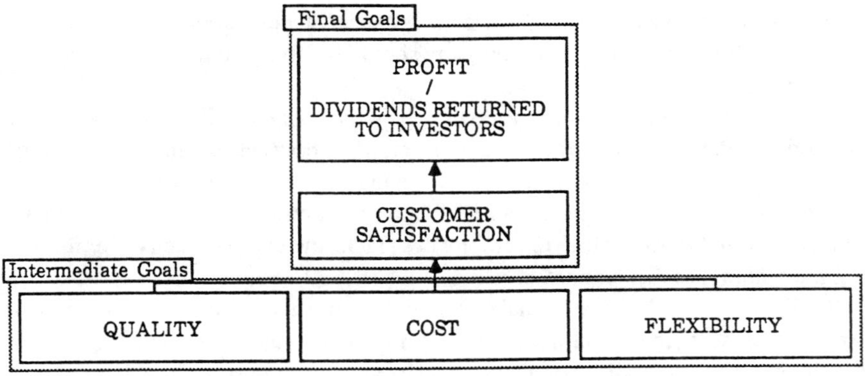

Figure 2.2. Company's Final and Intermediate Goals.

entire work force in the company's evolution is mandatory. But to what extent? And what are management's and the worker's responsibilities?

2.1.3 Scope and Responsibility for EI

Employment involvement can be defined as the participation of the entire firm's work force to improve the working environment, product quality, equipment productivity, and, eventually, company competitiveness. This participation occurs through cooperative relationships, open communication, and group problem solving and decision making. The employees are involved to continuously improve the current working environment and to help solve specific problems, such as determining the layout of a new metal-cutting cell or eliminating scrap generated by a given process. What is needed from all company employees is not only that they perform usual operations, but also that they use intellectual capabilities to determine how to improve these operations and the systems of which these operations are a part.

Clearly, the scope of EI is companywide. Employee Involvement should not start and end with production workers; involvement is required at all levels of the organization structure and also across these levels. Supervisors, managers, engineers, accountants, all have suggestions on how to improve the way their work is performed and should be given the opportunity to express and share these suggestions. Every individual should be interested in the company's future as part of his or her own future and, therefore, should participate in improving the conditions for preparing the company for the future. Some companies start quality circles composed of production workers and quality engineers but make the mistake of leaving aside the production supervisors (also accountable for product quality), thereby creating a frustrated group of people and depriving the quality circles from the experience and knowledge of these people who, incidentally, have often been promoted to the position of supervisors because of their good performance as workers. Every individual in the company should be involved in the improvement process.

The extent of the problems to be dealt with in the context of EI need to be clearly defined. A potential risk is to start the EI program without setting limits to the range of problems or to amounts of expenses for suggested improvements. If limits are not set, unrealistic suggestions can be made and then rejected by management, finally resulting in frustrating the employees. Thus, the entire objective for involving the employees can be missed.

Deciding what changes to make is clearly top management's responsibility. Among the most significant decisions of company managers are (1) the choice of the direction in which the company must be steered to ensure durable customer satisfaction and profits, and (2) the selection of the organizational and technical changes that enable the company to take this direction. However, defining how these changes will be introduced into the work place should be the joint responsibility of both management and the employees. For example, the choice of implementing manufacturing cells to increase quality and reduce cost is decided by management and not by any employee work team or by employee representatives; nonetheless, manufacturing engineers and foremen should be asked to help in assigning machines to cells, and input from production workers is necessary to identify how to position the machines in the cells, where to locate the tool racks, what materials handling system could be used, et cetera. The scope of employee participation is illustrated in Figure 2.3. Management and employees should work together to identify how to go from the current position to the position targeted by management.

After having announced the changes that will take place and the reasons for those changes, the following list of issues has to be discussed between management and the employees:

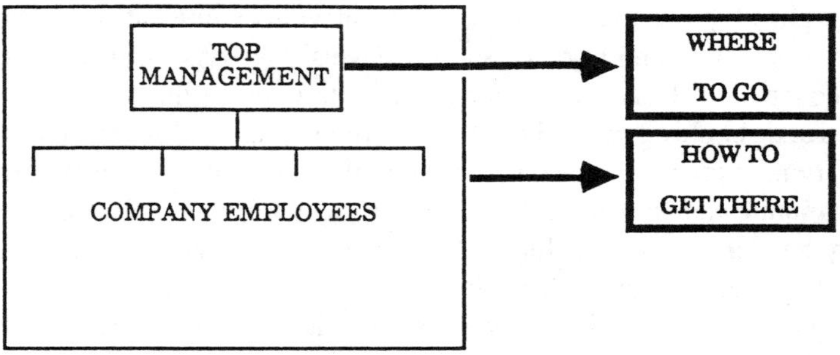

Figure 2.3. Scope and Limit of Employee Participation.

- The tasks and operations involved (for example, transferring tasks such as first-level maintenance, quality checks, and parts handling to machine operators).
- The organizational design that will support the new working conditions (for example, transferring material handlers to production assignments, or transferring quality inspectors to vendor counseling and certification jobs).
- The levels of qualification and the necessary training for implementing the new organization (for example, training former quality inspectors to manage relationships with vendors).
- The pay system that will be linked to the new organization (for example, replacing an individual piece-rate system by a team-incentive system).

When asking employees to share efforts, management should consider sharing benefits with them, too. Pay systems should evolve from the traditional system, which rewards merely turning out parts or products (such as is promoted by piece-rate systems), to a system that rewards workers for helping the company meet its end goals of increasing customer satisfaction and profits. In addition to production-team incentives, other innovative systems can be proposed, such as a "pay-for-knowledge" system, where employees obtain bonuses proportional to the number of tasks they master and, as a result, earn a salary that reflects their flexibility in job assignments.

The participatory approach for implementing organizational and technical changes is likely to evolve into nonconflicting situations and therefore yield positive results. There simply are no other good choices for either management or the employees: For management, because world-class manufacturing techniques cannot be implemented if the principal participants—the employees—take hostile or unsupportive attitudes; and for the employees, because participating in the implementation of the organizational and technical changes is a unique opportunity to influence their own way of working.

Employee involvement is essential for the company's management team because employee development in a common direction will facilitate the company's own development and improve its competitive advantage. The goal of modifying the company culture is that the entire company work force can help managers and project leaders to remove the obstacles on the way to world-class manufacturing and improved competitiveness. Figure 2.4 depicts first what happens when only a few individuals in the company strive to make the company progress with great pain because of the inertia caused by all the problems in the organization and the passivity of the rest of the work force: Very slow progress. It then shows that change and progress can happen faster when employees' own inertia disappear, and, in addition, the energy of the employees is synergistically combined with management efforts to remove the remaining problems.

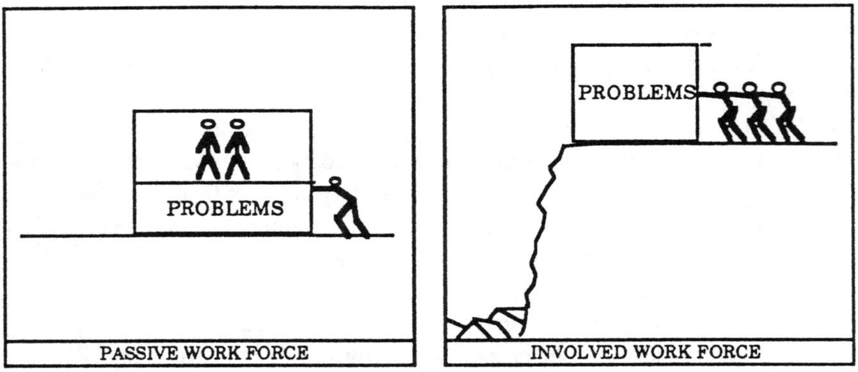

Figure 2.4. Involving the Work Force Facilitates Problem Elimination.

Using the participatory approach while implementing change enables the employees to gain a sense of ownership toward the organization and working environment being implemented, thereby promoting continuous improvement. After its change-initiator role, top management assumes a support role for employee participation while implementing change and afterwards to maintain full momentum toward continuous improvement.

2.2 HOW TO ACHIEVE EI

Employee involvement cannot be implemented for the sole purpose of having the employees "involved." Employee involvement is a foundation, it is a fertile ground that must be cultivated to allow changes to be firmly rooted and competitive manufacturing techniques to develop. The actions to be completed for achieving EI can be grouped into four successive phases:

1 Creating the conditions for EI.
2 Starting EI pilot programs.
3 Evaluating the results.
4 Extending the EI program to the rest of the company.

The actions to be undertaken in each of these four phases are presented in the following sections. In addition, a section on the pitfalls to avoid when implementing EI programs is also included. Finally, how to implement EI with teamwork and group approaches is addressed.

2.2.1 Creating the Conditions for EI

Once the company's top management is truly convinced that EI is crucial to the company's future, specific goals that will guide the EI program implementation must be developed and communicated to the company's work force.

Introducing EI in a company is in many ways similar to introducing democracy in a country. Individuals are asked to act with respect to the interest of the group. Employee involvement, as democracy, cannot be forced on the individuals. It occurs as the result of a set of initial conditions, such as mutual respect and trust between management and the work force, the existence of bidirectional communication channels, and the preparation—through training—of both parties to a new way of working. Without mutual trust, no common goals or effort can be shared; therefore, patience is needed on both sides in the process of building this mutual trust. Mutual trust can occur only as the result of a long-term effort (several months to one or two years is often necessary to establish mutual confidence) that requires informing the employees, management's consideration about employee concerns, and management's commitment to develop the work force through training. Employee involvement training includes ensuring that the entire company's personnel is aware of the importance of EI and training the people in charge about particular EI techniques (for example, training future team leaders to lead group discussions).

A structured approach to initiating EI is summarized by the diagram in Figure 2.5 and is briefly detailed thereafter.

- Establishing goals for a companywide plan is the long-term strategic view that management has concerning where the company should be in two, five, or ten years, and of the broad changes that will have to take place. The goal, for example, could be to become a world-class company within five years, and the necessary changes could be to switch from a functional layout to a cellular layout, to implement companywide quality control, total maintenance, and so on. Clearly, these goals should be established first without considering existing constraints, such as an aging work force, strong unions, or outdated equipment, which are considered later in consultation with the rest of the work force.
- Communicating goals (and necessary changes) to employees is prerequisite to work force participation. Initial education might be necessary for the work force to correctly understand and interpret the view of the management team.
- Involving the work force and reaching consensus consists of finding proposals for changes that are accepted by management, in terms of its final goals (customer satisfaction and profit), and by the work force, in terms of the future working environment. This phase's objective is to gain everyone's support for the changes. Consensus can be reached by adjusting existing conditions according to the desired changes (for example, by retraining the work force)

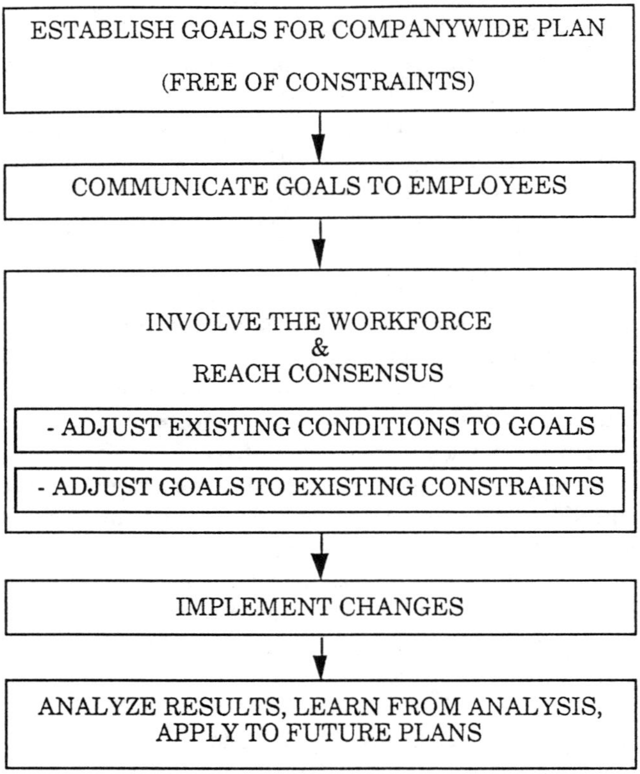

Figure 2.5. Plan for Initiating Employee Involvement.

and by adjusting the changes to existing constraints (for example, by delaying the introduction of state-of-the-art technologies such as computer-integrated manufacturing (CIM) because of a lack of in-house potential and outside success). Common efforts need to be shared among all company employees to enable the changes to take place. Crucial to worker participation development, this phase requires the circulating information between top management, employees, and employee representatives (unions). Communication channels need to be created for employees to provide their input to management and for management to communicate options and decisions to employees. Parallel communication channels can coexist through the conventional hierarchy, through negotiation with worker representatives, and through departmental team meetings. Educating the work force to be able to discuss management propositions and to give valuable feedback is a requirement for those companies without a tradition of employee consultation.

- Implementing changes is a natural consequence of having reached a consensus. At this point, management exerts little direct responsibility, and most of the changes are implemented by teams of employees associated with technical experts under the guidance of project leaders that report to the management team. Examples of change implementations are the design of manufacturing cells, the reduction of setup times, the reorganization of maintenance activities, and the implementation of statistical process control.
- Analyzing results, learning, and applying the conclusions to future plans is the last phase of the current plan in which management collects information from those involved about what went well or wrong during the change process, analyzes what could be improved, and applies the conclusion of these analyses to future plans.

A last but crucial condition for the success of EI is early union involvement in the EI program development. Unless they are asked to be involved, worker organizations might see the creation of employee teams and the request for worker participation by management as a threat to their authority and influence on employees and, thus, energetically combat the EI program. The approach to use with unions is similar to the approach to use with workers: Inform employee representatives as early as possible and involve them in the subsequent program-implementation phases. Company management should negotiate early with union leaders to gain their support. Since layoffs reduce union membership, it is in the union's interest to adhere to a program that aims at strengthening the company's position in the race against competitors and therefore also avoids the loss of business. Management objectives need to be precise and clearly communicated to avoid surprises and frustration. Some companies can afford to sign "no layoff" union contracts when they feel confident enough of their future position in the market. Signing such contracts, however, should not be mandatory in order to gain union support. In some companies, management and unions have worked together with the interesting result that the union organization has been the conduit to establish communication between management and employees: Union newsletters and information meetings have been a way for management to communicate with employees and involve its work force in proposed changes.

Clearly defining objectives, involving worker organizations, and preparing the work force through communication and training are preliminary steps in starting pilot programs.

2.2.2 Starting EI Pilot Programs

Starting the EI pilot programs consists of creating an initial structure that is able to support EI; that is, one that enables employees to analyze problems (such as current operational problems, proposed changes, et cetera) and communicate suggestions to management. The only way to involve the employees, to let them analyze problems, to achieve impressive improvement is through teamwork. No other approach can deliver significant and continuous results. A structure must be built based on a certain number of employee teams, which are able to communicate with management. Building this structure involves three steps:

1. Define responsibilities at the various levels of the EI structure.
2. Plan the implementation of the EI teams.
3. Implement the EI pilot teams.

These steps are detailed in the following paragraphs.

Define responsibilities at the various levels of the EI structure. Responsibilities regarding the functioning of the EI program must be defined at the company's top management level. Responsibilities must also be defined at the team level.

At the company's top management level, a group of individuals in charge of initiating and supporting the EI program is formed, as illustrated in Figure 2.6. This group, which is typically referred as the *steering committee*, is in charge of coordinating, managing, and fostering the EI program. This committee, which reports to top management, has the authority to make strategic decisions and to commit resources. In small companies, the CEO can play the role of the steering committee. In larger organizations, the steering committee includes operational managers, outside consultants, human resources facilitators (in charge of defining specific training needs), and technical advisors. Except for the technical advisors, committee members are permanent members. This committee can also be the same as—or part of—the committee in charge of monitoring the implementation of the other steps of the WCM strategic plan. The steering committee meets regularly (perhaps monthly) and remains available for additional meetings during the planning and implementation phases. The steering committee's role is to support, facilitate, and monitor action, but not necessarily to take action. As is also illustrated in Figure 2.6, an *EI program leader* should be appointed to carry out all the concrete actions necessary for the implementing the EI program. This is likely to be a full-time position during implementation; therefore, the EI program leader should be in charge of only the EI program implementation. The EI program leader, who is also responsible for forming groups, reports to the steering committee.

At the team level, *team leaders* must be designated first. Ideal team leaders are persons who enjoy communicating with other people, are open to new ideas, are able to lead group discussions and use problem-solving techniques, and are able to train team members in these techniques. They should also be able to communicate enthusiasm to the other members of the team. Team leaders play an important role in preparing meetings and in the transmitting information to the program leader

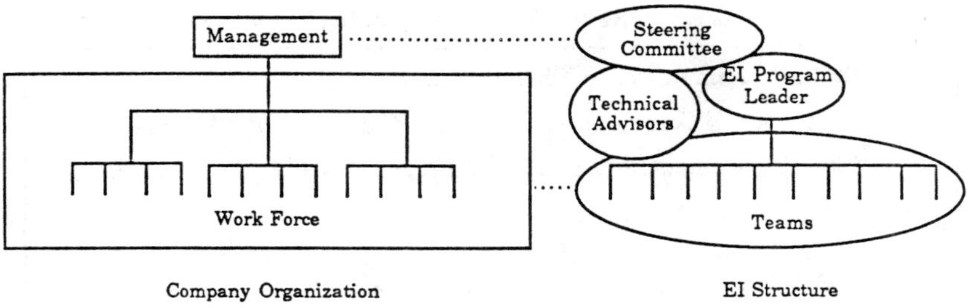

Figure 2.6. The Place of the EI Structure in the Company Organization.

and to the steering committee. Team leaders can be chosen by the steering committee from among staff employees and line supervisors. When preparing for and during team meetings, team leaders are assisted by *technical advisors* (who can be the same, or be different people from the technical advisors to the steering committee). Technical advisors are not permanent members of a specific team; they are called to a team meeting when a specific expertise is required for solving a problem. For example, a quality engineer could be called in to help a team that is improving an inspection procedure. A list of potential technical advisors for each domain of expertise should be established by the EI program leader in cooperation with department managers.

Steering committee members, the EI program leader, and—when designated—the team leaders are then involved in the next step.

Plan the implementation of the EI teams. First, the pilot area needs to be chosen. EI teams should not be created throughout the entire company at once. Smaller scale implementation problems are easier to solve if only a part of the company is initially concerned by the organizational changes, which minimizes the risk of failure. The mistakes made in the initial phases have limited consequences as they are noticed and corrected easily, and thus they are not repeated when the program is extended companywide. Success in the pilot area facilitates companywide implementation. The division or department chosen as the pilot area should be one where success is most likely to occur, where the manager is a strong supporter of the EI program, and where employees are most likely to actively accept the program. Evidently, if the pilot project approach is chosen for implementing the different steps to WCM (this approach is described in chapter 1), the pilot area for the EI program is the same as the pilot area chosen for the overall project.

The team leaders are then chosen and are involved in addressing the remaining issues of the implementation plan. These issues are listed in Table 2.3.

Company managers, steering committee members, and program leaders should be aware of the potential obstacles (due to the particular conditions of their

TABLE 2.3. Issues to be Addressed when Planning the Implementation of EI Teams.

- Define Communication Channels and
  Frequency of Communication between the Different Participants (Steering Committee and Teams, Program Leader and Teams, Between Teams, etc.)
- Define Level of Authority Given to Groups
  (Limited to Making Suggestions, Study of Solutions, Implementation of Solutions)
- Define How Suggestions or Solutions Will Be Evaluated and Implemented
- Define How Conflicts about Suggestions and Solutions Will Be Arbitrated and Resolved (Between Teams and Steering Committee, Between teams, etc.)
- Define What Resources Will Be Allocated to Groups
  (Time, Facilities, Supplies, Budget)

companies) that can thwart implementing EI teams and thus be prepared to act accordingly.

When there is a lack of employee motivation to participate in work teams, for example, more time should be dedicated to demonstrating the potential benefits of EI. In addition, sending employees to visit companies that have a history of successful EI programs, emphasizing training for EI, and the voluntary participation in work teams have all proved helpful in swaying the reluctant participant.

To avoid a potential rejection of the program by unions, union leaders should be involved as early as possible; in fact, even before the steering committee members are chosen. Objectives and the means to reach those objectives should be communicated to union leaders and discussed with them. Company management should carefully select skillful negotiators when appointing steering committee members and the EI program leader. In the context of the evolution of the company toward a new culture, it is important that a consensus is reached between the company and the union. (A major U.S. corporation famous for its synthetic fibers had to dissolve work teams and cancel EI plans in one of its plants after having lost a lawsuit filed by the unions because teams were formed in spite of union opposition.)

If there is a potential risk of lack of top-management support (for example, a risk that suggestions will not be timely reviewed or that employee input will not be considered when implementing changes in the work place), then the steering committee members and the program leader must be aware of this problem and confront the issue with company management.

Lack of resources or delays in resource allocation can result in a slow start for EI activities. The employees' and technical advisors' time, budgets, rooms, and supplies should be available when the first teams are formed and ready to meet.

These examples show various types of problems that result in suboptimal conditions for starting the EI program, which could result in frustrating the people in charge of the program implementation and the employees involved in pilot teams. These types of problems should be expected and resolved prior to implementation.

Implement the EI pilot teams. Implementation consists of the following four steps:

1. Informing the people in the area chosen as the pilot area.
2. Forming the teams.
3. Training the team members.
4. Allowing the teams to work and analyze team performance.

When informing future team members, the importance of EI as one of the steps to WCM should be emphasized. The functioning of the EI structure should be described, and the different people involved in the structure introduced. Teams are then formed, either inside or across departments. (Section 2.2.6 further details the formation of work teams and other practical issues such as the size of the teams and meeting frequency.) To foster the success of the pilot teams, it is crucial to select team members on the basis of a minimum level of competence and motivation. The pilot teams' role is to pave the way to a successful companywide EI program. During the first meeting, the team leader recalls the objectives of the EI program defined by company management. If necessary, time should be dedicated during this meeting to training team members about problem-solving techniques, possibly by working on an example. During the pilot phase, team leaders and the EI program leader meet regularly to discuss the project's progress and problems, which then are reported to the steering committee for evaluation.

2.2.3 Evaluating the Results

Evaluating the results of the EI program needs to occur during and at the conclusion of the pilot phase in order to guide the extension of the program to the rest of the company. Evaluation is also carried out in a continuous manner after extension. The EI process itself is difficult to evaluate because there is no precise method for evaluating how well the employees accept and support the EI program, which renders the evaluation not very significant in terms of end results for the company. More meaningful than the process used by the team to reach the results is the evaluation of the final results of a team's work. For example, if a team is in charge of a setup time reduction project, performance evaluation consists of looking at the percentage of overall setup time reduction, the time spent for the studies, and the types of proposed improvements, rather than focusing on an assessment of an individual's performance during the setup project. Similarly, if a team is asked to contribute to the continuous improvement of a department's operations, the performance of the team can be indirectly appreciated by looking at the department's overall performance improvement. When EI becomes a natural way of working, its results cannot be evaluated independently from operational results. If teamwork does not result in improving performance or does not enable the satisfactory completion of assigned projects, then the team behavior itself should be discussed by the steering committee, the EI program leader, and the specific team leaders. At the end of the pilot phase, when initial mistakes have been corrected and steering committee members judge that results are satisfactory, the EI program can be extended to the rest of the company.

2.2.4 Extending the EI Program to the Rest of the Company

Extending the EI program to the rest of the company occurs after a pilot period of several months and only after satisfactory results are obtained from the work of the pilot group. The way the EI program is extended to the rest of the company depends on the approach chosen by management for the implementation of the overall WCM program.

When the pilot-project approach is chosen for the WCM program, EI will be extended to other company departments or divisions as they enter into the scope of the WCM program, typically occurring a year to three years after the beginning of the WCM program. When EI is chosen as one of the most beneficial steps to implement before proceeding to other steps to WCM, then new groups are formed as recommended to top management by the steering committee, typically occurring six months to a year after the pilot group's formation.

After forming new groups, the steering committee continues to monitor the results of the EI program and to report its progress to management. Groups are formed or disbanded when judged appropriate; group composition can also be modified. The EI structure is a dynamic structure that evolves according to the company's needs. One of the steering committee's responsibilities is to verify that the EI structure and its functions correspond to these needs.

2.2.5 Common Pitfalls of EI Programs

Many companies that were truly convinced of the benefits of involving the work force, but did not know how to make this involvement happen, have suffered serious setbacks leading to frustration and results that oppose their initial goals. The following paragraphs review common pitfalls experienced by companies that did not realize they were playing with fire, or that confounded EI and paternalism, or did not give enough time for the team to meet and discuss, or that did not believe in the team approach.

Implementing EI is very much like playing with fire. Clear limits delineating the extent of the EI program have to be set to avoid expectations that cannot be fulfilled, which in turn, lead to both employee and management frustration. Budget limits given for specific projects, the type of improvements, or expected suggestions should be specified in advance. The role of employee teams is not to suggest investment in expensive technology or to outline a new organizational structure, but to propose ways to better implement these technological and organizational changes suggested by management and to suggest low-cost/no-cost solutions for improving existing operations.

Paternalism is not an appropriate approach to EI. The objective of EI is not to make the employees feel good, thereby inducing them to be more productive. This objective cannot lead to durable improvements. Rather, what is asked of employees is to contribute to the organization's improvement by sharing their knowledge of current operations and providing information that they are often the only ones to possess. As a result of the organization's improvement, it is expected that overall productivity will increase and perhaps that employees' will feel rewarded in using solutions that they have helped develop. The evolution of employee's personal feelings regarding their work is a probable result and not a planned means.

Teams do not have enough time to meet. Teamwork takes time and management should provide enough time (during work hours) so that team members can meet and work together. If teamwork is to become a new way of life, then it has to take place during the regular company life and adequate resources should be allocated to the teams to work effectively.

EI should not rely on individual-based approaches. More specifically, EI should not rely on traditional individual-based suggestion systems for continuous improvements. Suggestion systems have been used by companies for years. Some world-class companies like Toyota are still using them, but usually in combination with team-based approaches. Experience shows that individual-based suggestion systems go against employee involvement, as is demonstrated in the following arguments:

- Most of the existing suggestion systems reward individuals. This practice is contrary to the objective of having the individuals inside the company share common objectives and act as a team by sharing common efforts.
- Suggestion systems contribute to maintaining the idea that improvements are

not part of the employees' normal duties and that they should be specially rewarded.

- Flaws in the suggestion evaluation procedures can result in raising unrealistic expectations that lead to long-lasting frustration. In a major automotive company a worker suggested the replacement of a manual operation by a robot. The suggestion was rejected. Two years later a new engineer, not knowing of the suggestion, installs the robot. The employee still has the feeling that the suggestion was stolen from him.
- Suggestion systems can actually slow down the release of suggestions. Often, a yearly award limit is fixed per employee, and employees retain their second or third suggestion for the next year, rather than risking a "free suggestion."
- Finally, by putting the emphasis on personal gain, suggestion systems lead to damaging team spirit by provoking rivalry and jealousy among groups of employees.

Individual approaches cannot be successful unless the people place more priority on the interest of the firm than on their own. Suggestions should not require additional reward to regular remuneration. Continuous improvement and the suggestions that lead to it fall inside the definition of new job responsibilities. Suggestions should be made by individuals during regular group meetings, team approaches being the only way to foster employee involvement.

2.2.6 Team Approaches

As stated earlier, teamwork is the only serious approach to make EI happen. To develop effective teams and implement productive team approaches it is important to understand (1) the reason for working in teams, (2) what the different types of teams are, and (3) how to work in teams.

Why work in teams? For any organization, there exists many reasons why forming teams and working in teams can be beneficial. By grouping together persons who have different views of the same problems, more ideas can be proposed (for instance through brainstorming), more solutions can be compared, and consequently better solutions can be identified. In addition, a team can solve larger and more difficult problems by dividing the work load among its members, and problems that cross organizational boundaries can be more easily analyzed. When employees can propose changes, they are more likely to adopt these changes and further improve them when they are implemented in the work place. Furthermore, in organizations that are already composed of numerous functional groups of employees, it appears logical to provide these employees with an opportunity to regularly meet to discuss common problems and propose and implement solutions. Working in teams is already a common way of working for many companies.

The two types of teams. Clearly, there is a need to involve employees in (1) continuously improving the working environment, and (2) solving specific problems that management views as important for the company's future. To respond to these different needs, two types of teams must be created: *functional teams* and *project teams.*

Functional (or departmental) teams are formed within a department or a division, as illustrated in Figure 2.7. The mission of a departmental team is to continuously improve the working environment within the department. This type of team is permanent and usually led by the department manager, and its members belong to the same department and usually work together. Technical advisors external to the department can be invited to the meetings and bring expertise to solve specific problems. Functional teams should exist at different levels of the company hierarchy. Teams can be composed of production workers, engineers, purchasing agents, department managers, and so on. A department manager is likely to be part of more than one functional team: He or she will be a member of a department-managers team and will lead one or several teams within his or her own department.

Participation in functional teams can be voluntary, although it is highly desirable that everyone in a department be involved in a team. The best solution is to begin with voluntary participation, and based on the success of the initial teams, to form more teams when more employees volunteer.

Problems analyzed in functional team meetings can involve issues internal to the department or issues that cross department boundaries. Functional teams are in charge of implementing solutions internal to the department after management's approval. Problems crossing department boundaries are transmitted to a project team, which can be created specifically for dealing with that problem.

Contrary to functional teams, project teams are composed of individuals from different functional areas or departments and usually from different levels in the company hierarchy, as illustrated in Figure 2.8. The mission of project teams is to solve specific problems assigned by management. Project teams are temporary and

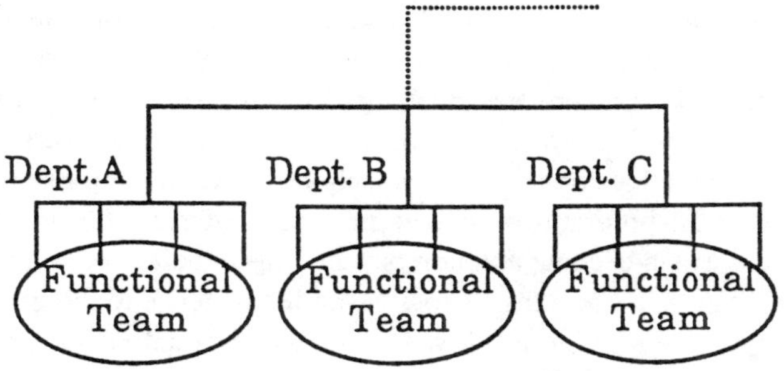

Figure 2.7. Formation of Functional Teams.

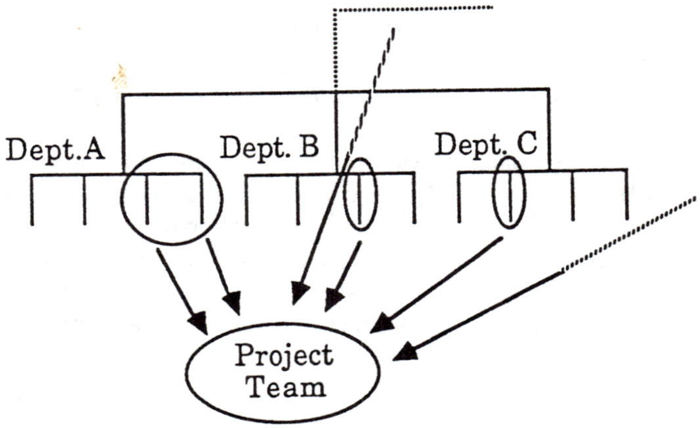

Figure 2.8. Formation of Project Teams.

disband when the solution to the problem is implemented. Members of temporary project teams are most likely to also be members of permanent functional teams.

Although team composition and the scope of the problems are different for both types of teams, the issues involved in team forming and functioning are similar.

How to work in teams? To answer this question, the following issues must be addressed and understood:

1. Team composition
2. Team resources
3. Team effectiveness
4. Team evolution

Team composition. Teams are usually composed of a core of permanent members to which temporary members are added. The permanent members are part of the team as long as it exists. The permanent core should not exceed six to eight members in addition to the team leader. Beyond this number teams usually become unmanageable. One of the team members is in charge of taking notes and helps the team leader prepare the meetings. This secretary role can be played successively by the different team members. Nonpermanent members are invited to one or several meetings by the team as technical advisors. For example, at the beginning of the EI program, an expert in teamwork can help the teams get started; then the teams become autonomous and this expert is no longer needed. Other advisors can then bring their expertise and experience to the group to help solve specific problems. An employee from the accounting department, for example, could explain to the team members how to cost justify a proposed investment.

The choice of team members is different depending on the type of team. Since a project team is formed to address a preassigned problem, its members are chosen from those best able to analyze the different issues involved in the problem. For example, a team in charge of simplifying and automating order entry procedures gathers people from the order entry, account receivables, central production planning, marketing, and customer service departments, each member bringing their original department perspective to the team discussions. Team member participation is usually subject to his or her immediate supervisor's approval.

To ensure that teams function effectively, sufficient resources must be allocated to them.

Team resources. The most important resources of a team are its members' technical skills and knowledge, which allow team members to contribute to solving problems. Training is essential for working in teams. Some of the team members are likely to have been part of a team. Training needs depend on the education level of team members. Some employees will need to be taught basic writing and math. Guidelines given in the next section on work force education will be helpful in planning these training programs. Specific techniques that team members need to be taught are problem-solving techniques (such as the use of Pareto charts, Ishikawa diagrams), data collection and analysis, report writing, preparing and delivering, oral presentations and group dynamics. The team leaders should be trained first, by external consultants for example, and then be actively involved in training their team members. These new skills can be acquired during the initial meetings or during specific training sessions. Other important resources that should be provided to work teams are (1) time, (2) budget, (3) external help, and (4) facilities for meeting.

Time is necessary for regular meetings. Since the management's goal is for EI to become a normal way of working, the meetings should take place during work hours. Meetings should last for about one hour, and their frequency should be one every week or every two weeks, with additional time given to team leaders and technical advisors for preparing the meetings. The team leader also must meet with the EI program leader and the steering committee members to report progress.

Budget should be allocated for trying and implementing solutions. Simple low-cost solutions should be tried by the team without the need for lengthy approval procedures.

External help such as outside consultants or experts from inside the company who are asked to participate as technical advisors to the teams also need to be planned.

Finally, appropriate facilities must be allocated for team meetings. Providing teams with sufficient resources will foster their effectiveness.

Team effectiveness. Once the team leader has formed his or her team and resources are allocated, the real work can start. First, teams need to set goals for their work, which is expressed in a formal mission statement. Examples of mission statements are (1) for a functional team: "Continuously improve the performance of our department by carefully analyzing and improving each of the operations involved in the department process," and (2) for a project team: "Study alternative workcell configurations for grouping the machines in department A, evaluate alternatives, and present evaluated alternatives to management." To reach these goals in the most effective manner, team members and the team leader must have a clear vision of the different steps involved; a structured approach must be defined and adopted by the team. An example of a structure approach that can be adopted for any process improvement is illustrated in Table 2.4.

By adopting a structured approach, team members are able to identify various obstacles to meeting an assigned objective and to propose suggestions to deal with each obstacle. In addition to the chosen approach, meeting preparation, record keeping, and the way conflicts are handled are crucial to a team's effectiveness.

Because meetings should always be prepared, the team leader needs to establish an agenda prior to the meeting. This agenda prevents the discussion from departing from the issues relevant to the problems at hand. When preparing the agenda, the assistance of external advisors—if invited to a meeting—might be helpful. A short informational meeting between the team leader and the technical advisors allows

TABLE 2.4. Team-Oriented Eight-Step Sequence to Process Improvement.

•1 Define the Process to Improve.
•2 Identify Inputs to the Process.
•3 Identify Customers Expectations Regarding Outputs from the Process.
•4 Define Objectives for the Process and Define how to Measure Them.
•5 Remove Obstacles on the Way to Objectives.
•6 Communicate the Improvement Plan to Management
•7 Implement Solutions.
•8 Measure Improvements and Adjust Solutions.

them to come prepared to the meeting and to be readily effective. Minutes are recorded during meetings, and the meeting minutes and team working papers are then classified and stored for future reference as this information is crucial to preparing the improvement plan.

Management expects the teams to make decisions. While making decisions, it is rare that conflicts do not emerge between team members over which decisions to take. Nevertheless, it is actually desirable that conflicts emerge; disagreement and conflict are a natural result when bringing together a group of people having different perspectives. The confrontation of ideas generally leads to better ideas, and this is one of the reasons for analyzing problems in groups. However, to be effective, team members must agree on a system that enables them to make a final choice. That system can be chosen among various existing systems, such as voting, either with unanimous vote or majority vote. The team leader should ensure that conflicts remain at the idea and issue level and do not become personal by involving team members' personalities.

These factors that influence team effectiveness are tangible factors. However, team effectiveness also depends on less tangible factors such as at what evolutionary stage the team is currently.

Team evolution. How to form teams and how to work in teams has been the subject of much observation and research, which promotes understanding of how teams evolve and how their effectiveness is affected by this evolution. Before reaching maturity and optimum effectiveness, teams go through various stages. These stages must be understood and recognized by the EI program leader, the members of the steering committee, the team leader, and by the team members themselves in order to understand and foster the current level of effectiveness of the various teams. For example, a group at a critical conflict stage might need assistance to go beyond this stage, or a group might inadvertently be locked in the illusion stage and thus make unrealistic decisions. The successive stages and the corresponding levels of team effectiveness are listed in Table 2.5 and are subsequently detailed.

1. *Formation.* The team members get to know each other as members of a team with common objectives,defining and accepting the team mission, and participating in the initial teamwork training. At this initial stage effectiveness is generally low.
2. *Conflict.* The first difficulties appear in teamwork. Due to the team members' various backgrounds, experiences, and origins, conflicts over ideas evolve into personality conflicts. In addition, the team now has a clear vision of the amount of work with which it is confronted. To prevent team explosion at this crucial stage, the team leader must actively work to resolve these conflicts (for example, by personal intervention during meetings or

TABLE 2.5. Successive Team Evolutionary Stages and their Corresponding Levels of Effectiveness.

Evolutionary Stage	Team Effectiveness
•1 Formation	Low
•2 Conflict	Low
•3 Cohesion	High
•4 Illusion	Lower
•5 Disillusion	Low
•6 Performance	Optimum

Adapted from: Mamagement by Randall B. Dunham & Jon L. Pierce, Scott, Foresman and Company, Glenview, Illinois, pp 470–474.

by personal off-meeting contacts with the members in conflict). Contrary to conflicts over ideas, conflicts over personalities result in low team effectiveness.

3. *Cohesions.* The initial personality conflicts have been resolved, and the group starts to function as a real team. Brainstorming sessions are fruitful, arguments develop over alternative solutions, and realistic solutions to problems are chosen. Effectiveness improves. Members realize the benefits of working in teams for solving complex problems. It is common that short-lived teams, such as project teams, remain at this stage and disband before reaching the illusion stage.
4. *Illusion.* The team is now so coherent that it starts to function as a single body. Every team member seems to share the same opinion, and team members truly enjoy working together. Thus, alternative solutions are not expressed for fear of altering team cohesion, and the team rejects external advice that is contrary to the team's position. These irrational choices lead to lower effectiveness.
5. *Disillusion.* As a result of the irrational decisions made at the illusion stage, internal problems appear that lead the team into the disillusion stage. Team members are frustrated by these results and become disillusioned about the work performed in teams, thereby further decreasing team effectiveness. However, by analyzing the internal problems and solving them with the team, a skillful team leader can lead its team from disillusion back to performance.
6. *Performance.* At this stage the team is mature. Team members know each other well, and from having experienced the previous stages they have learned to know their limits as a team and are again confident about the quality of the work done in teams. The team remains open to external advice and alternative solutions are confronted during meetings. Team effective-

ness reaches its optimum. Going through the five previous stages is necessary for the team to become mature and eventually reach the performance stage. The team leader has to be careful to avoid allowing the team to fall back from the performance stage to the illusion and disillusion stages by encouraging conflicts over ideas, not personalities, and by letting external advisors review and influence the work done by the team.

This section on EI has shown that the contribution of every individual—through teamwork—is a necessity for companies that want to continuously improve their operations and continue on their way to becoming mean, lean, and world-class competitive. Improving the working environment therefore must become an intrinsic part of everyone's responsibilities. Because managers alone cannot resolve all operational problems without the assistance of the rest of the work force, EI becomes essential to the long-term survival of companies competing on a global market. When starting an EI program, pilot teams are created first, and the results of the pilot phase analyzed before extending the program to the rest of the company. Early information and the involvement of union leaders are crucial to the success of the EI program. In addition, patience, mutual trust, and mutual respect are initial conditions for EI and prerequisites to the formation of the first teams. There is a distinction between types of teams: There are project teams that are temporary and cross organization boundaries, and there are functional teams, whose permanent members come from the same department or division. To enable the individuals in either type of team to contribute effectively to common efforts and thereby reach common goals, team composition, resources, effectiveness, and evolution are all issues that must be understood by managers and other employees in charge of the EI program.

2.3 WHY AND HOW TO EDUCATE THE WORK FORCE

Developing the ability and knowledge of a work force is another aspect of a continuous improvement strategy. Similar to EI which must take place at every level of the hierarchy, continuous education is needed for production employees as well as top managers. Education through regular training enables the company to acquire and retain the ability to compete in a continuously changing environment, for example, by familiarizing the production operators with new production processes or production managers with different accounting procedures. When a company decides to drastically alter its operating modes, employee education is more than ever a necessity. The entire work force must be prepared to adopt the new culture defined by top management and be willing to—and able to—actively be involved in its implementation. Educating the company's work force is indeed investing in the company's resources and must be envisioned with long-term objectives in addition to short-term results. Training programs must be carefully integrated with the company's strategic plan so that the company will experience desired results and will meet initial objectives while avoiding traditional pitfalls. This section presents a structured approach to educating the company's work force. It details how to plan and conduct training programs, and how resources already existing in the company can help top management and functional managers to educate their employees. It also presents the pitfalls that should be avoided in order to obtain the maximum return from the invested resources.

2.3.1 Investing in Human Resources

People working in a company are not merely an available supply of manpower for executing limited-scope, predefined tasks, but represent a resource that must be counted on to improve the working environment and thereby also improve the company's competitiveness. If EI is characterized by involving the work force in the elimination of problems, as was illustrated by Figure 2.4, educating the work force aims at enhancing each employee's ability to eliminate those problems, as illustrated in Figure 2.9.

A common approach used for improving employee education is internal training, including on-the-job training (OJT), as well as off-the-job training (OFFJT). Other approaches that affect the level of employee education when hiring are apprenticeship programs and the participation of companies in school or college programs. This section deals essentially with the internal training approach—educating employees that are already part of the company.

Many managers, unfortunately including top managers, who are requested to show evidence of short-term profits are not inclined to invest extensively in training because (1) the return on investment (ROI) of training programs is difficult

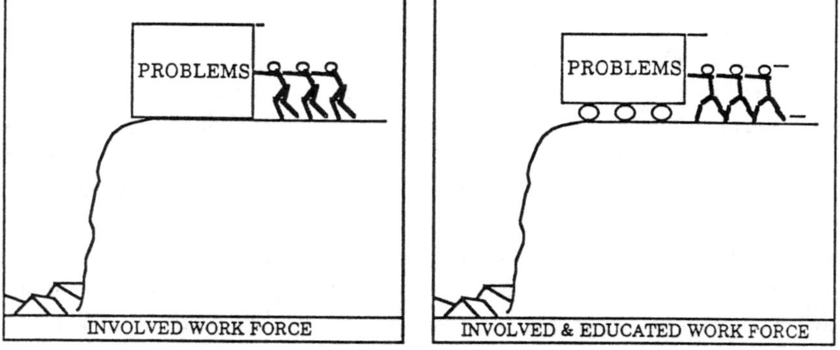

Figure 2.9. Improved Ability of the Employee to Eliminate Problems.

to measure and, when estimated, appears to exceed the company's standard ROI, and (2) these managers feel that there is a risk of losing the investment if the trained employees leave the company. This reasoning indicates that managers need to adopt a long-range viewpoint to plan work force education and, in addition, that educating company managers should have top priority. Frequently, in companies that are struggling to survive, a very low priority is given to training because the managers do not have time to plan the training and the employees are not available. Training is foreign to the "fire-fighting" mode of operations, and notably, the fire-fighting mode of operations is not encountered in world-class companies. Also struggling in highly competitive environments, these companies strive instead to implement working conditions such that they do not need to fight last minute fires: When a boat is designed to sail safety in rough weather, there is less risk of trying to rescue it from sinking. This is also the aim of training.

Education comes at a cost: The cost of paying employees while they are not working on the shop floor, the cost of the training programs themselves, the cost of the time spent by experienced employees to share their knowledge with less experienced ones, and the cost of planning the training. However, when adopting a long-range viewpoint it is clear that these costs are not merely expenses but are instead investments. A better trained, more self-confident work force is an additional competitive advantage. It is not uncommon for world-class companies to invest up to an equivalent of 10 per cent of their total salaries and wages in work force education. Competition requires the companies to adapt rapidly to changes. (For example, unprofitable product lines have to be stopped and replaced by new ones.) An educated work force understands that this sort of change is a normal way for a company to do business and does not fear the change, comfortable in the knowledge that it has the ability to be retrained for the new jobs. A company will not advance on the way to WCM if it leaves its personnel behind. Every step described in this book requires employee support—for which the employees must be prepared.

Occasional employee reluctance to training must be overcome. When training becomes strategic for a company, all employees must adhere to this strategy. Adequate information from management, as part of the implementation of a new culture for the company, should make clear what benefits employees will gain from the training programs they will follow. (The most tangible benefit could be increased wages, if a "pay-for-knowledge" paying system is implemented in the company.)

At this point, it is crucial that managers understand the importance of transforming the work force from a group opposed to changes and productivity improvements to an ally who sees the company's development as a way to enhance personal development. Employees becoming an ally of management against external assaults is the major benefit derived from work force education. A loyal work force is a by-product of this alliance, and as long as managers and employees are competitively rewarded and see potential for personal development, they will not leave the company. Human resource development is a process that benefits from itself (as illustrated in Figure 2.10), resulting in improved company competitiveness, which leads to a larger market share, which in turn, brings more revenues to the company—revenues that can be reinvested, in part, in improved training. The training investment, however, will pay off only if careful attention is paid to the training process itself: Who should receive what training, and How should training be provided? Fruitful training cannot be achieved without appropriate planning.

2.3.2 Planning Training Programs

It is natural for most people to reject that for which they are not prepared. Resistance to change is to be expected if employees are not well prepared to accept the changes and are not to be involved in their implementation. One might expect that people would be attracted by the novelty of working in a new environment or be eager to learn new techniques. However, if the employees do not clearly see what personal benefits they can obtain from the proposed change, they are likely to resist any alteration of the interests they currently hold. For example, before reallocating personnel from a centralized manufacturing department to decentralized product-line support groups, one of the first reasons for the employees to resist the change is the perception that the closer the people are located to the production equipment and personnel, the less is their prestige in the company. Preparation and

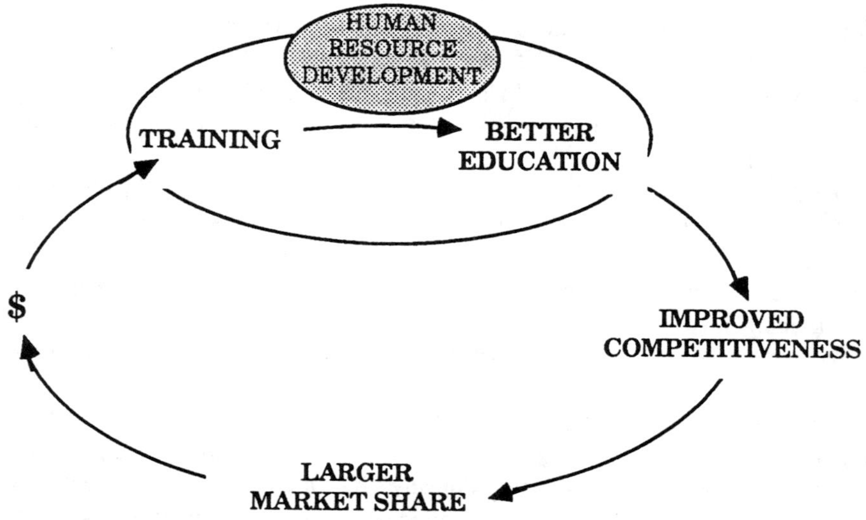

Figure 2.10. Training Cost-Benefits Cycle.

explanation of advantages, such as more varied and interesting job tasks, a more motivating work environment, and increased autonomy are necessary for these employees to accept and be involved in the changes. Training, by increasing the knowledge and skills of the employees and by modifying their attitude, prepares and provides people with the means to embrace such changes. Nevertheless, careful preparation for training itself is necessary to avoid facing partial or complete rejection by the trainees who may waste, in part or completely, the training investment. This preparation requires that management have a clear vision of the training's aim and to implement the conditions necessary for fruitful training. In addition, since training is a continuous process, the training's efficiency should be continuously monitored and improved.

Every company has some experience with employee training. In manufacturing companies, process planners have been trained to use computer-aided manufacturing (CAM) systems to prepare programs for computer numerical control (CNC) machines, and machine operators have been trained to use these CNC machines. Often employee training has a limited scope, such as learning a particular technique or process, which generally results in increased employee performance because the employee has to immediately use the learned technique. This type of training usually requires only minimum preparation. When the scope of change is broader, such as implementing a new accounting system with tens of users, then the training must be more carefully planned and prepared. Furthermore, when the objective is to implement the changes described in this book, which demand a radical cultural change across the company, then sound training organization is mandatory.

The following paragraphs describe how to plan the training programs prior to implementing the different steps to WCM. It is clear that this methodology can also be applied with success to training with a more limited scope. Before detailing the different steps of training programs, the necessary conditions for successful training programs must be understood.

The Conditions Necessary for Successful Training Programs

Implementing the necessary conditions for successful training programs is the responsibility of top management. Depending on the size of the company,top management can designate a training program coordinator. The training program coordinator must have identical authority as top management over division and department managers about all training matters, so that short-term department requirements do not prevent employees from participating in training programs—an essential element of the company's long-term strategy. Conditions necessary for successful training programs are listed in Table 2.6 and developed in the following paragraphs.

Define the overall training objective and benefits. The overall objective for

TABLE 2.6. Conditions for Successful Training Programs.

1) Define the Overall Training Objective and Benefits
2) Develop Program Steps
3) Adopt a Top-Down Approach
4) Give Managers Responsibility for Training
5) Evaluate the Overall Training Effectiveness

training and the resulting benefits should be that all the company employees gradually gain the knowledge, skills, and attitudes necessary for accepting and supporting the organizational changes toward WCM and eventually be actively involved in their implementation. In the larger context of fostering the company's long-term development, training should be considered as one of the ways to develop the company's human resources. Attaining world-class status is only one of the phases of the company's development. Top management should establish training plans according to its vision of what the company should be in the future.

Develop program steps. Once the training program's overall objective has been defined, the successive steps to reach this objective are developed. These steps cover (1) informing personnel regarding the planned training, (2) assessing the training needs, (3) developing specific objectives for training, and (4) selecting training methods and training the employees. The different steps, which are detailed in the next section, can be applied to any training program in the company, from learning how to use a new machine to learning effective leadership of team discussions.

Adopt a top-down approach. Managers are trained first—a condition that is *essential* in the context of the evolution to adopt a world-class culture. It is wasteful to train production operators to work in teams if their managers are not prepared to delegate authority to these teams. The decentralization of activities for each product line will yield full benefits only if the new decentralized department managers are trained to act as small independent enterprise managers. Training starts with the company's top management and progresses downward to hit each subsequent level when that level immediately above has acquired the potential to support the new company culture.

Give managers responsibility for training. Once the managers have acquired the new attitude to foster employee development, the authority to decide the training needs should be given to managers. At each level of the company organization, managers should be made responsible for training the people working under them: The foremen for the operators' training; the production managers for training foremen, production support staff, and so on. In companies where this system is not in place, training is usually under the responsibility of a central department labeled "Human Resources" or "Personnel." If this is the case, the department's activities should be decentralized into different departments so that the specific training needs of those departments are better fulfilled. A minimal central structure, which is in charge of interdepartment coordination and counseling, should remain active, but the responsibility for deciding the needs for employee training and the choice of training programs and schedules is delegated to the local structures under the responsibility of local managers. This organization is illustrated in Figure 2.11.

Evaluate the overall training effectiveness. This last condition is often neglected; however, when the equivalent of several percent of paid salaries and wages is invested into training, it seems reasonable to check whether the investment is used efficiently. Top management should verify whether the aim of the training program is being met, while lower level managers should check that specific objectives are met.

Checking for the acquisition of skills and knowledge is relatively straightforward and can usually be accomplished at the end of each training program by posttesting the trainees. When the trainees return to work, their manager also evaluates whether their performance has been improved. The evolution of an employee's attitude is evidently more difficult to measure than is the acquisition of new skills. How well the topics on the meeting agenda have been grasped can be assessed by asking the trainees to answer questionnaires once they have returned to their work site, by having them interviewed by their managers, or by

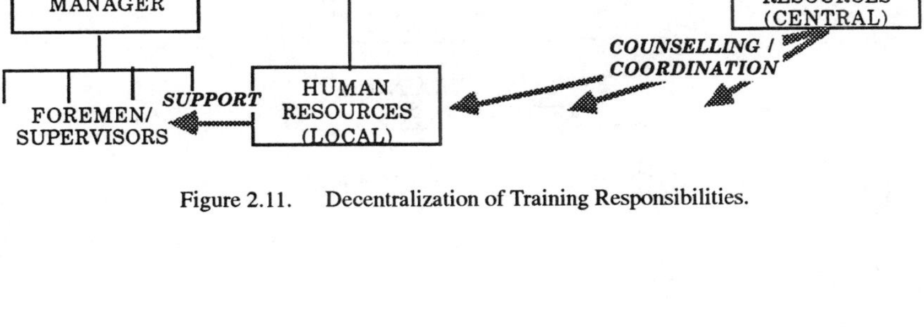

Figure 2.11. Decentralization of Training Responsibilities.

asking them to address these topics during one of the periodic work-group meetings. In addition, the trainees are asked to evaluate each of the programs in which they participate. One of the roles of the remaining central "Human Resources" structure is to consolidate the training evaluations, to guide the managers in the choice of future training, and to improve in-house training.

Once the conditions for successful training programs are met, the different steps involved in the training programs are taken.

The Different Steps of the Training Program

The sequence in which the different steps of the training program occur is illustrated in Figure 2.12, and each step is detailed thereafter.

Inform the future trainees. The goal of this first step is to communicate the overall training objective to the employees. Depending on the scope of the training,

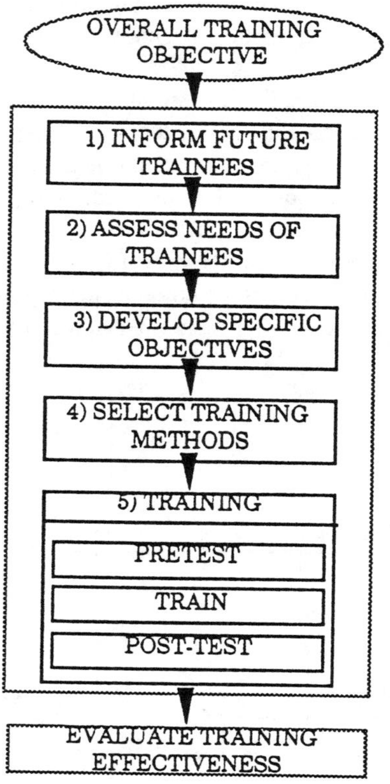

Figure 2.12. Training Program Sequence.

this information can be done by top management or by each manager in his or her department.

Assess the needs of the trainees. This assessment consists of analyzing the gap between what the employees need to know in order to meet the training's end goals and what they currently know. For example, when implementing TQ, production personnel often need to be trained in statistical process control (SPC) techniques, which require a minimum knowledge of mathematics that some operators might not have. Thus, needs assessment is to be conducted for each individual, preferably by his or her direct manager.

Develop specific objectives. Developing specific objectives is a direct outcome of the needs assessment step. It consists of identifying the means to reach the overall objectives. Examples of specific objectives are "Train all operators in department A to SPC techniques." And, "Train department managers in the Custom Products division in leading small-group activity meetings."

Select training methods. Training methods are selected according to the specific training objectives developed in the preceding step. Table 2.7 lists some of the training methods that can be applied.

These training methods should be combined so that the trainees are able to apply the theories they learn in classroom-type sessions. A session on problem-solving techniques, for example, can be closely followed by the actual analysis of job-related problems in small-group activities, or a personal study of materials could be a prerequisite to a classroom session. Job rotation, coaching by supervisors or more experienced colleagues, and temporary transfers to other departments are simple ways of enhancing the employees' knowledge and increasing their flexibility. Off-the-job classroom-type sessions preferably should be conducted by people from inside the company; for instance, training sessions on preventive maintenance can be conducted by maintenance workers or production workers from another department that are familiar with preventive maintenance. Training materials can be prepared with the support of human resource personnel.

Pretest, train, and posttest the trainees. Pretesting and posttesting the trainees is recommended. The aim of these tests is to allow the evaluation of the training program's efficiency. Pretesting can also help to check whether the trainees have the appropriate level of knowledge to follow a particular course of training. In addition, the instructors should monitor the trainees' progress during the training programs.

Careful organization and continuous assessment will ensure that the training programs followed by the employees aim toward the overall objective chosen by the company's top management. Management must clearly understand that an important training effort has to be made when taking the successive steps toward WCM. For companies without a tradition of excellence in training and human resources development, the development of training programs is one of the first of

TABLE 2.7. Training Methods.

In-House Training

- On-the-job training (OJT)
  - Job rotation
  - Coaching
  - Temporary transfers

- Off-the-job training (OFFJT)
  - In-house training sessions conducted by:
    - Production personnel
    - Company instructors
    - Outside Instructors (consultants, specialized instructors, etc.)
  - Personal study

Outside Training

- Visits to companies having experienced similar changes (Get on the road. Avoid the "not-invented-here" syndrome.)

- Seminars (equipment manufacturers, professional organizations, universities, etc.)

- Technical school

- Evening classes at university (within a degree program or not) these steps. These companies should be cautious to avoid the pitfalls commonly experienced that can lead to fruitless training efforts.

2.3.3 Common Pitfalls of Training Programs

Table 2.8 lists the most common pitfalls that companies experience when implementing WCM training programs. The consequences of these pitfalls and ways to avoid them are presented in the subsequent paragraphs.

Not training top and intermediate managers. One of the major mistakes of companies is to assume that they do not need to train or retrain the people in their management structure. When a company changes culture, its evolution can only develop from the top of the hierarchy to the bottom. Starting training at the middle of the hierarchy, or starting at the top and by-passing intermediate levels, results in creating groups with different beliefs and opinions on how the company should evolve. Top management, for example, might have the objective of reducing inventory so work teams might be actively working in that direction by analyzing setups and reducing setup times; meanwhile, department managers might be concurrently pushing for increased machine efficiencies by supporting the purchase of faster machines, possibly contributing to erratic production flows with increased buffer inventories. In addition, if managers are not familiar with the new concepts and methods presented to their subordinates, there is little incentive and support to apply these new methods and concepts on the job when the training is completed. As illustrated by Figure 2.13, training and education for WCM starts at the top of the hierarchy pyramid and progresses toward the bottom without interruption at any level. It is top management's responsibility to enforce the top-down approach.

Forgetting the fundamentals. Needs assessment should identify the individuals'

TABLE 2.8. Common WCM Training Program Pitfalls.

1) Not training top and intermediate managers.
2) Forgetting the fundamentals.
3) Training everyone in everything.
4) Using a shotgun approach.
5) Providing irrelevant training.
6) Not explaining "how to."

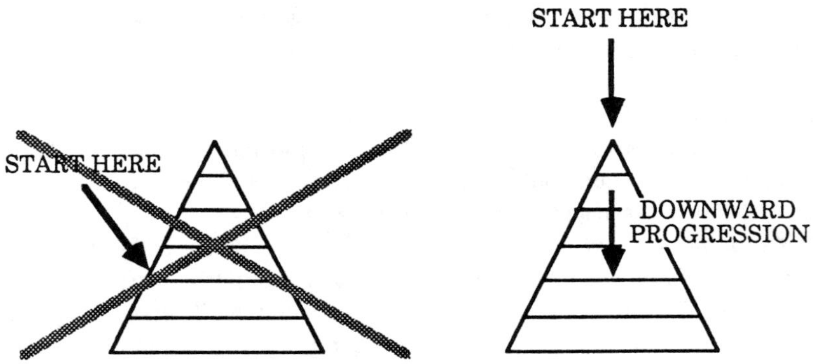

Figure 2.13. Training Programs should Start at the Top of the Hierarchy Pyramid.

deficiencies in the fundamental knowledge required to apply such techniques as performing quality checks, plotting SPC charts, and preparing written materials for group meetings. Calculus and English are often among the most needed training programs for those people who did not have the opportunity to complete high school.

Training everyone in everything. Another pitfall, which is an effect of poor planning, is to request that large groups follow identical sets of WCM training. For example, sending all the personnel on the shop floor to a set of SPC, total maintenance, setup time reduction, and teamwork training programs is not the correct approach to use because not every worker has the same training needs. Therefore, training programs should be tailored to individual needs. In addition, since it takes from several months to several years to fully implement a WCM program in a plant, workers should receive the training that they need only at the time when they need it in order to be able to rapidly and efficiently apply the new concepts they acquire. Thus, the training schedule must be matched to the company reorganization schedule.

Using a shotgun approach. Using a shotgun approach is another effect of poor planning or a lack of planning. It consists of choosing particular employee trainings on the basis of isolated conditions rather than by following a predefined plan (for example, sending a group of foremen to a setup time reduction seminar after having read an article on setup time reduction in a professional magazine). When work force development becomes one of the company's strategic objectives, the shotgun approach must be prohibited. Policies and guidelines concerning work force training should be developed by top management and issued to individual managers who then establish individual training programs with the help, if necessary, of the local or central human resource structures (refer to Figure 2.11).

Providing irrelevant training. Lack of planning also results in irrelevant training programs. Lack of coordination between the training schedule and the actual reorganization schedule can also result in irrelevant training programs (for example, teaching the Kanban system to machine operators while the actual Kanban system is still weeks from implementation). Because training program descriptions at times are not precise enough, employees are also inadvertently sent to trainings irrelevant to their current or future assignments. The role of the central and local human resource structures is also to reduce the risk of sending employees to trainings that do not correspond to their needs. These structures accomplish this by keeping the managers informed of the various existing outside training programs and, particularly, of who will benefit from these trainings. In addition, the information concerning in-house training programs should be reviewed to avoid ambiguity.

Not explaining "how to". Training sessions, even if followed according to a well-planned schedule, still might not increase the trainees' performance when the sessions give solutions to problems without describing how to practically implement those solutions. For example, a program on teamwork and small-group activities might do an excellent job at demonstrating the virtues of sharing and solving problems in groups and fail to explain how to practice problem-solving techniques, how to lead a group discussion, and how to communicate results to top management. Such a program might effectively modify the attitude of people such as machine operators, foremen, or production supervisors but will not provide them with the skills to hold or participate in small-group activity meetings. Both in-house training and outside training programs should also be checked for providing ways to implement the described solutions to problems.

In addition to these common pitfalls, companies will be facing numerous problems in developing their human resources similar to the problems they face in reorganizing their shop-floor operations. Recognizing the value of the final objective (human resource development), organizing for reaching this objective to avoid the largest number of problems, and being prepared to eliminate the remaining problems will enable people's energy to be focused on successful training programs.

2.4 SUMMARY AND MANAGEMENT GUIDELINES

To enter and durably remain in the limited circle of world-class companies, to compete in the selective markets of the nineties, the next step after having defined strategic goals is for top management to communicate these goals to the entire work force and involve the work force in attaining these common goals through companywide teamwork.

Communication channels need to be established or reinforced that (1) enable management to communicate the long-term objectives as well as shorter term goals to the work force, and (2) enable the work force to provide feedback and input to management regarding the means to reach these objectives and goals. Company managers should first assess the company's present culture and then, by comparing current culture to desired goals, identify the prerequisites for implementing change. (A new attitude often has to be acquired by both the managers and the employees.) Managers then plan a list of actions that will enable the sharing of common goals and efforts, which can occur only if the ultimate objective for the company (customer satisfaction and resulting profits) and intermediate goals (such as improved product quality, reduced costs, and increased company flexibility) are clearly stated and are understood by everyone. Management has to specify the nature and extent of the information it expects from employees to help build the communication effort, which is a basis for establishing mutual trust and respect between management and employees. This can be a long process that requires patience, especially on the management side.

An involved work force helps management solve problems rather than representing an additional obstacle on the way to improved competitiveness. Both managers and employees can find benefits in EI programs. In unionized plants, worker representatives should also be involved, and as early as possible. If management wants union support, the requirements are the same as for the rest of the work force: Early information, listening, negotiating, communicating, and recommunicating. Employee involvement cannot be successful if the employees or the unions do not know what management is trying to accomplish. Employee involvement is a never ending process; it is a new way of life within the company, where change becomes natural and is not feared.

Employee involvement can only happen through a team-based approach. Management's initial step is to create an EI program steering committee, which monitors the implementation of the EI program and reports to top management. An EI program leader is designated, and he or she becomes the main engine for driving the EI program. A pilot area is chosen in which the first teams are formed, with team leaders often being chosen by the steering committee. Technical advisors can bring their expertise to the various teams on a temporary basis. Channels of communication between the teams and management, the level of team authority, methods for evaluating the teams' suggestions, and resources to be allocated to the teams are determined at this stage. Once the initial pilot team results are judged satisfactory (commonly after several months to a year), the EI program can progressively be extended to the rest of the company. Two types of teams are distinguished: (1) Functional teams, whose permanent members belong to the same department and are in charge of improving the department's performance, and (2) Project teams, which are cross-departmental teams that solve specific problems assigned by management. Functional teams are permanent; project teams disband once the assigned project is completed. Work teams should adopt structured approaches to perform effectively as teams. Understanding whether the current evolutionary stage of a team is formation, conflict, cohesion, illusion, disillusion, or performance is useful to the persons in charge of the EI program for assessing and improving the team's effectiveness.

In addition to EI, work force development through continuous employee education enables companies to sharpen their competitive edge. Educating employees provides the work force with the skills, knowledge, and new attitude that enable them to share common efforts with management. Training programs should be planned with the perspective of the company's long-term goals. Managers must be trained first, and when line managers have been trained and understand the company's long-term goals, they should be responsible for training their subordinates. (In general, this new responsibility requires the decentralization of personnel previously part of the central human resource department.) Training efficiency has to be evaluated when employees return to the work place, and training programs selected or adjusted accordingly. When training employees, the use of internal experience should be combined with the discovery of external expertise. Managers need to be aware of traditional training program pitfalls, such as training that is not applicable in the work place or training that ignores fundamental bases because of a lack of needs assessment. Since moving toward WCM requires training and retraining an entire work force, it is particularly important to carefully plan the training program. World-class companies have clearly understood that their work force is their most important asset, helping to continuously improve their competitive advantage. These companies are equally convinced that continuously educating their work force is crucial to remaining competitive in the nineties and beyond.

3

Total Quality

In the last decade, international competition has strongly challenged most manufacturing industries to improve the quality of their products and processes. The effects have been detrimental to many companies that failed to recognize that better-quality goods and services fosters improved productivity, lower costs, and increased profitability. Today, most manufacturing companies are beginning to realize that a commitment to quality excellence is crucial for competing successfully in the global marketplace. Paramount for successful global competition is the knowledge of how to develop and implement a management-led, customer-driven total quality (TQ) program.

The slogans associated with total quality such as "make it right the first time," and "quality is everyone's job," are now quite familiar—to both manufacturing managers and hourly personnel alike. The basic concepts of placing the responsibility for quality in the hands of the people who make the parts—not with quality control inspectors—and emphasizing defect prevention—thereby reducing or eliminating the need for post processing inspections—have likewise been the guiding lights for many companies in their quest for a competitive edge. Although the concepts and slogans are easy to comprehend, the process of implementing total quality is both difficult and continuous.

This chapter addresses how to implement quality-management systems that simultaneously improve quality, increase productivity, and—through continuous improvement—reduce costs. The emphasis here is not on "selling" the concepts and benefits of implementing a TQ program, but to present some practical requirements and methods for achieving successful implementation. In particular, this chapter discusses management's role in reassigning responsibilities for quality and how to implement the key elements of the five integrated components of a TQ program. This chapter also addresses the use of statistical methods for total quality and how to organize and manage employee team-based programs for continuous quality improvement. In essence, this chapter emphasizes *what to do* and *how to do it* when using quality as a component for becoming world-class competitive in the nineties.

3.1 THE ROLE OF SENIOR MANAGEMENT IN QUALITY

Excellence begins with leadership. This phrase appropriately states the primary requirement for the successful implementation of world-class quality systems and procedures. It can also be stated that quality is the one cornerstone of WCM that does not require so-called management support—rather, it requires management's involvement, commitment, and leadership to cultivate and develop a new culture where quality is an operational strategy for the company and not just a departmental function. The key management issues that are crucial for making "quality" and "world-class manufacturing" synonymous are addressed in this section. Highlighted is management's leadership role both in implementing systems that define quality and conformance to requirements and in bringing about the required cultural changes. In essence, this section addresses management's role in reorganizing and reassigning responsibilities for quality.

3.1.1 Redefining Responsibilities for Quality

In most manufacturing companies, quality is the responsibility of the quality assurance (QA) department. This department is responsible for establishing systems and procedures that will ensure process quality and product quality. The production department, on the other hand, typically has the responsibility of seeing that the products are manufactured and shipped in time to meet the customers' delivery requirements. Specifically, QA people are responsible for quality, and manufacturing people are responsible for producing the product. Although it can be argued that the preceding statement is too general because everyone in the company is concerned with quality, the fact remains that most companies are still basically organized in this fashion.

Which is more important, quality or production? When asked this age-old question, management usually responds that both are important, but that quality comes first. Even though quality is said to be very important and a major concern of management, today in most companies, the reality of getting the product out the door to meet a deadline often takes priority over quality. For example, consider what really happens when it is mandatory that a half-million-dollar export unit be at the dock by Friday at 4:00 p.m. because the ship will leave as scheduled, with or without the unit; and there is no other ship available for five weeks; and, furthermore, the unit has a $5,000-per-day late-arrival penalty. The reality is that the unit ships on time—even if certain quality control (QC) evaluation and inspection procedures are skipped. The issue of quality versus production, however, is *not* one that needs to be addressed in order to achieve world-class-quality excellence. The real issue is how to organize and operate a company so that quality and production are complementary and not dichotomous.

Implementing management-led, customer-driven TQ programs requires fundamental changes in the way a company is organized, managed, and motivated. One necessary and fundamental change is to make quality a shared responsibility among everyone. This change is much more than a philosophical change. It is a cultural change, a change in how people think about and view quality in their jobs. If quality is primarily departmentalized, then there is little motivation or incentive to strive to continuously improve the company's products.

Many quality problems occur on the shop floor, often by either omission or commission of activities by production personnel. Although QC people may be fairly effective in identifying quality problems, they also often have difficulty in instituting the changes required to correct and prevent similar problems on the shop floor. Manufacturing, therefore, needs to be made more responsible for quality.

Traditionally, this idea has met inherent resistance since it often seemed like having the fox guard the hen house. Rather than signifying a conflict of interest, however, manufacturing's involvement in quality is well-suited for achieving problem correction and prevention. Manufacturing is most effective in maintaining real-time control at the point of manufacture; thereby eliminating the waste associated with quality problems. Through EI on the shop floor, an awareness of the factors that affect quality can be established and the responsibility for quality can then be transferred to the point of action. Many companies have already discovered that EI in quality and a greater emphasis on statistical process control can lead to significant improvements in the cost and the quality of manufactured products. (The logistics of implementing these activities will be discussed later in this chapter.)

Both design engineering, and sales and marketing also need to be made more responsible for quality. Being customer driven means being concerned with a customer's requirements for product quality, for example, by meeting the customer's defined product characteristics for quality performance and fitness for use. In addition, the issues of product safety, reliability, and maintainability are crucial in developing product designs and specifications. Quality must not only be designed into the product, but manufacturing must also be capable of building quality into the product. For instance, are the product design tolerances necessary and reasonable in light of both the customer's needs and the processing capability for manufacturing the product? How fail-safe are the designs in avoiding problems in fabrication, assembly, and customer use? Addressing these questions in the design process requires coordinated efforts by personnel from sales and marketing, design engineering, and manufacturing. The key concept is to use a teamwork approach to produce a product design that consistently meets or exceeds the customer's requirements while still being profitable to manufacture and to market. These coordinated efforts will not occur unless management effectively provides the leadership, recognition, and rewards that will foster this interdisciplinary structure. Required for the formulation of effective design reviews, this type of structure enables manufacturability to be judged in terms of how well the designs work together in preventing quality problems and in satisfying the customers' needs.

Materials management also has an impact on quality. The importance of receiving the right materials, in the right quantity, at the right time is fundamental to JIT manufacturing. In addition, production planning and control needs to develop and maintain a realistic plan for the master production schedule and capacity requirements, because imbalances between the scheduled work load demand and the available production capacity is typically the underlying reason for the conflict between quality and production.

It can be argued that every department in a manufacturing company has some impact on quality. A manufacturing company is best viewed and managed as a system of interrelated departments that function together to satisfy the customers' requirements for product performance, quality, and delivery. If the maintenance department fails to do its job, then the performance of the entire production system will be affected adversely. Likewise, the human resources department can inhibit overall system performance by failing to hire, train, support, and reward people properly. In essence, no one department or individual is responsible for quality—everyone affects quality. Thus, everyone needs to be made responsible for quality.

Making quality everyone's job requires bold leadership that can incite a change in culture: A change in the way people view their jobs and responsibilities, a change in the way people work together, a change in the way people are recognized and rewarded for performance. Only senior management can incite and foster such a change.

3.1.2 Quality Management by Committee

Some leading companies are now beginning to realize that quality management for WCM can best be accomplished by a committee. Upon initial consideration, this idea may seem somewhat ineffective or even ridiculous because committees are typically thought of as the organizational structure of bureaucratic inefficiency and ineffectiveness.

Much has been said and written about the need to bring about the cultural changes in a manufacturing company whereby quality becomes everyone's job; however, little has been suggested about how to actually accomplish this task most effectively. While the concept of "quality management by committee" may initially seem radical, the cultural changes required for successfully implementing total quality are, for most companies, likewise radical.

Total quality requires an integrated team approach and a sharing of the responsibility for quality. Forming the quality management (QM) committee is thus an important first step in establishing quality as a strategic issue for competitiveness. The following is a common representative membership for the QM committee:

- Sales and marketing
- Design engineering
- Production departments
- Materials & purchasing
- Production/inventory control
- Human resources
- Quality control
- Plant facilities & maintenance

The committee is thus a vocationally diverse group of people who share the responsibility for quality management in the company. In many companies, the existence of the QM committee makes it unnecessary for the existence of a separate department for quality assurance. Quality control personnel and activities are then incorporated in the production departments under the direction and support of production supervisors. This organizational structure is effective, but requires knowledgeable and quality-conscious production supervisors

Figure 3.1 displays the relationships and interactions of the QM committee with other organization levels in the company. Note that Figure 3.1 indicates that senior management clearly needs to provide the leadership and direction for defining the strategic goals and objectives associated with total quality. The QM committee is responsible for both the program's development, which is based on the TQ program objectives, and the management of quality in the organization. In addition, the QM committee is also responsible for reporting to senior management on the status and effectiveness of the program. At the department level, supervisors and team leaders are responsible for planning, implementing, and controlling the specific activities required for assuring quality. The major responsibility for QC is placed directly at the department level and primarily at the point of production. Thus, the actual real-time execution of numerous daily tasks and activities for QC is performed by employees on the shop floors, on the receiving docks, at the testing areas, and so forth.

3.1.3 Defining Objectives for Total Quality

Defining specific objectives for a TQ program is often a personal endeavor for a company, because a TQ program represents such a major commitment of people and resources to quality. Consequently, most companies choose to develop their own set of objectives rather than use a generic set of objectives from

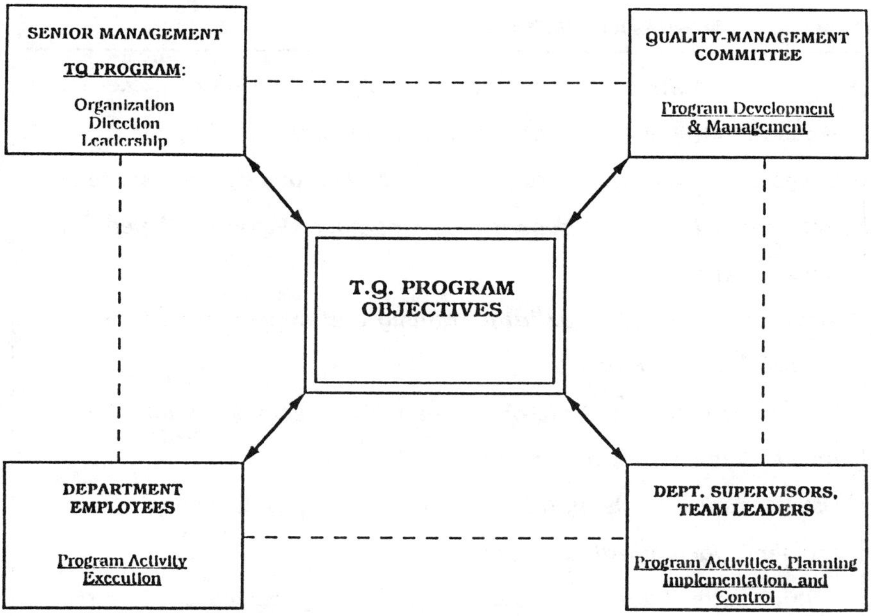

Figure 3.1. Quality Management by Committee & Employee Involvement.

a quality handbook, which results in the objectives being established after much consideration and discussion. Despite the personal flair, company objectives for achieving world-class quality excellence still share the common fundamental concepts and philosophies espoused by world-renown quality leaders, such as Feigenbaum, Demming, Juran, Crosby, and others. So while the wording and style may differ, the fundamentals are essentially the same. Table 3.1 shows a set of concepts and strategies that are fundamental in defining objectives for total quality.

The inherent challenge for management found in the objectives of Table 3.1 is simple to state, yet challenging and difficult to accomplish. Achieving quality excellence requires defining clear, meaningful objectives and implementing people-based quality systems for continuous quality improvement. Furthermore, achieving improvement requires understanding the manufacturing system and reducing the variation and the inconsistencies associated with the production processes. Only senior management can provide the leadership for cultural changes through new and innovative structures of responsibility, recognition, and rewards for quality. Many of the specific activities of a customer-driven, committee-managed, employee-executed TQ program are discussed later in this and other chapters.

TABLE 3.1. Objectives for Total Quality.

• *Provide products and services of a quality that meets or exceeds the requirements and expectations of the customer.*
• *Develop quality systems to foster continuous quality improvement.*
• *Define and implement quality programs based upon employee involvement.*
• *Provide all employees with the training and support needed to produce quality products.*
• *Develop and communicate objectives with all employees and assign responsibilities and accountability.*
• *Use statistical methods to monitor quality and isolate major problems for immediate solution.*
• *Acquire knowledge of production processes, focusing on defect prevention instead of defect detection.*
• *Develop and maintain a team approach that emphasizes enhancing the competitiveness of the company through increased quality and productivity.*
• *Provide an atmosphere that encourages every employee to achieve their full potential and pride in workmanship.*
• *Form partnerships with both vendors and customers that will improve quality in all aspects of purchased parts and product usage.*

3.2 ESTABLISHING A TQ PROGRAM

The previous section stated that achieving TQ through quality management requires leadership from senior management. A major part of that leadership involves providing a clear mission statement and a set of objectives that everyone can understand and to which they can subscribe. This section addresses the five integrated components of a TQ program (shown in Figure 3.2) and discusses the key elements and activities for each component. It is noteworthy that some of the

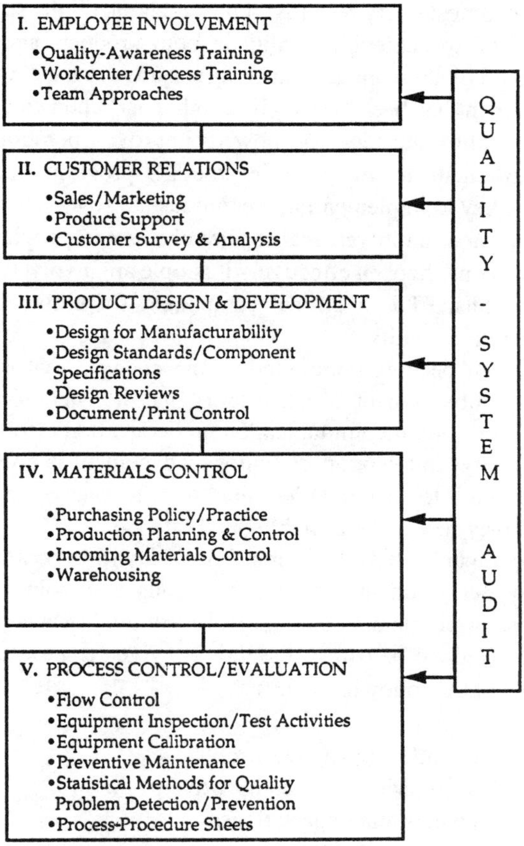

Figure 3.2. The Five Integrated Components of a Total Quality Program.

elements of the TQ program are fundamental steps for achieving WCM and, accordingly, are discussed in separate chapters in this book.

3.2.1 Employee Involvement

The most important element of any company's TQ program is people. People, not technologies, determine the success or failure of companies. Thus, the way people are hired, trained, treated, nurtured, recognized, rewarded, and involved is the key for achieving total quality. Without people's involvement and commitment, quality control will continue to be simply problem detection and correction, not problem prevention and quality improvement.

It is not surprising, then, that the first component of a total quality program is

EI—a concept that almost every recently published book on the subject of quality states is fundamental to achieving quality in both products and services. One company has defined EI as "management being sensitive to the employee's input." Employee involvement has been defined by another as "when all employees have an opportunity to contribute ideas on how to improve operations and product quality and to participate in implementing the TQ program." As discussed in chapter 2, EI is the key to implementing continuous improvement through change in an organization. Most improvements require changes in people's thinking and behavior, which can only happen effectively if people are involved and committed to improvement. Thus, EI is a critical component for eliminating waste and continuously improving quality.

Before people can become committed to the changes that are required for continuous quality improvement, it is necessary that they understand the importance of quality. Therefore, a communication and education effort, reflecting the commitment, sincerity, and support of senior management is absolutely critical. People will not commit to change unless they want to change, and they will not want to change if they feel uncomfortable or threatened.

Many WCM companies have implemented *quality-awareness training* for all employees. In one sense, quality-awareness training is ongoing and part of the continuous process of problem solution and prevention; however, quality-awareness training also needs to be part of the training given to new employees. This training should cover the following issues:

- Quality objectives and company philosophy;
- Company products, services, and competitive position;
- Customer requirements and expectations;
- Concepts of teamwork and EI;
- Employee's role and responsibilities for quality;
- Safety rules and good practices; and
- Overview of the TQ program: systems and procedures.

The introductory training can be presented most effectively by showing a professionally produced videotape, supplemented with an employee booklet that covers the content of the video for future reference. While preparing such materials requires effort and expense, the impact on the new employee is significant and long lasting—provided reality matches the philosophies presented in the tape. The following attitude has also been expressed frequently by both employees and customers: "If the company doesn't care enough to train their people, how serious can they really be about quality?"

One area where training and employee involvement overlap is in *work-center/process training*. In order to produce a quality product, people need initial training on how to perform the various process functions required by each particular job. Process training is most effective when it is "hands on" and supplemented with good work-center manuals.

A work-center manual is one effective approach to help ensure that proper instructions and procedures are available for preventing quality problems at the work-center. Work-center manuals are often developed and used by production employees, in conjunction with support from quality staff, and contain the following types of information:

- A description of the work center's function and how to operate it safely.
- A list of the tools, inspection devices, supporting materials, and other resources required for producing quality parts.
- A description of the types of QC activities required, such as first-piece inspection and SPC charting.
- Detailed instructions covering "how to" perform quality activities, including measurement procedures, data collection, recording and displaying data, interpretation, and corrective action.
- Specifications for performing required levels of equipment calibration and preventative-maintenance activities.

World-class manufacturing companies now recognize that providing employees with process training, along with complete and consistent instructions and informational support, is critical for performing work-center operations at high levels of quality. Furthermore, investing in employees through work-center/process training also provides the following benefits:

- Improves employee knowledge of job responsibilities and work-center processes.
- Fosters EI to achieve employees' full potential.
- Encourages employees to take ownership of their jobs and pride in their work.
- Provides people with the knowledge and skills for more effective EI in continuous quality improvement.
- Promotes good will and team spirit throughout the company.

Employee involvement is valuable in developing the work-center manuals. An experienced and knowledgeable employee is the expert on the work center, so let the expert help write the manual. Experienced employees are also a valuable asset in developing standards and procedures for certifying a worker as "quality capable" on a work center. In short, quality products cannot be produced by people who do not have sufficient training and information to perform the needed operations. Committing to improving quality means committing to improving people.

In addition to training, many companies have effectively used *team approaches* to achieve improvements in their quality. One form of the team approach is the *quality circle* in which a group of five to ten employees from the same work area or process meet regularly to address quality problems that affect them and their work performance. Part of the role of the team is to identify the causes of quality problems, investigate possible solutions, implement and monitor their corrective solutions, and report the results of their efforts to management. This role is primarily one of problem solving and prevention. Another role for quality circles is that of quality improvement. Significant quality improvements can often be made in an area where no specific problems are readily apparent. Both roles are important for improving quality.

Another form of the team approach is the *task team*. A task team is a group of people who function in a manner similar to the quality circle, but their emphasis is typically focused on solving a specific problem or a set of related problems. A task team's activities focus on a specific problem, not on a work area or particular process. Consequently, many companies use task teams for "quick kills" on specific problems and dissolve the team after a solution has been achieved. Some companies have stated that task teams maintain a level of enthusiasm and freshness often lost with quality circles after an extended period of time. However, any team approach can easily deteriorate into an ineffective use of time, unless management's commitment and support is both adequate and continuous.

One of the most common mistakes made by companies in using a team approach for quality improvement is overemphasizing the technological or mechanistic aspects of manufacturing and quality, and underestimating or ignoring the need for improving the problem analysis and thinking skills of the team members. Such skills are essential for quality improvement efforts, since they provide people with the analytical tools required to apply deductive reasoning, identify cost causes of quality problems, and apply their experience and knowledge in a structured problem-solving approach. Before a corrective action for quality improvement should be taken, the root cause of the problem needs to be identified and analyzed and alternative hypotheses about the possible causes tested. Most production workers are unskilled at effective problem analysis and solving, and tend to either apply the same standard solution to all problems or use a trial and error approach, both of which are usually ineffective and always more costly. Thus, a key for successfully implementing quality improvement is to initially provide the team members with approximately eight to ten hours of training on situation analysis, problem analysis, and decision analysis techniques, incorporating a hands-on workshop approach. Such training gives the people the skills they need to apply their knowledge and experience, and eliminate root causes of quality problems and waste in manufacturing.

3.2.2 Customer Relations

Customer relations is the second of the five integrated components of a TQ program. In today's competitive environment, providing the customer with a quality product and quality service is the determinator of success for a company. Thus, being able to identify and satisfy customers' needs is essential to achieving high levels of customer satisfaction—the driving force for any quality program.

The customer-relations component of a TQ program consists of the following three elements:

1. Sales and marketing
2. Product support
3. Customer survey and analysis

Sales and marketing is the first line of communication with existing customers concerning new products and product upgrades. In effect, sales and marketing acts as a customer liaison in determining product requirements that are based on the customer's needs and expectations. By working with other departments in the company, such as engineering, manufacturing, materials, and so forth, sales and marketing is initially and ultimately accountable for communicating the customer's requirements and helping to ensure that those requirements are met. Sales and marketing, thus, has a frontline position with the customer. This position is strategic not only in meeting the customer's needs but also in improving the company's products by enabling continuous communications with the customers. Because sales and marketing has this up-to-date information about the customers, it is crucial that a representative from sales and marketing be on the QM committee.

A company's dedication to providing the customer world-class products needs to be supported by an equally good commitment to product support. The product-support function helps ensure that the customer receives the necessary after-factory support for good product performance or, in the event of a product problem, facilitates the appropriate corrective action.

A world-class product-support function provides service to the customer in the following areas:

1. Field service:
    a. Provides consultation for field service and after-market installation support.
    b. Reviews product manuals and installation, service, and other procedures for accuracy and completeness.
2. Warranty disposition and analysis:
    a. Coordinates the warranty function, including the receipt of returned goods and claims.

b. Verifies and determines the cause of customers' warranty claims.
c. Provides feedback and communication to appropriate company employees regarding customer claims for corrective actions.

3. Service parts:
   a. Provides support to customers with a complete service-parts program and twenty-four-hour phone service.
   b. Furnishes customers with complete service-parts information, including, exploded-view service drawings, specifications, and service instructions.
4. Service training: Provides support in developing service and maintenance training programs for the customer.

Many companies' success can be partially attributed to the efficiency and effectiveness of their product support. For example, Caterpillar is a tough competitor in the off-road vehicle industry because it is well known worldwide for having excellent field service for its products.

The use of survey instruments to ascertain customer satisfaction is becoming more common as companies recognize that customer satisfaction truly is the driving force behind total quality. The objectives of these customer surveys are:

- To obtain information pertinent to the performance of the company's product and the effectiveness of their service program.
- To gain further insight into the customer's needs, both current and future.
- To have a formal communication vehicle between the customer and the company concerning product performance.

The analysis of survey information provides input from the customers, which is essential not only for satisfying the customer's individual needs, but also for providing useful ideas and directions for the continuous improvement of product quality and service.

3.2.3 Product Design and Development

Product design and development is the third of the five integrated components of a TQ program. This component focuses on improving design quality by reducing variation and by detecting and solving quality problems during the designing stages rather than in the manufacturing stages of the life of a product. The emphasis is to achieve quality improvement through the coordinated efforts of sales and marketing, engineering, and manufacturing by focusing on product- and process-design improvement.

Many WCM companies address quality improvement in product and process design under different names. Commonly used names are design for manufacturability, simultaneous engineering, concurrent engineering, design for assembly, design for competitive advantage, and design for excellence. Regardless of the specific name, the strategy is that of an integrative approach that addresses in the design and development process all factors that can impact production, quality, and customer satisfaction.

The objectives and associated benefits of *design for manufacturability* (DFM) are the following:

- To develop products that meet or exceed customer requirements and quality expectations.
- To design products in light of their manufacturing capability; thereby facilitating trouble-free production at a low and competitive cost.
- To reduce the product-development cycle in order to bring products to the market sooner.
- To improve productivity and lower manufacturing costs by releasing complete, manufacturable designs that do not require performance-related redesign and/or engineering changes.
- To provide a quality-competitive product that will not only provide the best return on investment today for the customer, but will continue to do so in the future.

Improving the quality of the product design and development process provides good leverage for a company's investment in total quality. Studies have shown that during the four stages of product design and development (vision, investigation, development, and manufacturing) approximately 85–95 percent of the development effort is expended on only 25–35 percent of the final product cost. This situation is further illustrated in Figure 3.3. For example, often 65–75 percent of the product's cost to manufacture is determined and fixed during the vision stage of product development, which only accounts for 5–15 percent of the total effort expended for the development of the project. Without a DFM strategy in place, too often 90 percent of the product's manufacturing product cost is already fixed before manufacturing becomes involved.

The impact of DFM has also been illustrated in recent studies on the relative costs of making a change for quality improvement during the four stages of product development. For example, the cost of making a change during stage three (development, involving product building and testing) is roughly twenty times the cost of making a change during the vision or early investigation stages where product concepts and performance can be evaluated through simulation. Similarly, the cost of a change during the production stage can often be 100 times the cost of a change made in simulation. Improving the quality of the design process provides good leverage on a company's investment in total quality. The investment of time

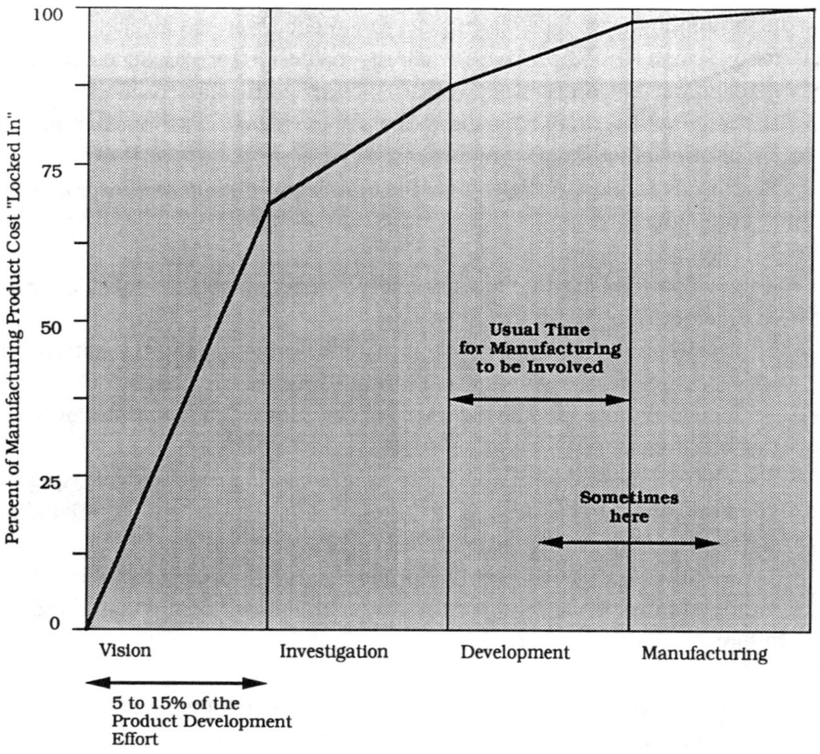

Figure 3.3. Four Stages of Product Development.

and effort to improve the product design and development process pays significant dividends later in manufacturing by eliminating quality problems and the need for product redesign and engineering changes.

Implementing a DFM strategy often involves various quality tools. One tool that can assist in the DFM effort is quality function deployment (QFD), which is a methodology for ensuring that customer requirements are translated and communicated into relevant technical specifications during the four stages of product development. By means of detailed instructions and specifications for product performance, QFD provides a matrix-structured procedure for defining and communicating customer requirements and expectations from early concept through the manufacturing and operating levels.

Whether or not a formal QFD methodology is used, all effective DFM programs have essentially the following elements:

1. A multifunctional team of all key players, which is formed before the product architecture is defined.
2. Design standards, involving both proprietary and industry standards, that are defined and used as guidelines for both designing and testing components and product systems.
3. A complete set of product specifications, developed and agreed on by the team, that cover such issues as manufacturing costs by subassembly, performance specifications, option structures, and quality requirements in terms of mean time between failure and other defined measures. These product specifications are based on the customer's requirements or "voice."
4. The performance of a complete design review, including a competitive analysis and benchmark of alternate products in the marketplace. This design-review process helps ensure that cost-competitive products meet customer specifications by providing realistic cost figures and making certain that quality and serviceability have been designed into the product.
5. A review and finalization of the product specifications before its release to production, which is based on the benchmarking and analysis in step 4.
6. Each element of the product design and development project being scheduled by the team. The project is managed according to this schedule, and necessary scheduled changes are made only with full team involvement and agreement. This schedule also provides documentation of the product-design and development process and all related changes for quality improvement.
7. The promotion of the DFM strategy by top management, plus the restructuring of the corporate system for accountability and rewards. The standard for excellent performance in a team-oriented DFM strategy is significantly different from the traditional department-oriented focus in most companies.

Many WCM companies have used DFM with impressive results. For example, DFM enabled Hewlett-Packard at its Disk Memory Division to dramatically improve the quality of an existing computer mass-storage device. Improvements relevant to the existing product were a sixfold improvement in reliability, an 87 percent reduction in warranty costs, a 50 percent reduction in workmanship errors, an increase of 3.6 times in production yield, a 57 percent reduction in product cycle time, and a bottom line, 23 percent reduction in production costs. Many other companies have reported similar gains relating to existing products and have also reported very significant improvements in bringing new products to the market sooner and, consequently, achieving major gains in market share.

Document control and print control are also important elements for ensuring quality in the product-design and development process because they are able to disseminate accurate and timely product-design information to those involved. This information is typically embodied in three types of documents: (1) blueprint drawings, (2) bills and materials, and (3) engineering change orders. Lack of accurate and timely information in these essential documents can be a source of great frustration and frequent quality errors and problems on the shop floor. Implementing and maintaining a good document-and-blueprint control system requires effort and commitment by management. Nevertheless, it is absolutely essential to ensure quality. There are no good excuses for the lack of a good document-and-print control system.

3.2.4 Materials Control

Materials control is the fourth component of a TQ program. Materials obviously play an important role in both product quality and cost. It is therefore essential to consider the impact and operation of the following elements in total quality:

1. Purchasing policy and practice
2. Production planning and control
3. Incoming-materials control
4. Warehousing

The purchasing policy and practice effects the success of developing and maintaining win–win relationships with vendors who are capable of supplying quality products and service. Effective policies for WCM often involve the use of blanket orders and the consideration of coordinated planning and scheduling for JIT operations. As quality and delivery assume greater importance, it is common for price negotiations to be based on cost information and cost savings that can be shared jointly between customers and vendors. A greater degree of sharing is also beneficial, in many cases, regarding technical resources. In general, purchasing's policies and practices in WCM companies are focused on developing ongoing win–win programs with certified suppliers. (The vendor certification process is discussed in detail in chapter 10.)

Production planning and control is one module of a Manufacturing Resource Planning (MRP II) system and is used to control the flow and scheduling of production lot orders in terms of order release, shop-floor activity update, and order close-out activities. If shop-floor production personnel are rushed to meet a customer's delivery requirements,then shortcuts in quality control naturally follow. In addition, fabrication and assembly errors are more likely to be made when workers are under pressured conditions. With good production planning and control, the age-old question of which is more important, quality or production, is no longer relevant. The role of production planning and control in a mean, lean, and world-class competitive company is addressed in chapter 8.

Incoming materials control addresses the inspection process used on all incoming materials, both for rough materials and finished parts. Receiving inspection activities often include the following procedures in order to perform this QA process:

- First-article inspection is performed on all dimensions when a new part or vendor is involved.
- A sampling plan with zero defects (C–O) is applied for determining the acceptance of all parts on subsequent orders.
- Discrepant material is transferred immediately to a QC hold area to avoid comingling with good parts or products.
- Inspection records are periodically reviewed to check for vendor feedback on potential problem areas regarding part quality.

While incoming-materials control will always have a function in a manufacturing company, its role will continue to decrease as total quality is implemented to a greater degree.

The last element of the materials-control component is warehousing. Even in JIT environments, modern warehousing facilities play a role to ensure a high level of customer service by recognizing the importance of being able to satisfy changes in the customers' schedules. The challenge is to achieve a balance between reducing warehouse space yet maintaining a flexible reaction capability to the changing demands of customers. Many of the issues discussed in this book—such as setup-time reduction programs, cellular manufacturing, a closer cooperation with vendors for JIT shipment, and better production planning and scheduling—facilitate the reduction of warehouse space without sacrificing customer service. Quite simply, the implementation of total quality involves coordinated activities with other WCM components.

3.2.5 Process Control and Evaluation

Process control and evaluation, the last of the five integrated components of a TQ program, encompasses the systems and procedures for ensuring quality in the many activities involved in shop-floor manufacturing. This component of total quality has traditionally received the greatest amount of attention by management and consists of the following six elements:

1. Flow control
2. Inspection and test activities
3. Equipment and gauge calibration

4. Preventative maintenance
5. Statistical methods for detecting and/or preventing quality problems
6. Process-procedure sheets

Flow control is the control of material movement between and within departments during processing; it also provides knowledge of and accountability for the status of any part on the shop floor. A good flow-control system should allow any person, at any time, anywhere in the plant to determine where the material came from, where the material is going, what operations have been performed and completed on the material, and what operations have not yet been completed. It should also provide documentation showing that all QC activities and inspections have been completed up to that stage of manufacturing. Flow control can be accomplished effectively through the use of move tickets for controlling the transfer of materials between drop zones—areas designated for receiving lots of material in each department. Flow control is essential for good quality by ensuring that all materials are clearly identified, which minimizes the possibility of lost parts or the inadvertent use of incorrect or poor-quality parts in any manufacturing operation.

Inspection and test activities involve the use of various mechanical and statistical tools to ensure that both the in-process components and the final products conform to the customers' functional requirements and corresponding dimensional specifications. Inspection activities encompass both first-piece and first-article inspection, in-process inspection, and final-assembly test activities. Although one of the goals of total quality is to minimize the amount of inspection activities by implementing more real-time and prevention-oriented process controls, a well-defined inspection program is still necessary to monitor and control the quality of parts and products produced at the process level.

A simple test of the completeness and effectiveness of an inspection and test system is to verify that answers to the following six questions can be found in the existing documentation for all quality-checking procedures:

1. Why is the procedure performed?
2. How should the procedure be performed, including such issues as the location for quality checking, the characteristics to be measured, and the equipment and gauging to be used?
3. Who is to perform the procedure?
4. When should the quality check be performed?
5. Where are the results of the quality check documented, and how long is the documentation maintained?
6. What procedures should be followed for components and products when they either pass or fail the quality check?

Equipment and gauge calibration programs are a fundamental part of total quality because monitoring quality in production requires the capability to measure quality. A world-class quality program requires formal, documented procedures and schedules for regularly certifying the accuracy and repeatability of all inspection instruments used to make quality decisions in the manufacturing process. Effective calibration programs must include the following elements:

- Complete specifications of calibration requirements and procedures for all applicable tools, gauges, machinery, and test equipment used in measuring parts, processes, and products.
- Explicit timetables for calibration with accelerated frequencies for less-stable equipment.
- Documentation showing that all required calibrations were performed (such as putting calibration tags on the equipment).
- Calibration requirements on test equipment being traceable to national standards.
- Defined procedures for preventing the use of equipment that does not pass calibration standards.
- Procedures for notification and rectification should calibration activities disclose the possibility that discrepant materials were produced or even shipped.

Preventative maintenance (PM) is a key element to achieving total quality in WCM because maintaining a high level of product quality and customer service requires reliable and dependable equipment. Today, PM is becoming a program with a team approach involving both commitment and cooperation between operations and maintenance department personnel to implement effective programs of systematic inspection, detection, and prevention of unforeseen equipment failures.

PM is an essential element of a company's approach to WCM because it provides the following benefits:

- Increases employee safety.
- Minimizes costly equipment breakdowns and possible further damage.
- Helps ensure product quality.
- Minimizes production delays and possible customer-service problems.
- Lowers operating costs.

Because of the importance of PM in WCM, the strategies for defining and implementing total PM programs are addressed in detail in chapter 9.

Statistical methods for detecting and/or preventing quality problems represent more than just another element of the process-control and evaluation components of total quality—it is a cornerstone of total quality. Because of its importance, this subject is also addressed in greater detail in section 3.3.

Process-procedure sheets are essential in providing employees with complete and consistent instructions and support information needed to perform their tasks in a manner that will help ensure quality. With the continuing deterioration of both process-related skills and communication skills in the workforce today, the use of visually based instructions is becoming more widely used because of its greater effectiveness in communicating process requirements. For example, the use of a computer-based, desktop publishing system, often coupled with a digital, still-video camera, provides a simple yet effective way to disseminate product-specific information and process instructions.

Process-procedure sheets are often issued with each shop order and contain the following types of information:

1. Visual displays (laser-printed photo images) linked to specific process instructions.
2. Customer feedback on previous product performance and quality problems.
3. Complete lists of all tools and supplies needed to perform the work correctly.
4. Detailed descriptions of sequential activities and procedures that are critical to ensure quality.
5. Cross-references and checks for current engineering-revision levels.

In order to manufacture and ship a quality product, it is critical that the people performing the work have the knowledge and the direction to perform their jobs properly. Process-procedure sheets address this need by providing a means to both capture and communicate process knowledge.

3.2.6 Quality-Systems Audit

The quality-systems audit is a useful management tool for implementing TQ systems. A comprehensive audit can provide an objective evaluation of the capability of a company's quality system to produce and deliver conforming products to its customers. The audit can also show the quality system's effectiveness in achieving this capability. Audits are beneficial in that they provide identification, and induce management's awareness, of existing quality problems, thereby forming a basis for corrective action and continuous quality improvement.

The purpose of the internal self-audit is to measure the effectiveness of given quality procedures compared to the overall objectives of total quality. In effect, the audit needs to address the following questions: What is the value of a given quality procedure for achieving total quality? Is the procedure necessary, redundant, too complex, clearly defined, or valuable?

Many companies conduct audits on a semiannual or annual basis. To be effective, the auditor (or audit team) must have wide knowledge and experience in quality systems and an ability to communicate openly and impartially with people of all levels in the company. Many companies use the services of an outside quality consultant to satisfy these requirements, then report back to management on the findings and needs for corrective action and document the results. It is also essential that senior management be involved with and responsive to the audit process and ensure that the results are reported back to all the people who are affected by it so that they may take appropriate actions. A properly conducted audit is a valuable tool for identifying opportunities for continuous quality improvement.

3.3 STATISTICAL METHODS FOR PROCESS CONTROL AND IMPROVEMENT

Statistical methods are an essential ingredient of and strategy for a WCM company to achieve both continuous quality improvement and conformance quality—the degree to which a product performs to its intended design requirements. Ishikawa's Seven Tools for quality improvement (check sheets, Pareto analysis, cause-and-effect diagrams, histograms, scatter plots, data stratification, and control charts/graphs) involve the use of both statistical methods and statistical thinking. The value of statistics is that they provide a means for measuring and understanding manufacturing processes. The value of statistics was already recognized back in 1894, as reflected in the following quote by Lord Kelvin:

> I often say that when you can measure what you are speaking about, and express it in numbers, you know something about it; but when you cannot measure it, when you cannot express it in numbers, your knowledge is of a very meager and unsatisfactory kind.

Statistical methods, when properly applied, can be effective in both finding the capability of a process to produce a part to given specifications and in controlling and reducing the variation of a process to improve part quality. Statistical methods are also effective in defining quality-related problems so that effective methods for problem solving and prevention can be implemented. The next section addresses some key statistical methods for process improvement and control.

3.3.1 Problem Identification

A primary role of management in total quality is to focus the employees' energies and efforts on the task of continuously improving quality. Providing the proper focus and direction requires knowledge of the most critical problems to be addressed and information regarding the causes that lead to these problems. Pareto analysis and cause-and-effect diagrams are widely recognized and used as effective tools for addressing these two questions. To capture the data required for subsequent Pareto analysis, many companies have developed numerical schemes to code and classify quality problems that occur throughout the manufacturing, distribution, and product-use cycle.

Figure 3.4 shows an example of a failure coding scheme, which facilitates the simple application of Pareto principles for identifying and prioritizing existing quality problems. The code consists of three digits. The first digit describes where the failure was detected. For example, the quality failure could have been detected at receiving, during fabrication, during assembly, during final testing, at shipping, in the field, or at the original equipment manufacture (OEM). The second digit describes why the failure occurred. For example, the quality failure could have been caused because of a vendor or materials error, an operator error, a print-design error, and so forth. The third digit of the code provides details on the cause of the quality failure already defined in the second digit. As Figure 3.4 shows, there are seven defined causes relating to a vendor or materials error, eight causes associated with operator errors, and so forth.

Whenever a problem with quality occurs within the given manufacturing, distribution, or product-use cycle, the failure is coded by a designated employee, and the code is then recorded. Problems with scrap, rework, customer returns, vendor parts, and so forth can be recorded with an appropriate code to describe the failure. For example, parts scrapped during fabrication because of an operator error in reading the blueprint or process instructions would be coded as 224 or 234, depending on whether the operator error was due to carelessness or lack of knowledge. If the error was because of lack of knowledge, the code would point out the need for additional operator training. Coding all quality failures in this manner provides data that can be stored, sorted, and displayed as Pareto charts through the use of computer data base and graphics software. Figure 3.5 shows an example of a Pareto chart developed from quality-failure coding information.

The use of a quality-failure coding system is beneficial in several ways:

- It identifies major or recurring quality problems that are in need of corrective action either by solution or prevention or both.
- It provides a communication tool between the employees who detected the problem and the management/employee teams formed for problem correction.
- It provides a means to measure the progress of corrective-action efforts.
- It identifies weaknesses in the quality system and provides direction for continuous improvement activities.

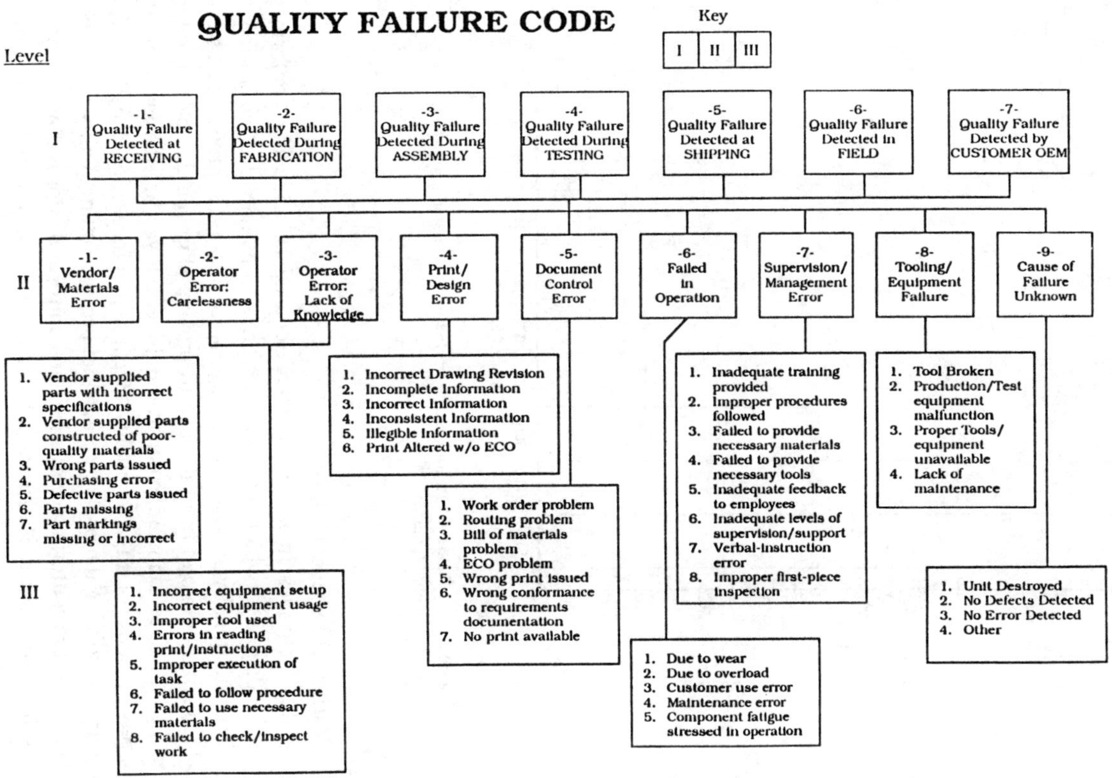

Figure 3.4. Quality Failure Coding System.

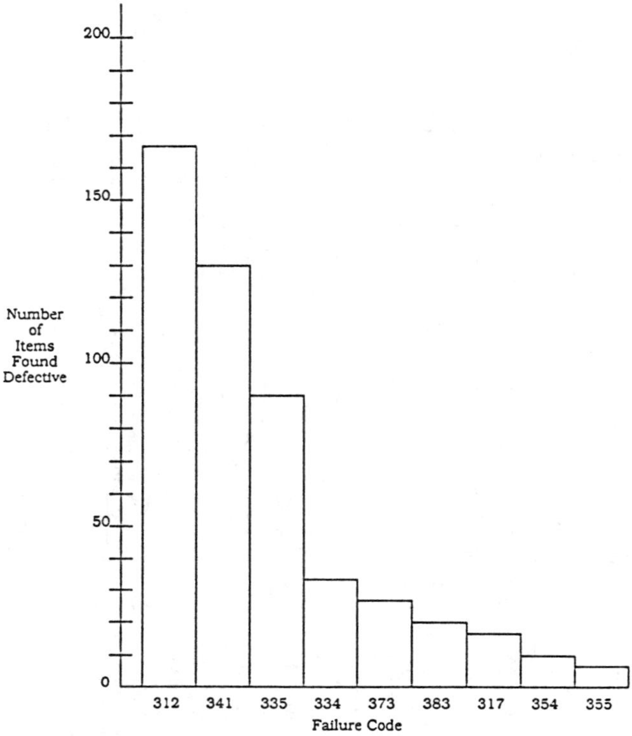

Figure 3.5. Pareto Chart for Quality Problems Detected in Assembly—First Quarter.

3.3.2 Statistical Process Control

Many quality activities in a company are focused at achieving conformance quality—the degree to which a product performs to its intended design requirements. One traditional way to ensure conformance quality is through part inspection, which sorts the good from the bad to prevent shipping defective products. Unfortunately, inspection is an expensive and ineffective after-the-fact technique and, moreover, does not improve quality. Because part inspection does not involve identifying and solving the problem, it cannot prevent future quality problems. A better way to ensure conformance quality is to control and improve the process that produces the desired parts. Statistical process control is now recognized by manufacturing managers as an important tool for improving quality and productivity.

Statistical process control is a simple charting procedure used for real-time monitoring and control of key product characteristics at a production process. Statistical process control is a key strategy for defect prevention because monitoring a production process while it is operating is the best way to prevent bad products from ever being made. Statistical process control uses statistical methods (data sampling, charting, and analysis) so that real-time decisions can be made concerning how a process is performing in relation to the required standards of that process, with the result that nonconformance can be avoided.

All manufacturing processes have some variability. Since some variation in a manufacturing process is expected, the value of SPC is that it shows whether the observed variation is normal or abnormal. Abnormal (that is, special-cause) variation is usually associated with an unexpected change in the operating conditions that adversely effects product quality. This unexpected variability may be caused by changes in material consistency, machine or tooling conditions, operator performance, or other factors. The goal of SPC is to detect abnormal variability when it occurs, correct or eliminate the cause of the abnormal variability, and thereby prevent the production of defective products. SPC can be used to determine how a process is performing and when an adjustment is necessary.

Failure to use SPC when it is appropriate increases the cost of manufacturing a product. Additional costs typically are incurred through the need for final-product inspection activities (component inspection, sorting, grading, scrap, and rework). Without SPC, it is also common for operators to make unnecessary adjustments to a process (overcontrol) and, consequently, induce additional variability into the process. Although performing SPC takes time and effort, it is usually a good investment. One study, for example, showed that the cost of sampling and SPC charting was approximately $1.20 per occurrence; the cost of detecting and correcting the same quality problem in the field was approximately $2,500 per unit.

The way in which SPC is applied varies depending on the type of product characteristic that is measured and on a number of other factors relating to cost, production rate, and so forth, as shown in Figure 3.6. Although different SPC chart types exist, the way in which SPC is used is generally the same. Typically, SPC is applied to a manufacturing process (machinery) that processes discrete parts. A random sample of parts is taken periodically from the stream of parts coming off the process. Some key characteristic or attribute of each part is measured or observed, and the outcome of this sample is plotted on a control chart over time. If no pattern exists on the chart and all the plotted points are within a band (the upper and lower control limits), then the variation is normal and the process is judged to be "in control." Thus, no adjustment is necessary. On the other hand, if patterns exist or a plotted point falls outside the control limit, then abnormal variation is present and the process is "out of control," which can adversely affect part quality. When this occurs, the process is stopped and the cause of the abnormal variation is corrected, thereby preventing the continued production of poor-quality parts. When a manufacturing process is sampled periodically, one can say that the

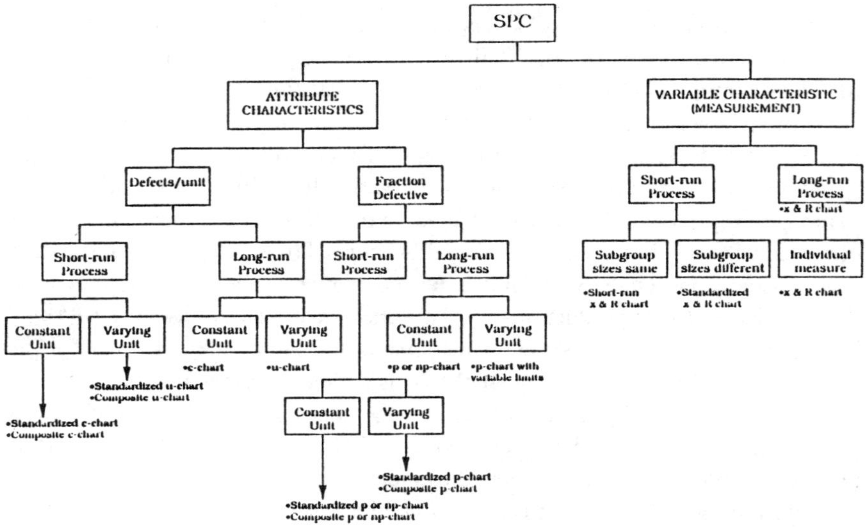

Figure 3.6. Guidelines for Choosing SPC Technique.

process is "in control" if its mean level and level of variability remains stable and predictable over time. Under these conditions, nearly all the plotted points in the chart (over time) should fall within the band between the lower and upper control limits if the variation in the process is because of normal or expected causes.

Figure 3.7 shows an X-bar chart and an R chart for controlling a turning process by measuring the actual diameter of a shaft that has a nominal size of 1.25 inches. The X-bar chart is used to detect a shift in the mean level of the process, whereas the R chart is used to detect a change in the degree of variability in the process. In this example, a sample of five shafts is measured (each hour). The average (X-bar) of the five shafts is calculated along with the range R (highest value minus lowest value). Both values are then plotted on their respective charts. Note that in this example the process was "out of control" at samples number 6 and 16 on the R chart.

Statistical process control is fundamental to total quality because it incorporates the key strategies of both defect prevention and EI. With SPC, the operator performs the sampling and charting, and then determines if the process is "in control"; thus, it is the operator's responsibility, not the QC department's responsibility, to maintain control over the process. Through statistical process control, the process operators are better able to know how the process is performing and when adjustments are necessary so that they can prevent the production of defective parts. Furthermore, SPC is cost-effective by being a real-time strategy and by reducing waste in all aspects of the production process. Statistical process control has become an essential element of total quality in WCM companies.

$\overline{X}$ AND R CHART EXAMPLE

PART NO.	6839999
CHART NO.	
PART NAME (PRODUCT)	B-UNIT HS Shaft
OPERATION (PROCESS)	Seal Diameter
SPECIFICATION LIMITS	± .010
OPERATOR	
MACHINE	CNC LATHE
GAGE	1-2" MICROMETER
UNIT OF MEASURE	.001
ZERO EQUALS	1.250

	1	2	3	4	5	6	7	8	9	10	11	12	13	14	15	16	17	18	19	20	21	22	23	24	25
DATE	11/5																								
TIME	8:00	9:00	10:00																						
SAMPLE MEASUREMENTS 1	5	2	1	5	-1	3	2	-3	-1	2	6	0	5	2	0	1	2	1	7	0					
2	2	-4	2	3	-4	-2	2	-4	0	-4	4	1	-4	-1	0	-13	2	2	-2	6					
3	4	1	-1	-4	1	-8	-2	-4	4	4	2	-1	1	0	1	9	4	4	-6	-2					
4	9	2	-3	1	0	17	3	0	-4	-5	10	0	-3	-2	5	2	-2	1	-3	-1					
5	5	2	-4	-3	-5	-4	2	4	2	-3	5	1	-2	0	1	0	3	1	1	-1					
SUM																									
AVERAGE, X	5.0	-.2	-1.0	.4	-1.8	1.2	1.4	-1.4	.2	-1.2	5.4	.2	-.6	-.2	1.4	-.2	1.8	1.0	1.0	0					
RANGE, R	7	6	6	9	6	25	5	8	8	7	8	2	9	4	5	22	6	6	13	6					

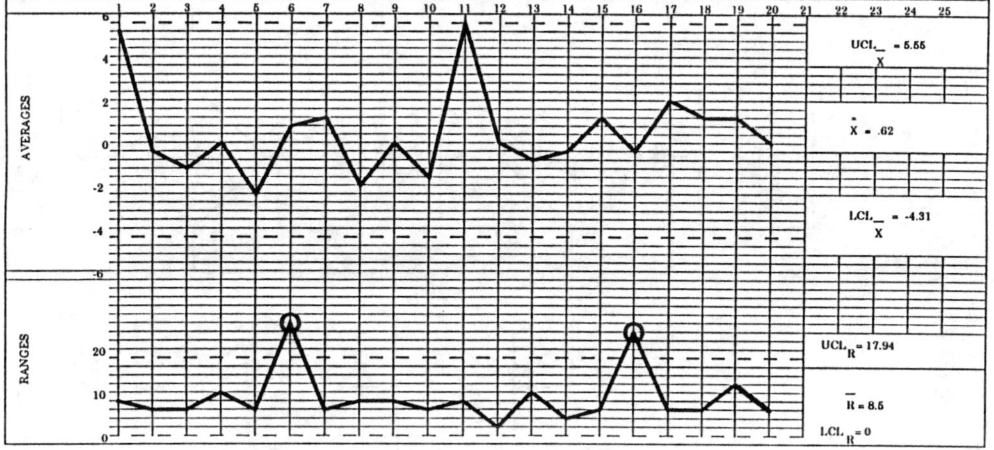

Figure 3.7. X-Bar and R Chart for Turning Shaft #6839999.

Figure 3.8 shows an overview of the steps and statistical methods that are typically used (along with SPC) to ensure process control. It is important to note that process control must be applied where it is most needed in terms of the characteristics that most affect final product quality. Having a team work and focus on every product characteristic for SPC is neither practical nor necessary. Once the process is in control for a key characteristic, the next step is to determine the capability of the process.

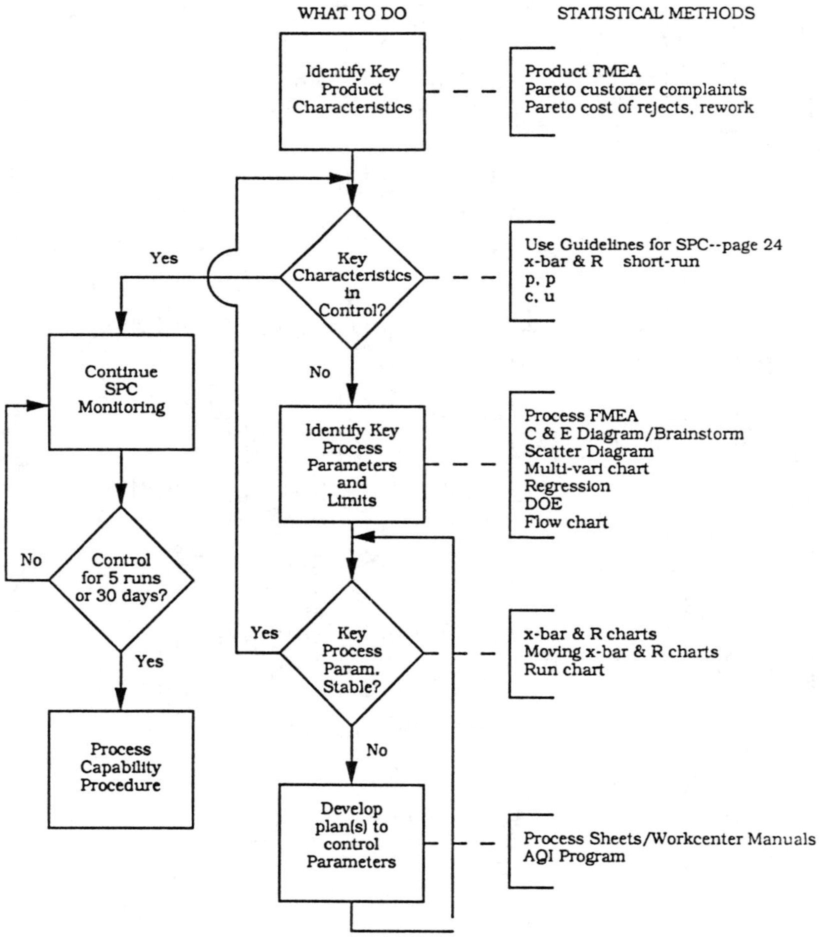

Figure 3.8. Process Control Procedure.

3.3.3 Process Capability

When a key characteristic is "in control" on a process, it does not necessarily mean that all the parts are being produced within specifications. One simple yet effective way to prevent quality problems is to ensure that parts are routed to and run on capable equipment. Capability, in this sense, is the ability of a stable (in control) process to manufacture a part that meets engineering (customer) specifications. Process-capability analysis is a statistical comparison of natural process variations to customer and engineering specifications. The natural process variation is defined as six standard deviations (6σ), calculated from sampled data on a part characteristic. The customer and engineering specifications are stated as an upper specification limit (USL) and a lower specification limit (LSL), respectively above and below the nominal or target specification level.

Process-capability analysis provides numerical indices that describe the ability of a machine or process to produce a part within given specification limits. One common capability index is the C_p ratio, which is the ratio of the engineering tolerances to the natural tolerances, as shown in the following equation:

$$C_p = \frac{USL - LSL}{6\sigma}$$

As the formula indicates, if a machine (or process) is capable, then (USL - LSL) must be greater than the 6σ natural tolerance of the machine (or process).

Since it is also important that a process be centered within the upper and lower specifications limits, a more stringent capability index is the C_{pk} ratio, defined as the minimum of the ratios of the difference between the specification limits and the process average over 3σ. This can be expressed in an algebraic format as follows:

$$C_{pk} = \text{minimum } \{Z_u, Z_l\}$$

where

$$Z_u = \frac{USL - \bar{x}}{3\sigma}, \quad Z_l = \frac{\bar{x} - LSL}{3\sigma}$$

It is common practice in industry today to strive for C values greater than 1.33 to ensure that a part will run both economically and without quality problems on the given process. (Note that C_{pk} can, in fact, be less than zero if the process average is either above the USL or below the LSL.

As an example, consider the situation shown in Figure 3.9 where the process average is at a value of 10 with a process standard deviation of 2 units. If the LSL and USL of a part were respectively 0 and 14, then the C_{pk} value would equal

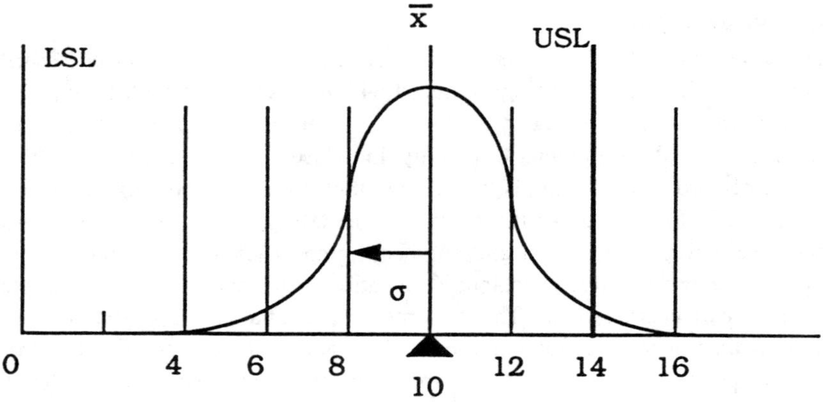

Figure 3.9. Illustration of Capability Indexes.

unity, but the C_{pk} value would equal only one-third, as shown in the following calculations:

$$\overline{x} = 10, \sigma = 2, LSL = 0, USL = 14$$

$$Z_u = \frac{14 - 10}{3\,(2)} = .667$$

$$Z_l = \frac{10 - 0}{3\,(2)} = 1.667$$

$$C_{pk} = \min\{.667, 1.667\} = .667$$

$$C_p = \frac{14 - 0}{6\,(2)} = 1.667$$

In this case, it would be essential to center the process in order to bring the C_{pk} value to a value of 1.167 as well. In any case, however, the process is only marginally capable for producing this part based on an ideal C_{pk} value of 1.67.

Process-capability analysis, relatively simple to perform, follows these five steps:

1. Record all process conditions, such as the tooling, equipment settings, materials, and any other factors present at the machine during the time in which the data is collected.
2. Collect the sample of parts and record the statistic or statistics of interest. Ideally eighty to one hundred parts should be measured if technically and

economically feasible. However, in no case should less than thirty-six samples be collected.

3. Perform the capability-analysis calculations and make plots or histogram displays of the data for visual presentation.
4. Interpret and communicate the results of the analysis with all the people whom it will affect—typically, manufacturing engineering, design engineering, production, QA, and shop-floor personnel.
5. Take appropriate action if the results of the analysis indicate that the process is not capable on the part or family of parts evaluated.

Process-capability analysis is a valuable tool because there are many activities that a company can do with the results, such as the following:

- Determine the capability of a machine or process with a given set of tooling.
- Determine the need for maintenance or replacement of equipment and tooling.
- Schedule and load jobs on machines that are capable of producing quality parts without difficulty.
- Determine the effects of repairs on equipment and tooling.
- Evaluate the quality improvement of new equipment, and use process capability indices as specifications for purchasing and accepting new equipment.
- Evaluate the effects of time on the quality of a product (part shrinkage, for example).

Process-capability analysis has proven to be a valuable tool for achieving better quality in many companies. When first applied, it provides new insight into not only the capabilities and limitations of existing equipment and processes, but also indicates how to improve process performance and prevent the quality problems that, all too often, seem to happen at random without identifiable causes.

Most processes are not in statistical control when the process control team begins to work on them. In fact, it often requires a considerable amount of time and effort by the team to bring the process into control. A formal capability index of the process cannot be determined until the process is in control. During this time, if the process is not capable of producing parts within spec 100 percent of the time, the team can work immediately on improving the process capability, rather than waiting until process control is achieved.

A process-potential study can be used to understand both the variability of the process and its potential for producing products that will meet specification. If the process-potential study shows poor capability, the team then has an early indication to work on improving process capability in addition to process control. In process-potential studies, the capability index calculation procedure is followed even though the process is not in statistical control. However, the C_{pk} calculated is not a true capability index, and therefore a PP (process potential) should be placed before the C_{pk}.

Figure 3.10 shows the steps and statistical methods appropriate for conducting process-potential studies, process-capability analyses, and process improvement. If the process-capability analysis shows that the process is not capable at a level of $C_{pk} > 1.67$, then a detailed analysis may be required to understand the relationships between key process parameters and product characteristics. The objective of this analysis would be to determine the process-parameter limits and targets that would reduce variability and/or center the process. Advanced statistical methods such as *design of experiments* (DOE) and *regression* may be required at this step. (The team leader and training coordinator should plan training and seek outside expertise to assist the team if necessary.) An example of a team working to determine key process-parameter limits might procede as follows:

> A process control team determined that ambient temperature greatly affected the thermoforming process. Upon further investigation, it was determined that the limits should be 100°F–115°F (the process runs best in this range with respect to the key product characteristic of interest). The team now knows that if the ambient temperature is controlled between 100°–115°F, the key product characteristic will be in control and within specification.

Once the process-parameter limits have been determined, the team can check to see if the process is running within these limits. Statistical process control charting techniques can be used to check the process (these charts may already exist from following the process control procedure and, therefore, may not have to be done at this step). If the key process parameters are not within the limits determined from the previous step, some action plans need to be developed to address this situation. These plans should enable the process parameters to be controlled and kept within their allowable range. These plans may be part of an annual quality improvement (AQI) program.

After the team has improved the process to the point that it is capable ($C_{pk} > 1.67$), a plan for holding this gain and auditing the process should be developed. This is an important step so as not to let the process improvements degrade over time.

3.4 ANNUAL QUALITY IMPROVEMENT

Many manufacturing companies are becoming fully aware of the necessity to improve quality, increase productivity, eliminate waste, and—through continuous improvement—reduce costs. Organizing for AQI is a strategy that many companies are using to accomplish these objectives. Dr. Joseph Juran defines *quality*

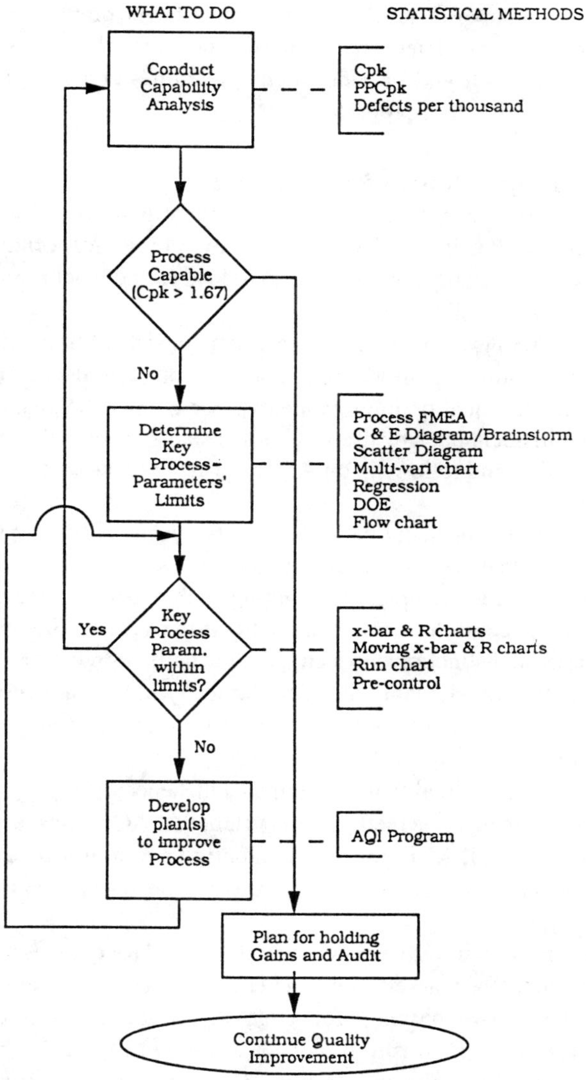

Figure 3.10. Process Potential, Capability & Improvement Procedure.

improvement as "a superior level of performance never previously attained and resulting from the human determination to set a new record." This section defines the components of AQI and provides some guidelines and suggestions for implementing an effective AQI program.

3.4.1 Establishing a Strategy for Success

Annual quality improvement is a key strategy used by many WCM companies. The strategy is really quite simple: By implementing prevention-oriented, process-focused AQI projects, companies are able to reduce variation, eliminate waste, and improve process and product quality.

The AQI program is a structured process that consists of implementing particular projects that result in quality improvements throughout the company. In order to be effective, this quality improvement process must be managed. Quality improvement will not happen by itself. It is the responsibility of the QM committee to initiate, appraise, and follow-up on AQI projects to make certain that quality is improved.

It is important to understand the difference between quality improvement and QC. Quality control centers around maintaining a level of quality. When a process goes out of control, for example, "firefighting" strategies are used to try to bring the process back in control. In contrast, quality improvement changes the level of quality. This critical distinction between QC and quality improvement is illustrated in Figure 3.11. Improved quality allows a company to increase productivity and reduce costs, thereby enabling them to compete more effectively in an international marketplace.

The opportunities for quality improvement and, hence, AQI projects come from all areas of the company. It is therefore important that AQI projects not be defined solely by QA personnel; AQI projects should originate from people in all areas of the company. Since quality is everyone's job, everyone must have a role in defining AQI projects.

Because continuous improvement is a strategy used for competing successfully in today's international marketplace, it is essential that a company have a well-structured and effective program for achieving continuous improvement. The purpose of an AQI program is to provide a well-planned, well-managed process through which specific goals, time frames, and measurable objectives can both be set and monitored. Furthermore, an AQI program helps ensure that the organizational structure and the momentum required for continuous improvement are in place.

3.4.2 Defining AQI Objectives and Projects

An AQI program consists of various AQI objectives, each of which defines a measurable goal for quality improvement in a process. Many companies initially

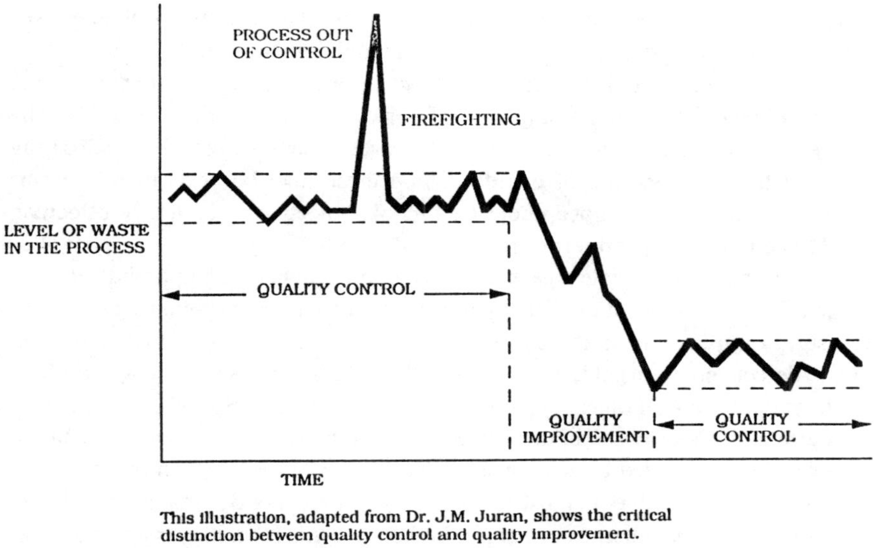

Figure 3.11. Juran's Definition of Quality Improvement.

have difficulty in defining and executing an effective AQI program. One of the reasons is that AQI objectives often have too large of a scope. For example, someone might say that they want to improve productivity in their machine shop, or reduce the defect rate at final-product testing. While these are certainly admirable objectives, they are too broad to be effectively planned, managed, and measured. To avoid this problem, it is useful to insist that every AQI objective must be:

- Process focused
- Measurable (now and in the future)
- Able to be completed in one year (or less)
- Project(s) oriented/executed
- Prevention oriented (reduce chronic levels of waste)

The first requirement is to identify an AQI objective that is process focused, such as reducing the level of waste from a spot-welding operation or reducing parts that need to be repainted due to a high number of defects or blemishes in the painted surface. Not all processes, however, need to be directly related to manufacturing machines. For example, a drafting process with high levels of print errors that consequently causes quality problems on the manufacturing floor and in the field should have a drafting-process AQI objective. Thus, while the interpretation of process can be widely defined, it is useful and essential to define AQI objectives directly related to specific processes that yield high levels of waste.

Another requirement is that the AQI objectives must be statistically measurable now and in the future. In defining an AQI objective, it is not adequate to state the desire to "improve the process," "eliminate waste," "achieve zero defects," or any other such broad measure of quality. While these may be very appropriate as long-term goals, they are not appropriate for AQI objectives, which to be effective, need to be relatively short term (from several months to a year).

Determining appropriate measures of accomplishment is an important yet often difficult part of establishing AQI objectives. One or more appropriate statistics for measuring improvements in quality must be selected. Such statistics as the C_{pk} value, defects per thousand, percent rejected, and scrap rates may be appropriate. At the start of the AQI objective, it is necessary to state, in terms of these statistics, the current levels of waste in the process. Once this base level is known, a target goal can be established by management based on what level of improvement is necessary and justifiable. In most cases, the goal should not be set at the maximum level of improvement that can be achieved because this level may not be the most cost-effective or practical. In any case, however, some measures of accomplishment are essential not only for defining AQI objectives and its related goals, but also for defining the various activities involved in each project.

Not only must AQI objectives be process focused and measurable, but they should also be prevention oriented. The AQI objectives should address actions that will better control the process and reduce the variability of products emanating from the process. In this case, there is again considerable value in identifying causes and effects so that predictive and preventive measures can be determined and implemented.

Each AQI objective must also be project oriented and executed. For each objective, one or more projects are defined (often over a period of time through Pareto analysis, cause-effect analysis, et cetera) to achieve the objective. Ideally, the cause-effect analysis is performed with both supporting data and team brainstorming so that the major contributors to chronic waste can be identified and corrective action projects can be established. In essence, the projects define "how to accomplish" the objective.

It is common to identify and establish two or more projects for each AQI objective, as shown in Figure 3.12. These projects can be handled by a project team that has the knowledge and capabilities to bring about the desired improvement. For example, when addressing the issue of reducing spot-welding defects, it may be appropriate to define three projects: One project can address reducing the variability of the incoming spot-welding material; another project can be focused on training to improve the skill levels of the machine operators; and a third project can be defined to model the relationships between the settings of the machine that

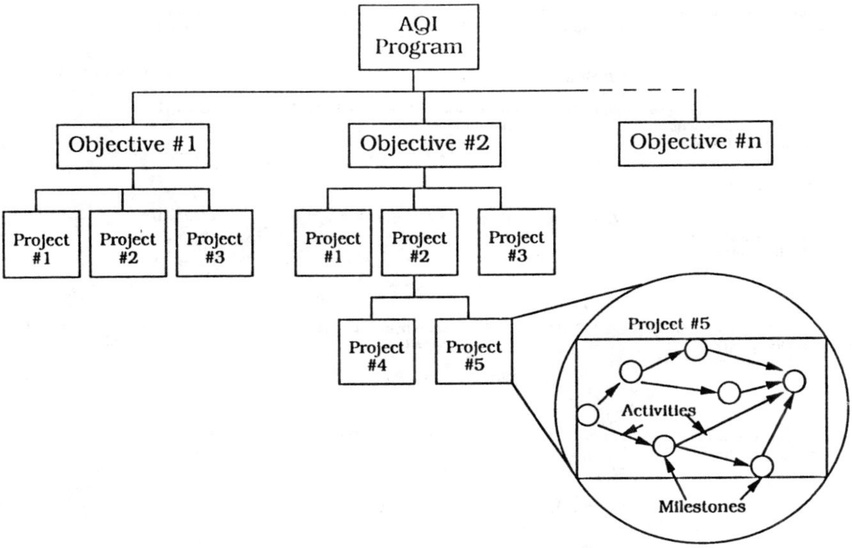

Figure 3.12. Structure of AQI.

is controlling the pressure, the temperature, and the tensile strength of the welded area. It is important to note that projects do not necessarily need to be process focused, but can also be focused on systems, policies, and procedures required to achieve the AQI objective.

Annual quality improvement projects can be managed and controlled more effectively if specific milestones and time frames for those milestones are established. Furthermore, specific responsibilities can be delegated to members of the team concerning the designated time frames. After responsibilities are delegated, the final issue to address concerns how to monitor progress and measure accomplishments for AQI. Figure 3.13 shows a portion of AQI objective number 2 covering project number 2.

In summary, an AQI objective is "what we want to accomplish" in quality improvement, and the associated projects define "how to accomplish" it.

3.4.3 Implementing AQI Projects and Measuring Results

Once an AQI objective is defined, then specific AQI projects can be established and assigned to a team as previously discussed. The team strategy of the AQI project is the most effective way for implementing continuous quality improvement. Most frequently a team is established to focus on solving a specific problem or implementing a set of changes that are necessary to bring about quality improvement. Regardless of what the team is called—task team, quality-circle team, or AQI team—the important issue is that a team approach is taken to address

AQI OBJECTIVE #2: Reduce assembly rejects by 50% of the Sept.-Dec. levels for serialized units (as measured at final test) CONTINUED

PROJECT	TIME FRAME	MEASURE OF ACCOMPLISHMENT	RESPONSIBLE PERSONNEL
2. Improve Current Process Procedures/Instructions with increased use of visual aids	3/90 - 3/91	Have Process Sheets on 80% of Assembly Activities	TEAM 1B
MILESTONES:			
• Determine Process--Procedure needs and Establish Formats	3/90 - 6/90	Process Sheet Formats	TEAM 1B
• Select, Purchase, and Install Computer-Assisted Process Sheets (CAPS) System	4/90 - 7/90	CAPS System Installed and Training Initiated	TEAM 1B above, supported by: A. Smith (Engr.) P. Burns (Q.A.)
• Implement Process Procedure Sheets	12/90	Process Sheets on 50% of Assembly Activities	J. Collins T. Few

Figure 3.13. Example of AQI Program Plan.

problems and bring about quality improvements. A team approach emphasizes that quality is everyone's job and that everyone has certain responsibilities in accomplishing that job.

One of the components of AQI is that the quality improvement must be measurable with the milestones and time frames that are established for the various elements of project activity. Measuring and tracking progress not only provides effective feedback to management on how the AQI project is proceeding, but also provides valuable feedback both to those people who are involved in the project and to those who are affected by the quality problems and chronic levels of waste.

It is often effective to post project performance on some type of chart, whereby everyone can see how the team is progressing. Such charts can be posted at the machine or in work areas close to the area of concern. It is often difficult to foster a worker's enthusiasm for a project if there is no specific feedback detailing his or her achievements. Posting progress also generates a sense that not only are people concerned about monitoring progress, but also that they care enough to take the time to present and display this information. Can you imagine going to a basketball game in which there was no scoreboard? Quite simply, keeping score is important to people: It challenges them, it motivates them, it creates a sense of excitement. Keeping score on AQI projects is equally as important as selecting and defining the AQI objectives. Failing to keep everyone involved in and aware of progress can cause the commitment and support of the team's members to slowly deteriorate.

Many companies are implementing *cost-of-quality* programs in order to provide a basis for defining AQI objectives and for measuring their progress in quality improvement projects. These accounting-based programs typically address the following four cost elements:

1. Prevention costs (employee training, design reviews, and so on)
2. Appraisal costs (inspection, testing, and so on)
3. Internal failure costs (scrap, rework, and so on)
4. External failure costs (warranty, recalls, repairs, and so on)

As the implementation of total quality progresses, the total cost of quality decreases, with prevention costs representing an increasing percentage of total quality costs. While it certainly is important to be able to measure quality costs for good total quality management, it is crucial to maintain the primary emphasis of total quality on quality improvement, not on accounting activities.

3.5 CHAPTER SUMMARY

Developing and implementing a management-led, employee-involved, customer-oriented TQ program is paramount for competing successfully today. The success of a company's TQ program is primarily determined by senior management's commitment to and leadership in changing the company's culture, which affects the organization, attitudes, and rewards systems pertaining to quality.

Articulating and communicating the company's philosophies toward quality, through the company's mission statement and quality objectives, is a key first step. Implementing systems for continuous quality improvement through employee involvement is similarly paramount for success. Employee involvement can be effectively implemented through team approaches for identifying and solving quality problems and through a committee approach to managing the quality systems at a company. Only with senior management's involvement, commitment, and leadership can such an approach to quality take place and succeed.

4

Standardization and Simplification Via Group Technology

Standardizing and simplifying all parts and processes is a fundamental step in the journey to becoming mean, lean, and world-class competitive. Standardization and simplification is a key strategy for the nineties because it is a low-cost low-technology approach for identifying waste, reducing costs, and ultimately, increasing profitability. This chapter addresses the role of group technology in standardizing and simplifying design and manufacturing operations; thereby facilitating the production of quality products with the highest levels of efficiency and profitability. The major emphasis of this chapter is the application of the group technology philosophy for identifying and forming focused operations and facilities—often referred to as cellular manufacturing.

4.1 THE ROLE OF GROUP TECHNOLOGY IN DISCRETE PARTS MANUFACTURING

Complexity is truly a curse upon manufacturing. The typical company makes thousands of different parts, in many different batch sizes, using a variety of different manufacturing operations, processes, and technologies. It is beyond the capability of the human mind to comprehend and manipulate such vast amounts of detailed data; yet, people still need to make decisions regarding how to run a manufacturing company and succeed in today's competitive environment. The pressures on management are continuing to escalate as global competition drives the need for producing a greater variety of high-quality products, in smaller lot sizes, and at lower costs. These ongoing demands continuously increase the level of complexity present in a manufacturing environment. What is needed is both a strategy and a tool that can be used by managers to achieve greater understanding, order, and simplicity regarding manufacturing operations.

4.1.1 Definition of Group Technology

Group Technology (GT) is a manufacturing philosophy and strategy that assists a company in understanding what it manufactures and how those products are then manufactured. Group Technology provides a means to identify and exploit the "underlying sameness" or similarities of parts and processes. Once identified, it is possible to capitalize on these similarities by processing together groups of similar parts (families), by standardizing and simplifying closely related and repetitive activities to avoid unnecessary duplication of effort, and by efficiently storing and retrieving information related to recurring problems; thereby avoiding solving the same problem again and again. A vice president of manufacturing once described GT as a means to standardize and simplify manufacturing parts and operations so that they come within the bounds of human comprehension. Once comprehension is achieved, the means exist to eliminate waste and achieve improved efficiencies and productivity that create value for the customer.

4.1.2 Applications in Design, Manufacturing, and Management

The applications of GT in manufacturing companies are quite broad; however, the greatest opportunities for improving operations are in processing discrete parts in small to medium batches—an ongoing trend in today's environment of JIT manufacturing.

In product design, GT helps to classify and code parts on the basis of their geometric similarities, with emphasis on part families with similar or equivalent function, shape, and size. Group technology is often implemented in design engineering through the use of a formal classification and coding procedure that is incorporated into a computerized design-retrieval system. Thus, when a new part is needed, the design engineer can search the data base for existing parts that have similar functionality and geometric features as the new required part. In some cases, it is possible to use an existing part and thereby avoid generating another unique part number. In other cases, a new part can be designed to have the same manufacturing requirements as a family of existing parts. Group technology thus provides cost savings and simplification through reductions in engineering design time, new-part designs, bills of materials, and part proliferation.

In manufacturing engineering, GT focuses on similar machining operations, similar tooling and machine setup procedures, and similar methods for transporting and storing materials. By identifying similarities in manufacturing, similar piece parts can be grouped into distinct families and processed together in a dedicated workcell comprised of dissimilar machines and processes. For example, a company that produces 12,000 different component parts in batch sizes of 10–1,500 pieces on 300 different machines can obtain major efficiencies by grouping the majority of these parts into forty to sixty distinct families, each having similar manufacturing requirements. In this manner, the company can process any part in a given family using a focused and standardized set of operations. In addition, certain efficiencies can be gained by simply being able to view and manage the resulting families of parts as an entity. This standardization provides economies of scale that justify expenditures on reducing the costs of design and manufacturing for an entire part family. Quite simply, group technology via cellular manufacturing provides a strategy to gain the advantages normally associated with high-volume (repetitive flow) production, but doing so in a small- to medium-volume batch environment. This strategy has fundamental importance to batch manufacturing operations in a JIT environment.

Group technology can also be a building block to computer-integrated manufacturing (CIM), because GT provides a link between computer-aided design (CAD) and computer-aided manufacturing (CAM)—typically manifested through computer-aided process planning (CAPP) systems. Computer-aided process planning systems provide the benefits of greater consistency, accuracy, and timeliness in translating part design specifications into manufacturing operation instructions when converting a part from a rough to a finished state. In general, GT provides a means to organize and structure data about manufacturing parts and processes and subsequently to store and retrieve that information using computers. Because CIM's foundation is integrating, retrieving, and analyzing information through computer technologies, GT is, therefore, a fundamental building block for CAD/CAM integration and CIM.

Although GT has many significant applications, it is most successfully employed when viewed as a philosophy for achieving order and simplification in manufacturing operations. It is evident that GT is a fundamental strategy in achieving WCM. What is not evident is where, when, and how GT can be applied to achieve WCM given the size, type, and current environment of a company. These issues are addressed in this chapter.

4.1.3 Part Families

The application of GT is primarily based on the concept of part families. A part family is a group of parts that have either design similarities or manufacturing (geometric shape and size) or manufacturing similarities (machines, tooling, process sequences, et cetera). Some parts may look similar to each other, but because of differences in materials, tolerances, or other production requirements, they have different manufacturing requirements and thus do not constitute a "manufacturing family of parts." Additionally, parts that look different from a design perspective may in fact be processed in a similar fashion. In most cases, the emphasis is to identify families of parts that can be manufactured in a standardized and simplified manner; however, since all the parts in a family are certainly not identical, what set of similar attributes need to exist in order to group parts into the same family? How does one identify the characteristics of a part that warrant it being placed into one family and not another? Identifying and forming part families is fundamental for achieving high levels of standardization and simplification. The following sections of this chapter address various techniques for establishing part families and discuss the applicability of GT in companies of various types and sizes.

4.2 CLASSIFICATION AND CODING

One method used to establish part families is to examine the parts and then classify and code each part into a group with similar attributes. Classification is the process of identifying and establishing the various classes or divisions that exist for a set of parts based on relevant attributes. For example, the ratio of length to the diameter of a cylindrical part might be one attribute that distinguishes a family of "shaft type" parts (requiring turning between centers in a lathe) from a family of "hubs" (requiring turning in a chucking lathe). This type of information is embodied in a code: A string (set) of characters or symbols that determines and indicates the class (family) to which a part belongs. In addition, codes provide a means to describe part attributes in a format that is suitable for data analysis and processing. The coding process involves assigning characters and/or symbols that characterize a part's attributes, thereby also determining the family to which it belongs.

4.2.1 Code Structures

The characters or symbols used in a GT code can be either all numerical, all alphabetical, or a combination of both. Although most of the early GT codes were numerical and only six to twelve digits in length, modern commercial coding and classification systems often emphasize CIM integration and, therefore, use larger computer-based, non numeric codes to increase user-friendliness and meet the needs of both design and manufacturing engineering. All codes can be classified into one of three basic structures:[1]

1. Hierarchical code or monocode
2. Chain-type code or polycode
3. Hybrid or mixed code

---

[1]Group Technology Oriented Coding Systems: Structures, Applications, and Implementation, Nancy Lea Hyer and Urban Wemmerlöv, Production and Inventory Management, Second Quater 1985.

The hierarchical or monocode consists of dependent digits; that is, the meaning of each character or symbol is dependent on the meaning of the previous character in the code string. The common postal zip code is an example of a monocode. Consider the zip code 53706. The value 53 indicates that the postal region is the southern part of the state of Wisconsin. The value 7 denotes the city of Madison, and the value of 06 represents the region of the city occupied by the University of Wisconsin-Madison. In the zip code 53206, the value 2 indicates the city of Milwaukee and the value of 06, in this case, refers to a postal region in Milwaukee. Thus, the meaning of the value 06 depends on the value of the city code designator. Figure 4.1 shows an example of a hypothetical monocode for a particular manufacturing environment. The advantage of the monocode is that it has a very compact structure; thereby allowing a large amount of information to be packed into a relatively small number of digits. Defining and later interpreting the meaning of each digit, however, is more difficult with a monocode.

The polycode has a chain-type structure with independent digits so that each characteristic or attribute of the part is assigned a fixed position in the code. The meaning of each character does not depend on the meaning of the previous character. An example of a polycode is 10/20/51—the slash form of expressing dates, where the meaning of the value 20 clearly does not depend on the meaning of the value 10. Polycodes have the advantage that they are easy to read and interpret, and also provide a quick identification of parts with similar attributes that often have similar processing requirements.

The hybrid code is an attempt to combine the advantages of both the monocode and the polycode, and is the most common structure for commercial coding systems. Typically, the code structure consists of several groupings of short segments of independent digits. For example, the first digit in the hybrid code may denote the type of part (shaft, gear, hub, housing). The next few digits in that group are a polycode describing particular attributes for the part type denoted by digit one. The remaining groups of digits in the hybrid code represent dimensional characteristics, material types, tolerance specifications, and so forth. Again, within each group, there could be a given number of digits in a chain-type structure to describe the particular attributes. Hybrid codes are the most versatile and, therefore, are often used to describe parts from both a design and a manufacturing point of view. They are relatively compact in structure, yet they provide for ease in identifying parts with specific characteristics and related processing requirements.

Currently, formal GT codes are being replaced, in some applications, by relational data base systems. A relational data base allows actual data to be stored, thereby eliminating the formal classification and coding process. Natural or query languages can be used to sort, select, and essentially classify parts on the basis of whatever attributes may be needed for part analysis. This type of system has the added advantage that report generation and graphical presentation are readily

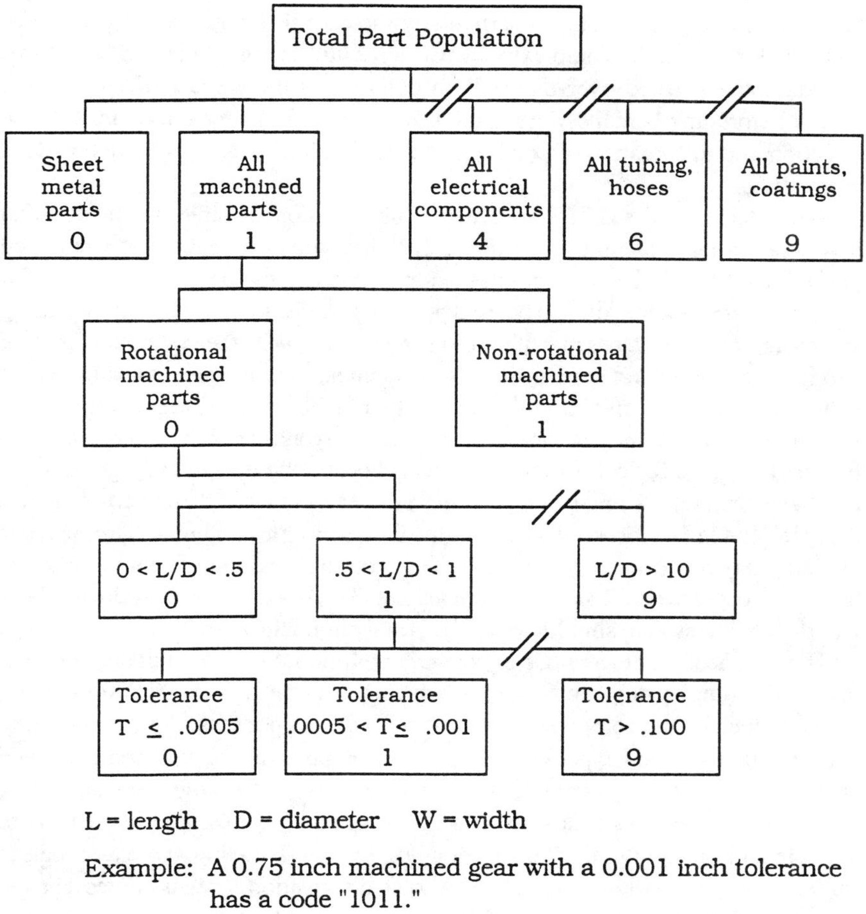

Figure 4.1. Diagram of a Portion of a Hypothetical Monocode.

available and easy to use. In addition, such data base systems are particularly advantageous in companies where networked computer structures and CAD systems are used.

4.2.2 Selection and Justification of a GT Coding System

While all GT coding schemes involve the use of logic, attribute definition, and symbols, ironically, there are no standardized procedures for generating codes. Furthermore, purchasing and installing a GT coding system can represent a sizable investment of both capital and personnel, with no guarantee that commensurate benefits will be gained. This is particularly the case if the number of parts to be coded is large, and the complexity of part attributes is high. One U.S. company, for example, invested over one million dollars and three years of engineering efforts before any benefits were gained. It is, therefore, very important that the selection and justification of coding and classification (C & C) systems be done correctly.

When deciding on a GT coding scheme, there is a considerable amount of initial investigation and planning to be done. The first step is to determine what need exists for a formal group technology coding or relational data base system. For example, is the system going to be used primarily by design engineering for storing and retrieving parts? Does manufacturing engineering need the system for forming part families, or only for storing and retrieving manufacturing process plans? Some coding systems are only useful for variety reduction and design information storage and retrieval and do not have the capability as stand-alone procedures to find part families for cell formation. This is because codes typically group parts that have geometric similarities but differ in tolerances and/or batch sizes; hence, the parts should be processed on different types of machines. Similarly, some types of codes may not group together parts having dissimilar design features, although they are processed on the same set of machines. The key question to address is how will the coding system simplify operations and eliminate waste?

The justification process has to consider both initial costs and operating costs of a GT coding system. Initial costs, which can often total $100,000 or more, typically include the computer hardware, software license fees, user training, and part coding and entry expenses. Ongoing operating costs include computer and data base maintenance, computer time and supplies, part changes and additions, and training. These costs have to be weighed against the expected benefits, both tangible and intangible. Because most commercial GT coding and classification software is now marketed as part of a total information system addressing the issues of CIM integration, the justification process is both broader in scope and more strategic in nature than in the past.

In a recent study on the economic justification of C & C systems, the following reasons were given as to why more companies do not implement C & C into their organizations—in spite of the benefits reported by both users and vendors.[2]

1. The justification process is time and resource consuming because, typically, many departments in a company are affected and each anticipates different benefits. Unless top management takes a leadership position, the project usually loses momentum and is abandoned.

---

[2]Technical Report by Jerry W. McKinney, Manufacturing System Engineering—University of Wisconsin-Madison, May 1988.

2. Traditional justification techniques do not consider the intangible or strategic benefits and, thus, do not look attractive given the high initial costs.
3. The cost of "not implementing" C & C is not considered; therefore, the benefits associated with being flexible and responsive to changes in the marketplace are overlooked.
4. Many people do not understand group technology and C & C systems and, therefore, do not feel comfortable in recommending their use. Essentially this is a fear of change.
5. The initial investment in capital and personnel is quite high, and the reported or expected benefits are not guaranteed nor immediately attainable.

These reported observations provide several implications for management to implement C & C systems successfully. Education is, once again, the primary and most important step in implementing new technology into an organization. The fear of the unknown and the resulting inertia or resistance to change is formidable. Since the expertise for education most likely does not exist within the company, it is essential for management to identify knowledgeable resources and bring them into the organization (via seminars, courses, consultancy, et cetera) to foster growth and advancement. It is also essential to recognize that education and technology are often long-term investments and typically not visible in next quarter's profit statement. However, WCM is not to be achieved by, or viewed as, a "quick fix."

4.2.3 Group Technology in Small Companies

On the other hand, the use of GT coding can be both inexpensive and simple to implement, thus making it easily utilized by small companies. Figure 4.2 shows a polycode that was developed by a company that needed to write routing/process sheets for the approximately 800 parts manufactured in their machine shop. Although around 250 operation sheets had already been written, problems with inconsistency and inaccuracy were present—largely because the sheets were written over a period of time by different process planners. Consequently, the company developed a simple C & C procedure, reviewed all existing routings/process sheets, made revisions when needed, and coded each revised process plan accordingly. The part number and GT code number were then stored on a personal computer using inexpensive data base software. Now, when a process sheet needs to be written, the part is coded and the data base is searched to identify the part numbers of any and all existing process sheets. The process planner then goes to the file cabinet and withdraws the process sheets, which are used for reference. Although this system is unsophisticated, it works well and provides improved standardization and consistency in their manufacturing process plans and operations. This company later used the code to identify families of parts that could be processed together, and thereby determined how to reorganize the layout of their

PROCESS SHEET CODING SYSTEM

	DIGIT					
VALUE	FIRST	SECOND	THIRD	FOURTH	FIFTH	SIXTH
1	Rotational L/D ≤ 2	No holes	No keyway	L.D. ≤ 1"	S.L.D. ≤ 1"	Tol. ≤ .0005
2	Rotational 2 < L/D ≤ 15	One hole at center	External keyway	1"<L.D.≤6"	1"<S.L.D.≤6"	.0005<Tol. ≤.001
3	Rotational L/D >15	One hole off-center	Internal keyway	6"<L.D.≤20"	6"<S.L.D.≤20"	.001<Tol. ≤.005
4	Flat Part	Two or more holes	Surface slots, grooves, angles	L.D. > 20"	S.L.D. > 20"	.005<Tol. ≤.010
5	Box-shaped part					.010<Tol. ≤.016
6	Other main shapes					.016<Tol. ≤.030
7						.030<Tol. ≤.060
8						.060<Tol. ≤.125
9						Tol. > .125

L.D. = Largest Dimension
S.L.D. = Second Largest Dimension
L = Length of Rotational Part .
D = Diameter of Rotational Part

Figure 4.2. Polycode for Coding and Classifying Routing Sheets.

facilities into a cellular configuration composed of distinct workcells. Clearly, the philosophy of group technology has applications in any size manufacturing company.

4.3 FOCUSED FACILITIES AND OPERATIONS

The layout design of a manufacturing facility is one of the most important factors affecting product quality and cost. The manner in which the equipment is configured on the shop floor affects material flow; manufacturing leadtimes and work-in-process inventories; in-process quality control; and the manner in which work is scheduled, processed, and controlled through production. Layout configuration is not only a long-term strategic decision, but is also a determinant for achieving WCM.

The GT philosophy has a fundamental role in determining layout configuration, particularly for companies involved in producing discrete parts in small-to-medium batch sizes. Since these types of operations intrinsically involve high levels of complexity in terms of a multitude of different parts, machines, and other processing requirements, the need and opportunities to standardize and simplify are great. In particular, GT provides a means to group parts into families and group machines into work centers, and then combine those work centers into workcells that are dedicated to processing a family of parts. By focusing operations in workcells, operational complexity is reduced and greater control of (and opportunities for) ongoing improvements in quality and productivity are possible. This section discusses the various types of layout configurations and provides an overview of the advantages and limitations of manufacturing workcells.

4.3.1 Types of Layout Configurations

There are basically four ways machines can be configured on a factory shop floor. The layout configuration that is most appropriate depends on the volume and variety of different parts to be processed. The relationship between part volume and part variety is shown in Figure 4.3 for each of the four types of layout configurations, which are typically called:

- Product line layout
- GT flow line workcell
- GT manufacturing workcell
- Functional (job-shop) layout

The four configurations differ in the way they process parts and in the pattern that the material (work) flows through the machines.

The product line layout, shown in Figure 4.4a, is used for continuous-flow fabrication (or assembly) of large quantities of a part or in process industries. Consisting of special purpose machines and material handling equipment, the line

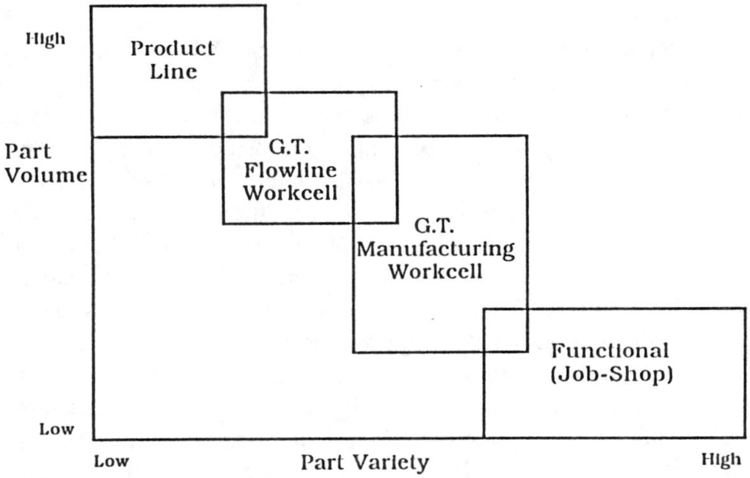

Figure 4.3. Part Volume/Variety Relationships with Manufacturing Systems Configurations.

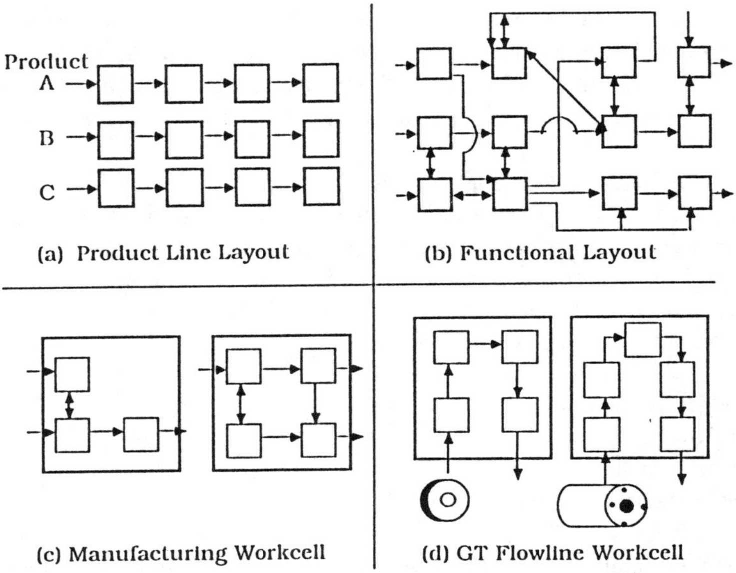

Figure 4.4. Four Configurations for Machine Layout and Part Processing.

layout is arranged in a balanced line to accommodate continuous-flow processing. Although the line is capable of high-volume production with high throughput rates, low in-process inventories, and high degrees of focus and control, the specialized nature of the line makes it very inflexible and incapable of producing a variety of different parts.

In the functional or job-shop layout, shown in Figure 4.4b, machines are grouped together by similar functional type, and they are often designated and controlled as different departments. The functional layout is the most flexible because it consists of general purpose machines that are capable of processing dissimilar parts in a wide variety of different volumes. However, functional layout configurations have numerous disadvantages: Low throughput rates, high levels of work in-process inventories, significant amounts of material handling, and spaghetti-like patterns of material flow, which in turn, makes scheduling and control of work difficult to manage. In essence, functional layouts lack focus, standardization, and simplicity in terms of work flow and control. Although the functional layout is still the most common layout configuration in the United States, many companies are now reorganizing their facilities into cellular configurations.

In a manufacturing workcell, machines of dissimilar functional type are grouped together and dedicated to processing a family of (similar) parts. A GT manufacturing workcell (Figure 4.4c) provides greater focus and control than the functional layout, while still allowing for the processing of different parts (in the same family) in small to medium batch sizes. In essence, each workcell is a small dedicated job-shop. The disadvantage of this cell configuration is that continuous-flow processing is often not possible because the flow of work is not unidirectional. Although the analysis of existing routings will typically suggest the use of such workcells, it is usually a mistake to implement workcells that have backtracking flow patterns.

The GT flow-line workcell, shown in Figure 4.4d, is the configuration that best realizes the full benefits of cellular manufacturing. This type of workcell provides the advantages and efficiencies of high-volume production (product lines) in a low- or medium-volume (flexible manufacturing) environment. Specifically, each flow-line workcell can process a GT-based family of parts in low-to-medium batch sizes with piece-by-piece continuous flow through a line of general purpose, dissimilar machine types. Although backtracking is not allowed, machine skipping (leap frogging) is typical. The GT flow-line workcell represents the necessary and ideal configuration to help a company achieve the objectives and advantages associated with WCM. Hereafter, the GT flow-line workcell will be simply referred to as a "workcell."

4.3.2 Advantages of Workcells

Reorganizing a factory from a functional to a cellular configuration is a major undertaking requiring the involvement and dedicated commitment of both management and shop-floor personnel. Therefore, it is particularly important that everyone knows and believes in the advantages of cellular manufacturing and also understands how and why these advantages are possible. The following details the major advantages of cellular manufacturing and the rational for each.

1. *A reduction of 70–90 percent in production lead times and work-in-process inventories has been reported by many companies of various sizes.* These magnitudes of reductions result primarily from changing from a lot (batch) mode of processing to a continuous-flow mode of processing. The difference between processing modes is illustrated in Figure 4.5 for the particular case of processing a single job of 100 pieces through two consecutive operations, resulting in a 50 percent reduction in the production lead time for continuous processing. The amount of lead time reduction is also dependent on the number of production operations—as Figure 4.6 shows. Since many jobs in a factory have five or more operations, it is not surprising that continuous-flow processing has significantly reduced production lead times and their associated levels of work-in-process. Such reductions, however, are only possible with flow-line workcells that have reasonable balance and reliable system components (issues which are addressed in chapters 5 and 9 respectively).

Given:

Job	Batch	MACHINE 1		MACHINE 2	
		Setup	Process	Setup	Process
1	100 pcs	60 min	1 min/pc	70 min	0.9 min/pc

CASE A: LOT PROCESSING

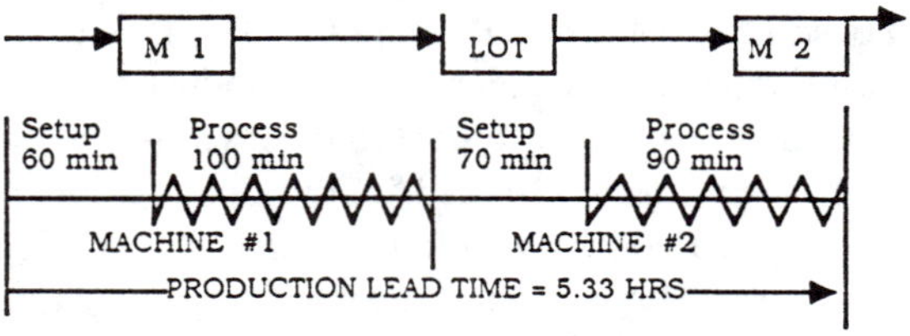

CASE B: CONTINUOUS FLOW

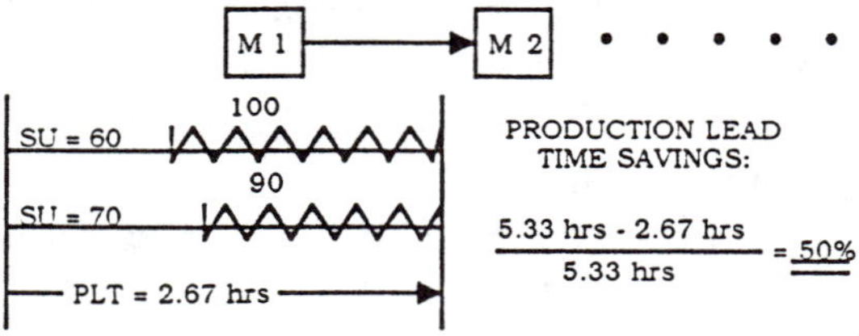

Figure 4.5. Lot Versus Continuous Flow Processing.

2. *A reduction of 75–90 percent in material handling is commonly achieved, along with a 20–45 percent reduction in the amount of factory floor space required to produce the same number of products.* Since material handling adds nothing to the product except cost, this reduction is an effective elimination of waste. Furthermore, by reducing both the amount of handling and the levels of in-process inventories, considerable floor space is gained for productive activities.
3. *Considerable advantages related to direct labor are also achieved with workcells.* By implementing cross-training programs, operators can learn to attend and run more than one machine at a time. Not only does this increase worker productivity, but it also provides management with the flexibility to run the cell at different capacity levels by simply adjusting the number of operators and assignments in the cell. (Design strategies for operator levels and assignments are discussed in chapter 5.) Remember to structure win-win situations to assure labor that these changes are not a threat to job security or job quality. Most companies have been able to reduce the direct

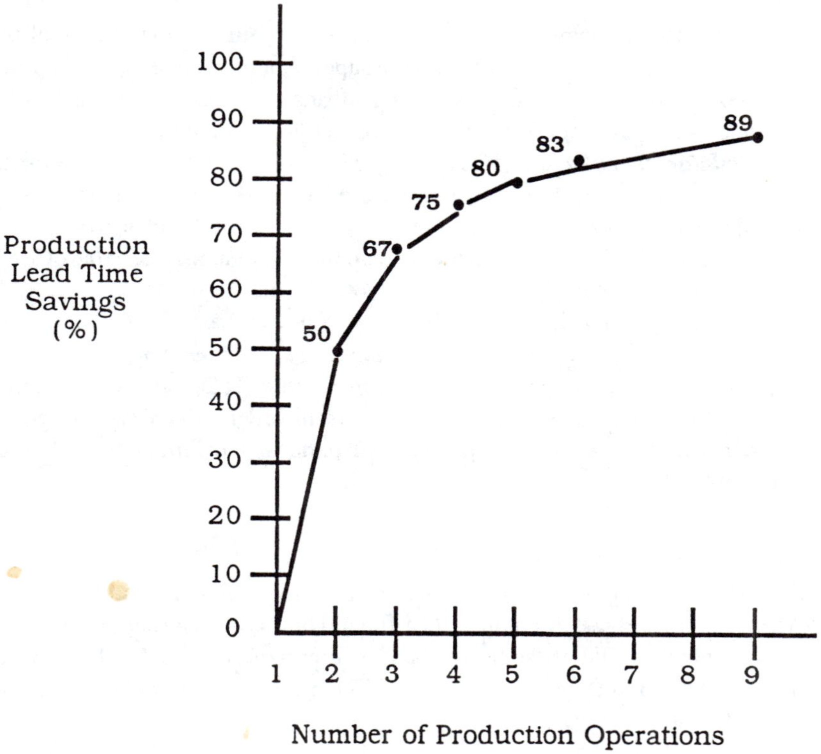

Figure 4.6. Production Lead Time Reduction Versus the Number of Production Operations.

labor content in a part by 15-55 percent, yet still provide higher levels of employment with greater degrees of worker involvement and job satisfaction. Harley Davidson, for example, has reported productivity gains of 33 percent and labor grievance reductions of 35 percent—a nice combination!

4. *The standardization that results from focusing on part families also helps decrease machine setup times by 65–80 percent.* While part of this reduction is attributable to standardized tooling and fixturing, a significant reduction is also due to the improved organization and planning that comes from having a better understanding and focus of what is being manufactured. Since setup-time reduction is another component of WCM, the subject is addressed in detail in chapter 6.
5. *Quality-related problems are reported to decrease 50–80 percent in work-cell environments.* These reductions are attainable primarily from increased machine operator involvement in problem prevention. The combination of clustered operations, continuous-material flow, and statistical tools such as control charts, provide both an awareness and a means to respond quickly

to quality problems. Increased worker involvement in the quality of parts produced in the cells also allows the supervisor to assume more of a role as "coach"—supporting the cell team's efforts, and less of a role as "policeman"—controlling to ensure that production goals are met.

6. *Cellular manufacturing simplifies shop-floor controls and decreases paperwork.* Control by paper can be replaced by control by sight. Companies have reported corresponding reductions in indirect labor of 25-55 percent. The issue of production planning and shop-floor control is discussed in chapter 8.
7. *An indirect benefit that often results from the process of planning and implementing cellular manufacturing is better communication between product design and manufacturing engineering.* This often occurs from the need to redesign or modify some parts in order to achieve the product consistency necessary to produce all parts in the family in a flow-line workcell.

4.3.3 Limitations of Cells

Although cellular layout configurations have many advantages, they also have some disadvantages when compared to functional layouts. When assessing the overall economic value of the tradeoffs, it is important to consider the impact of these limitations in different environments. The most common limitations of cells are the following.

1. *Reduced machine utilization is one of the biggest limitations of organizing a factory in a cellular layout.* Consider a company that has one vertical boring mill with only a 78 percent utilization. The cell formation analysis suggests that two separate workcells should be established and that each requires a vertical boring mill. If another vertical boring machine is purchased to satisfy the objective of continuous flow operations in the two workcells, then the individual boring machines will have only 39% utilization on the average. This dilemma raises the fundamental question of what is the overall objective? If the company wants to produce high-quality parts to satisfy the customer's needs while keeping inventories at a minimum, then high machine utilization—though traditionally held sacred—should not be a primary objective. While high levels of machine utilization are desirable, making machine utilization a primary objective has all too often filled warehouses and stockrooms with materials for which no customer demand exists. Reduced machine utilization may be a justifiable cost if they are outweighed by the overall returns and benefits from cellular manufacturing.
2. *In spite of the term "flexible" manufacturing cells, workcells are, in fact,*

less flexible than functional layouts. Flow-line workcells are only more flexible compared to product line systems. Consequently, cellular layout configurations are more prone to obsolescence, particularly in environments with relatively short product life cycles. While changing the layout of facilities to best meet the needs of the product is not intrinsically undesirable—in fact just the opposite—there is a cost associated with planning and implementing new layouts.

3. *In a cellular layout, equipment failure has more costly and damaging effects in terms of decreasing equipment utilization and lowering the throughput capacity of the production facilities.* In an environment of continuous-flow processing and minimal in-process inventories, a breakdown of any machine in the flow line can quickly bring the entire line to a stop. This unacceptable situation can be greatly minimized, however, through the use of "Total Preventive Maintenance," a topic discussed in chapter 9.

Besides these particular disadvantages, cellular manufacturing also has associated requirements that demand the investment of costs and efforts. One of these requirements is a training program to cross-train the operators on different machines in the cell and teach statistical problem solving and QC techniques. It is also necessary to develop and build a teamwork atmosphere and mode of operating in the workcells. It is important to understand that simply implementing workcells is not sufficient to reap the benefits of cellular manufacturing: Management must also provide the education and training required to achieve a team approach and employee commitment to creating value for the customer.

4.4 CELL FORMATION

One of the first steps to cellular manufacturing is to identify or "form" suitable workcells. The cell formation problem can be stated quite simply. Given a set of parts (with known routings, batch sizes, machine setup and processing times, and estimated annual requirements) and a set of machines, group the parts into "families" and organize the machines into "machine groups," so that the families can be assigned to machine groups to form cells in a way that satisfies the constraints of the problem and minimizes the costs of production. The process of cell formation identifies and exploits the linkages that exist among parts and machines. For example, parts are linked or related if they are processed on the same set of machines, and machines are related if they process the same set of parts.

The cell formation process is illustrated in Figure 4.7. Given an initial set of parts, part families can be identified either on the basis of part attributes (often through the use of a GT code) or by examining part routings for sets of parts with

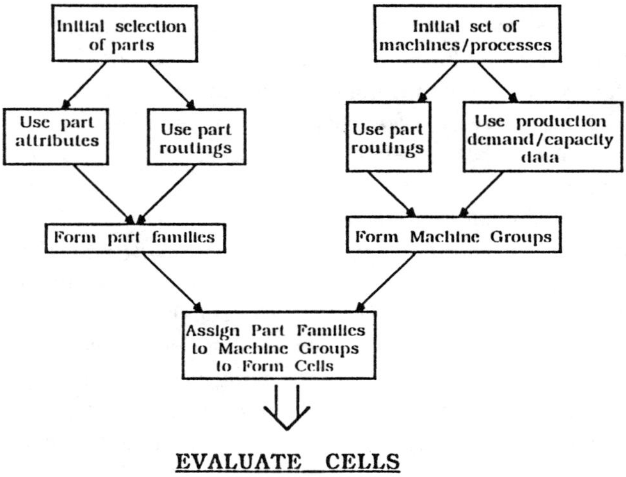

Figure 4.7. Cell Formation Process.

production similarities. Similarly, given an initial set of machines, key groups or clusters of machines can be identified either on the basis of routings or by examining production data, capacities, and utilizations. The part families are then assigned to the machine groups to form workcells, often through a process requiring both judgments and adjustments. A description of the various procedures for the identification of part families/machine groups is given in the following sections.

4.4.1 Visual Methods

Visual search methods involve looking at physical parts, drawings, and routings to identify families of parts that have similar processing requirements. This approach has the advantages of being quick and relatively inexpensive; moreover, it also incorporates the knowledge of personnel experienced with machine capabilities and processes. In many instances, however, this informal GT approach provides inaccurate and inconsistent results, particularly if a large number of parts and machines is being considered. At best, visual methods provide a "rough cut" determination of possible workcells, which then can be refined into flow-line workcells that have good consistency and continuous-flow processing. The quality of the results, to a large extent, depends on the knowledge and initiative of the personnel involved.

4.4.2 Classification and Coding

The identification of manufacturing workcells by using a GT C & C system involves three steps. The first step is classification, that is, establishing sets of categories of relevant attributes, characteristics, and class sizes for all the parts being considered. In step two, an alphanumeric code is assigned to each part; the code then serves as a means to represent classified information. The third step is to identify part families that have part processing similarities by using the coded data of part characteristics. It is essential that the codes used to identify "cell families" be capable of capturing and representing traits that are related to manufacturing processing.

This approach to cell formation has the advantages of eliminating inconsistencies and inaccuracies associated with human judgment and providing an opportunity to select and group parts in different configurations depending on the number and types of machines to be used in the proposed workcells. The disadvantage of this approach is that considerable time and expense are often required to develop and implement a data base of coded parts. Furthermore, because the codes often provide limited insight into design and processing alternatives, the judgments and initiatives of personnel from design and manufacturing engineering are still needed. If a company already has a classification and coding system, or needs one for additional reasons, then it is reasonable and advantageous to use it for cell formation. On the other hand, implementing a classification and coding system solely for cell formation is difficult to justify because other approaches can be used to determine workcells in a far more timely and cost-effective manner.

4.4.3 Production Flow Analysis

One of the more effective approaches to forming cells in facilities design is production flow analysis (PFA), developed by Burbridge in 1975. Production flow analysis is a structured technique used for determining part families and machine groups simultaneously by analyzing route sheets for parts (or subassemblies) fabricated in the shop. Production flow analysis groups into families parts that have similar operational sequences and machine routings; grouping the machines that perform these similar operations into cells. Figure 4.8 illustrates the comparison of an initial machine-part matrix and a restructured matrix. Notice how the grouping procedure was performed by reordering the rows and columns of the matrix to form the cells.

The PFA technique has several advantages as a means to identify potential workcells. First, this technique can be used when the shape of the parts has little or no relation with the manufacturing methods needed to produce them; thus, the tendency to identify part families solely on the basis of similar function, part names, or physical appearance is avoided. Second, PFA can identify workcells more quickly and with much less effort than can the classification and coding system. Third, because PFA is based on routing sheets, the technique focuses solely on current manufacturing methods and uses existing processing equipment and

Mach \ Part	1	2	3	4	5	6	7	8	9	10	11	12	13	14	15	16	17	18
1		1		1	1												1	
2	1		1		1											1		
3						1					1				1		1	1
4		1						1	1	1			1	1	1		1	
5						1			1	1				1			1	1
6						1			1	1				1			1	
7	1				1							1				1		
8	1				1							1				1		
9		1		1				1					1					
10		1		1				1					1					

(a) Initial Machine-Part Matrix

Mach \ Part	6	9	10	11	14	15	17	18	2	4	7	8	13	1	3	5	12	16
3	1			1		1	1	1										
5	1	1	1		1		1	1										
6	1	1	1		1		1											
4a		1	1		1	1	1											
4b									1		1	1	1					
1							1		1	1	1					1		
9									1	1		1	1					
10									1	1		1	1					
2														1	1	1		1
7														1		1	1	1
8														1		1	1	1

Cell 1 (machines 3, 5, 6, 4a); Cell 2 (machines 4b, 1, 9, 10); Cell 3 (machines 2, 7, 8)

(b) Reordered into Three Groups with Duplication of Machine #4

Figure 4.8. Production Flow Analysis with a Machine-Part Matrix Reordering.

tooling. Fourth, PFA offers a way to reorganize existing facilities and gain some advantages of cellular manufacturing with the least possible investment.

Using existing routings as a basis for cell formation, however, has some possible pitfalls. If accurate and consistent routings do not exist for all the parts, then it may not be possible to identify part families and/or machine groups. The identified workcells may also incorporate the limitations and weaknesses that are inherent in the current manufacturing methods and procedures. Production flow analysis also requires that the process planner and/or analyst address the problems associated with multiple routings, unique operations, subcontracted operations, and the situation of one or more "critical" machines being required for the majority of all parts. In the critical machine situation, it is often necessary to initially eliminate these machines from the machine-part matrix prior to sorting and then duplicate the machines so that each cell will contain at least one of the critical machine types. For example, note in Figure 4.8b that machine number 4 was duplicated for workcells 1 and 2. Furthermore, the PFA technique does not identify flow-line workcells unless they are already inherent in the routing structure. Consequently, considerable cooperative efforts and initiatives are usually required by design and manufacturing engineering.

4.4.4 Key-Machine (Clustering) Approach

The key-machine approach involves identifying machines that are considered very important (special capabilities, high investment cost, large work loads, et cetera) and then forming a workcell around one or more of these machines. The key machine processes parts that, in turn, are routed to other machines. These other machines then become candidates for inclusion in the cell containing the key machine. In Figure 4.9, machine number 2 is the key machine. It processes parts P1, P4, and P5, which therefore, are considered as belonging to that cell. These parts, in turn, are processed on machines M1, M3, M4, and M5. In order to determine which of these machines should also be included in the cell, it is appropriate to measure the similarity or "affinity" between machines in terms of some quantitative measure. One such measure is a similarity coefficient defined as follows

$$S_{ij} = \frac{X_{ij}}{X_{ii} + X_{jj} - X_{ij}}$$

X_{ij} = number of parts on BOTH Machine i and j
X_{ii} = number of parts processed on Machine i
X_{jj} = number of parts processed on Machine j

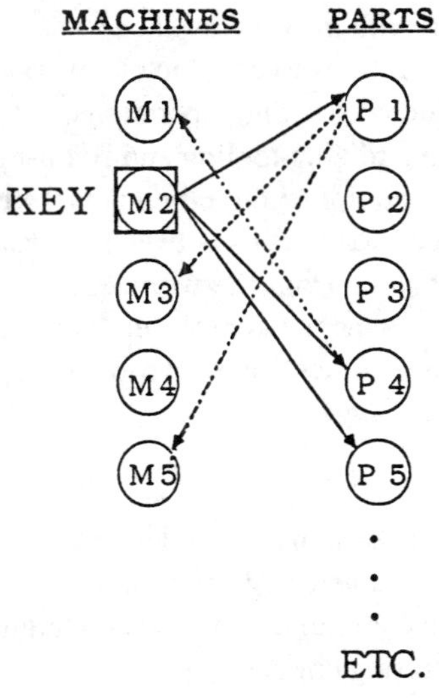

Figure 4.9. Key Machine Approach to Cell Formation.

If a candidate machine has a similarity coefficient with a key machine that is larger than some arbitrarily defined critical value, or some other machine already is being considered in the cell, then the candidate machine is entered into the cell. There are numerous measures of similarity and procedures for applying this approach to cell formation.

4.4.5 Developing Flow-Line Workcells

Because many cell-formation procedures are based on existing routing information, the initially identified workcell configuration usually lacks consistency in both processing methods and sequence. This lack of standardization will result in a workcell that, if implemented, will not achieve continuous-flow processing. The backtracking in material flow and the imbalance between the workstations in the cell will result in poor levels of equipment utilization and throughput. Many companies have learned this lesson through painful experience and the expense of reorganizing the workcell and its processing on the shop floor. It is therefore essential to invest a little additional time and effort before implementation to achieve flow-line workcell designs.

Developing flow-line workcells from the "rough cut" cell configuration often involves one or more of the following activities:

- Redesigning the parts in terms of either geometry or materials to make processing consistent with in-line flow.
- Combining different routings to develop one workable routing procedure.
- Resequencing operations or adding operations.
- Changing and/or standardizing tooling and fixturing.
- Purchasing additional equipment to eliminate interruptions to the flow, backtracking, or unbalanced work loads between workstations.
- Removing exception parts from the workcell, and subcontract or use a "remainder cell" to process the nonconforming parts.
- Reconfiguring a large workcell into two or more smaller cells that exhibit continuous-flow processing.

Because it is important to design and implement flow-line workcells so that the full benefits of cellular manufacturing are achieved, cell formation is usually an iteration process requiring communication and cooperative efforts between design engineering and manufacturing engineering.

4.5 EXAMPLE OF CELL FORMATION USING THE WUBC ALGORITHM

This section presents an example of an efficient cell formation technique based on the within-cell utilization based clustering (WUBC) algorithm.[3] This procedure utilizes routing and work load information to form part families and machine cells simultaneously. The algorithm assigns machines to cells based on work load and cell size restrictions and assigns parts to cells in a way that tries to ensure that a majority of the operations can be performed within the cell. In addition, the algorithm assigns the appropriate number of machines to each workstation (operation) in the cell based on within-cell utilization of the workstation, thereby providing good work load balance.

Machines are admitted into a cell if their work load fraction (WLF) is greater than an arbitrarily specified value called the cell admission factor (CAF). The workload fraction of a work center S_i in a cell C_j is defined as the ratio of within-cell work load on work center S_i, which is caused by those parts that have already been assigned to cell C_i to the total work load of all parts in the work center. Work load fraction can be expressed as follows:

$$\mathrm{WLF}[S_i] = \frac{\sum\limits_{k \in C_i} \omega[S_i, P_k]}{W[S_i]}$$

where w $[S_i, P_k]$ is work content of a part P_k (that has already been assigned to cell C_j) on work center S_i, and $W[S_i]$ is total work load of work center S_i due to all parts that are routed through it.

This heuristic ensures that the work load fraction of any work center that has been assigned to a cell is equal to or greater than a specified fraction—the CAF. For example, if CAF is specified to be 0.5, then any work center that is assigned to a cell, say C_j, has 50 percent or more of its total work load from parts that have already been assigned to the cell C_j.

The algorithm also uses a cell size upper limit (CSUL), which is the maximum number of work centers that will be allowed to be assigned to a cell. This value can be changed to generate different numbers and configurations of cells.

[3]Ballakur and Steudel, "A Within-Cell Utilization Based Heuristic for Designing Cellular Manufacturing Systems," INT.J.PROD.RES., V25:N5, 1987.

Consider a hypothetical shop with seven parts and five work centers, where work centers S1, S3, S4, and S5 have one machine each and work center S2 has two functionally identical machines denoted as S2A and S2B. The routing of the parts through the work centers along with the processing times are given in terms of a work center-part load matrix in Table 4.1. For this example, a CSUL value of 6 and a CAF value of 0.5 have been chosen.

The WUBC algorithm forms cell number 1 by initially choosing S1 as the key work center because it has the highest total work load per machine. The work center S1 is then admitted to cell number 1. Because parts P2, P5, and P7 are routed through work center S1, they are now assigned to this cell. Each operation of the parts P2, P5, and P7 is evaluated and a list of nonkey work centers (in this case, work centers S2, S3, and S5) is formed. This situation is illustrated in Figure 4.10. Notice how the addition of parts P2, P5, and P7 to cell number 1 establishes an affinity to also bring work centers S2, S3, and S5 into the cell.

Next the WLF of each of these work centers is calculated. For example, the WLF for work center S3 is the ratio of work load on S3 for parts already assigned to the cell (P2, P5, and P7) to the total hours of work load on S3 (6.5 hours, in this case, as shown in Table 4.1).

$$\text{WLF [S3/S1]} = \frac{2.5 + 3.0 + .5}{6.5} = \frac{6.0}{6.5} = .923 \text{ (ADMIT)}$$

Similar calculations of WLF can likewise be made for work centers S2 and S5. The list of work centers is then sorted in decreasing order of WLF as shown on the following page.

Since workcenter S3 has a WLF value of 0.923 (which is greater than the CAF value of 0.5), it justifies itself being admitted to cell number 1 along with work center S1. Work center S5 is also admitted to cell number 1 since it has a WLF

TABLE 4.1. Work Center-Part Load Matrix.

Work-centers	PARTS P1	P2	P3	P4	P5	P6	P7	Total Workcenter Workload (Hours)	Number of Machines
S1		0.5			5.0		2.0	7.5	1
S2	2.5	2.0		4.5		1.5		10.5	2
S3		2.5			3.0	0.5	0.5	6.5	1
S4	2.5			1.0		0.5		4.0	1
S5			3.0		1.0		2.0	6.0	1

NOTE: Work center S2 has two machines. The rest have one machine each. All the machines work for eight hours a day and the capacity of each machine per day is also assumed to be eight hours.

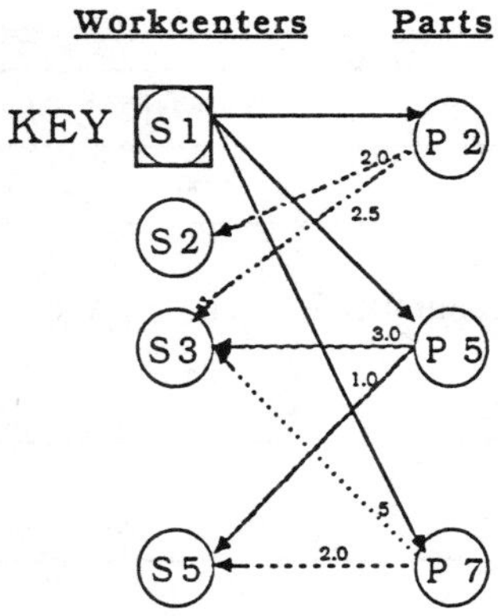

Figure 4.10. Initial Formation of Cell 1 Based Upon Key Work Center S1.

Cell No. 1	Key Work Center No. S1		CAF = 0.50	CSUL = 6
Work Center	No. of Machines	Cell Work	Total Work	WLF
S3	1	6.50	7.00	0.923
S5	1	3.00	6.00	0.50
S2	2	2.00	10.50	0.19

value equal to the value of CAF. Work center S2 has only one part, P2, in common with work center S1 as shown in Table 4.1. S2's WLF value of 0.19 is less than the CAF value of 0.5; consequently, it is not admitted to cell number 1.

Work center S3, the one with the longest value of WLF, is now chosen as the key work center for expanding cell number 1. P6 is the only assigned part that is routed through S3 and is, therefore, assigned to cell number 1. The list of nonkey work centers and their respective WLFs with S3 as the key work center is as follows:

Cell No. 1	Key Work Center No. S3		CAF = 0.50	CSUL = 6
Work Center	No. of Machines	Cell Work	Total Work	WLF
S4	1	0.50	4.00	0.13

Even though work centers S1, S5, and S2 share common parts with S3, they are not considered in this stage because they were either admitted or rejected in the earlier stage (with S1 as key work center).

The work load fraction of the nonkey work center S4 is calculated as shown in the chart below. S4 is admitted to the cell based on its WLF value of 1.0.

Cell No. 2	Key Work Center No. S2		CAF = 0.50	CSUL = 6
Work Center	No. of Machines	Cell Work	Total Work	WLF
S4	1	4.00	4.00	1.00

The final assignments for cell number 2 are as follows:

Cell No. 2	Key Work Center No. S2				CAF = 0.50	CSUL = 6
Work Center	No. of Machines	Cell Work	Capacity	WCU	No. Admitted	Additional Required
S2	2	8.50	16.00	1.06	1	—
S4	1	4.00	8.00	0.50	1	—
Includes parts P1, P4, P6						

Work center S2 has a WCU value of 1.06. Because the value is so low, only one machine of the work center, namely S2A, is assigned to the cell. The other machine, S2B, is assigned to the "remainder" cell.

Attributes such as the number of intercell moves, the average percentage of parts processed in cells, and the maximum cell size are also calculated as follows:

- Number of intercell moves = two
- Number of cells (excluding remainder cell) = two
- Maximum cell size (excluding remainder cell) = three
- Average percentage of operations that can be processed within cells for all cells (excluding the remainder cell) = 91 percent
- Percentage of parts processed in all cells (excluding the remainder cell) = 100 percent

The WUBC algorithm is simple to apply and useful in the early stages of cell formation because it helps identify workcells based on routings, work loads, and other user-specified factors of practical significance. Furthermore, other criteria can be used to select the key machines in forming cells. The alternative workcells can then be analyzed and refined to develop flow-line workcells.

4.6 EVOLUTIONARY IMPLEMENTATION OF GT

Achieving standardization and simplification through the application of GT concepts is an evolutionary process and one that is governed by both the needs and size of the company. Applying supposedly good solutions to the wrong problem generally produces poor results and dissatisfaction with the entire process. This section addresses the four stages of evolution a company goes through and discusses where, when, and how GT can be applied to advance a company forward in the journey to achieving WCM status.

4.6.1 Phase one: Recognition and Organization

Phase one is the most primitive phase of the evolutionary application of GT concepts. During this phase, a company begins to recognize the need to standardize the ways in which they design and manufacture parts. Often this recognition is precipitated by management's feeling that too much time and expense is being expended solving incidental or "brush fire" problems, and management's observation that the solutions do not have any carryover benefits. In many cases, the problems are the result of growth in a company's sales and operations without sufficient management of the growth process.

In design engineering, management begins to recognize that not only have the number of different parts, drawings, bills of material, and types of purchased materials grown significantly; but that the costs to maintain this information continues to grow with little assurance that these costs will eventually be contained. This situation is further exacerbated by design engineering's efforts to develop new products at lower costs and of greater quality. To react to these situations, management often makes some simple changes that are an initial step to standardization, such as when design engineers are assigned to specific product lines, product line reviews are conducted to look for ways to standardize on component designs and materials, and "meaningful part number" schemes are sometimes employed. For example, a part number designation of 600-699 in a string XXXXXXX600 might be used to represent a given type of parts—possibly a family of pinion gears. Similarly, a print review process among engineers from different product line groups might be used to try to ensure that some levels of design consistency are met. Companies also attempt to standardize the "part names" given to parts. Many companies also purchase new filing systems and drawing storage equipment in order to improve the ability and ease of organizing and retrieving design documentation.

In the area of manufacturing planning, phase one involves the recognition that serious problems, involving poor accuracy and consistency, exist in the process plans that are used to convert a part from rough stock to a finished item. Furthermore, the costs associated with generating and maintaining these process plans continues to escalate. The extent of these problems is illustrated in an analysis conducted by one company that reviewed their existing process plans for a family of expander sleeves. It was determined that forty-two different routings, involving twenty different machines, were developed for only sixty-four different styles and sizes in this family of parts. The different routings were developed by eight or nine manufacturing engineers, two process planners, and twenty-five numerical control (NC) part programmers over a span of several years. Upon review, it was determined that these 64 parts could all be processed on four machines using one of two possible routings. Management's response in phase one to these types of problems typically involves reorganizing people's areas of job responsibility. For example, one company assigned their most experienced process planner to determine all the routings for any part to be processed in the shop. The detailed planning for each operation in the routing (involving cut planning, feed and/or speed selection, calculating standards, et cetera) was then delegated to an analyst who was responsible for one or possibly two types of processes. In effect, the goal was to gain some improvement in standardization through job-task specialization.

In phase one, shop-floor manufacturing usually recognizes a need to achieve higher equipment utilizations to reduce manufacturing costs. To reduce the time involved in machine setup activities, shop orders for parts that are viewed as a family are often collected and then processed together in each department. For example, the turning department may setup a chucking machine to run several orders of a "family" of hubs before sending these orders to the slotting department for further processing. Processing standardization is done primarily at the department level and is usually the result of individual initiatives and ingenuity. In the area of material handling, attempts to standardize storage and movement of materials are sporadic and not part-family oriented. Typical progress is characterized by the use of a standardized size and type of pallet, a system of "move tickets" to control when and where material should be moved, and designated areas for the storage of materials (sometimes according to type of part).

During phase one some of the more blatant problems and costs incurred by lack of consistency and standardization are recognized. Simple solutions are employed at a department level and typically involve reorganizing the way in which people's responsibilities are defined and assigned. At this point, GT is not recognized as a concept to improve manufacturing. Furthermore, there is no recognition that standardization and simplification are strategic issues for gaining a more competitive position in the marketplace.

4.6.2 Phase two: Planning and Implementation

Phase two is the period in which management realizes that standardization must be implemented in a more systematic way across the organizational structure of the company. During this period, technology is investigated—and often implemented—as the tool to deal with the now apparent magnitude and complexity of activities associated with the design and manufacturing operations. While the specific approach taken depends on the size and the environment of the company, the emphasis of this initial stage is on planning and implementing GT concepts and strategies.

Design engineering is often the leader during phase two, taking a primary role in using a GT classification and coding system or a relational data base system to gain reductions in part proliferation and design time. Larger companies typically purchase and may even customize a computerized system. Smaller companies often develop their own system, often based on published GT literature, and install it on a less costly computer. Sometimes the resulting system is viewed by engineering as their own system; thus, minimal efforts are made to extend its use to other departments. The implementation and use of CAD systems is common in phase two, along with some involvement with data base systems for storing and retrieving part attributes.

Phase two is also the period in which manufacturing engineering begins to investigate and install computerized systems to further improve the accuracy, consistency, and cost-effectiveness of the process planning task. Using a GT code to classify all routings, these systems provide the capability to store, retrieve, and manually edit existing routings when revisions are required, or when a routing for a new part belonging to an existing family is needed. While larger CAPP systems can become quite sophisticated and expensive, smaller companies can gain many of the same benefits by using a simple "homegrown" coding scheme and inexpensive computer hardware and software. With both systems the concept and objectives are essentially the same; the only difference is in its application and justification.

Along with the emphasis to gain better consistency in the area of process planning comes a recognized need to improve standardization of shop-floor manufacturing. Phase two is the period when the first efforts to develop and implement workcells are initiated. Quite often one or two pilot workcells are installed on the shop floor; that is, the commitment to cellular manufacturing is exploratory, and there is often insufficient analysis performed to design cells and parts that will yield good continuous-flow processing. Although this "try it out" initiation process is not necessary and should be avoided, it often occurs and can result in a poor start to an important strategy for manufacturing. The subject of how to design and implement workcells correctly is addressed in chapter 5.

Phase two is also the period when projects involving machine setup reduction are initiated. If pilot workcells are being installed, then setup-reduction activities follow in natural succession. Early projects are often characterized by an approach of trying to reduce setups by some "management dictated goal," say a 70 percent reduction, which is typically selected based on the results of a magazine article or a seminar. Currently, little, if any, analysis has been performed to determine the net effects of setup-time reductions on the overall system in terms of achieving work load balance and better material flow. As a result, the bottleneck in production can simply shift from the machine on which the setup-reduction analysis was performed to another machine. Similarly, phase two setup-reduction studies are often based on reductions in direct labor costs, and little emphasis is given to determining the balanced relationships between batch sizes and setup times for the entire sequence of processing operations. Nevertheless, these setup-reduction projects can yield significant results and clearly are important first steps for improvements in standardization and simplification.

4.6.3 Phase three: Refinements and Synergy

During phase three, management's strategies to design and manufacture a quality product in a cost-competitive manner evolve from primarily a department-oriented application of engineering and technology to one emphasizing greater interdepartmental planning and execution of activities. It is a period of refinement and continued progress of many of the standardization and simplification activities initiated in phase two. Greater emphasis is given to communication and synergy between departments, and EI is beginning to emerge as a means for achieving improvements in standardization and simplification improvements in shop-floor operations.

In design engineering, new partnerships are formed with manufacturing engineering to conceive and design products that are easier to fabricate and assemble. The emphasis is to involve manufacturing engineering early in the design process and consider and incorporate manufacturing capabilities and quality issues before the product design is determined. For example, when IBM designed their Proprinter™, it was dictated by management that the unit could not have any screws, springs, or adjustments required for assembly; thereby allowing for an automated assembly processes. The result was a better-quality, more competitive product with fewer parts. Hewlett-Packard has also emphasized design for manufacturing approaches and has achieved many significant reductions in production costs and solutions to quality and reliability problems.

Phase three is a very active and beneficial period for manufacturing. Manufacturing engineering and shop-floor manufacturing are becoming highly integrated and are focusing on achieving process-oriented and continuous work flows. Cellular manufacturing is a major strategy for achieving standardization and simplification in parts processing. Balanced flow-line workcells are now prevalent, and setup-time reduction is being planned and executed in a manner that facilitates continuous-flow processing. Also, numerically controlled machines are utilized to a greater degree, typically as an integral part of a workcell, thereby providing better flexibility and tooling standardization. The resulting simplicity of processing operations on the shop floor facilitates improvements and standardization in operator training, quality measurement methods and systems, and incoming materials from selected and qualified vendors. In essence, the majority of the benefits of cellular manufacturing (discussed in section 4.3.2) are finally realized in phase three.

4.6.4 Phase four: Strategic Operations

During phase four, management continues to refine and then utilize the concepts of standardization and simplification and the associated benefits of flexibility, responsiveness, and quality as strategic weapons against its competition. At this time, systems for information integration are implemented and utilized both internally throughout the organization and externally with customers and suppliers. The concept of "strategic partnerships" are also in place, based on development of win-win arrangements between all the players in the team.

During this phase, design engineering, marketing, and manufacturing engineering work together closely on new product concepts, designs, and development and, in some cases, as members of product-based teams. This is evident in that more emphasis is placed on collecting and analyzing customer data on product performance and on identifying ways to make better products with more standardized procedures. From the standpoint of technology, parts are designed using a CAD system that references a data base of similar parts. Once the part is designed, a CAPP system generates a standardized process plan and the control-stream information necessary to drive any computer numerical control (CNC) machine tools. This CAD/CAM system is integrated into the production planning and control, which in turn, schedules and controls the materials for JIT production.

Material flow is likewise well orchestrated with shipments of quality materials to the point of use in standardized containers for each type of part. The standardized containers and simplified flow-line processing allows for quick shop-floor control procedures based heavily on human and/or visual (not computerized) systems. In essence, phase four represents the point where a company works as an integrated team of people using simple and standardized systems and procedures to compete; thereby removing the curse of complexity on manufacturing.

5

Design and Implementation of Manufacturing Workcells

The reorganization of plant facilities from a functional (job-shop) layout to a cellular configuration is a major competitive strategy for many discrete-part manufacturing companies. However, simply grouping machines into workcells on the shop floor does not guarantee that the advantages associated with cellular manufacturing will be achieved. Improperly designed workcells will not yield continuous-flow processing, and even greater losses in productivity than with the functional layout can result. This chapter addresses five categories of design decisions that must be considered in order to achieve effective and efficient manufacturing workcells. These five decision areas are:

1. Cell layout and capacity measures.
2. Operator levels and assignments.
3. Buffer levels and transfer batch sizes.
4. System operating parameters.
5. Just-in-time variables and effects.

This chapter also discusses personnel and implementation issues pertinent to using manufacturing flow-line workcells as the strategy for becoming mean, lean, and world-class competitive.

5.1 CELL LAYOUT AND CAPACITY MEASURES

After the machines and parts have been assigned to a workcell (through cell formation), the next issues to address are cell layout configuration and workstation capacities. Both of these issues concern the amount of machine capacity needed to provide some measure of work load balance between workstations and the specific configuration to use in placing the workstations on the shop floor. Cell layout and machine capacities are important factors to address because they determine many future design decisions. Together, cell layout and machine capacity provide the bridge between cell formation and cell design.

5.1.1 Cell Layout Configurations

The typical flow-line workcell is composed of several functionally different workstations, which are clustered together to produce a group of similar parts with continuous-flow (piece-by-piece) processing and material movement. Each workstation consists of one or more functionally similar machines that will establish a work load balance between the operations necessary for continuous flow. To further minimize the effects of work load imbalance, variability in machine processing times, and machine breakdowns and repair times, buffer storage of some finite capacity is often allocated between the workstations.

Figure 5.1 shows a number of different configurations for flow-line workcells.

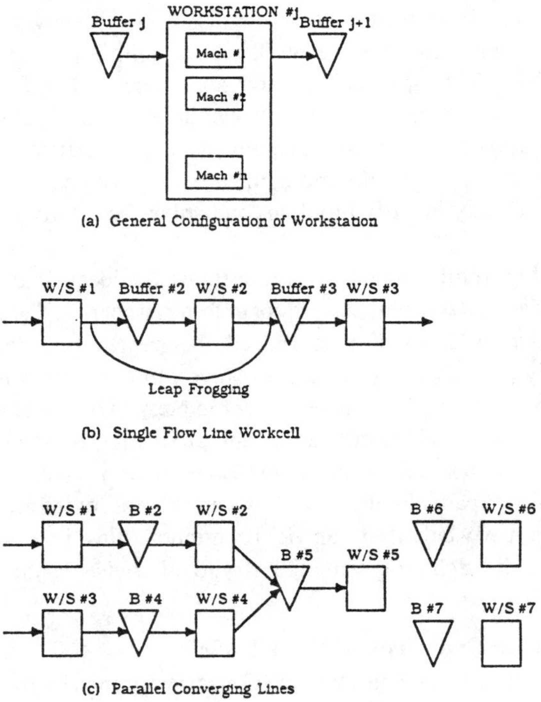

Figure 5.1. Flow-Line Workcell Configurations.

Figure 5.1a shows the general configuration of a single workstation consisting of one or more machines with similar processing capabilities. Work enters the workstation from an upstream buffer and is placed in the downstream buffer after being processed on a machine. The most appropriate buffer size to use is one of the design considerations that needs to be addressed in order to achieve good workflow processing. In a traditional, single flow-line workcell, depicted by Figure 5.1b, the flow of materials is from upstream to downstream workstations, however "leap frogging" is allowed. On the other hand, backtracking of flow from downstream workstations to upstream workstations is not allowed, because such flow would inhibit continuous-flow processing and the subsequent reductions in work-in-process and job throughput times. Figure 5.1c shows a variation where parallel lines converge to a common workstation and diverge again to parallel lines. This type of configuration is often appropriate in situations where a key machine exists but duplicating that machine is inappropriate.

The layout pattern of the machines in a workcell on the shop floor is also an issue of significance because it affects both operator assignments and production rates. One of the most popular layout patterns is the U-shape since it provides for flexible staffing, improved operator utilization, and better operator awareness and control. As an example, consider the six workstation cell shown in Figure 5.2a. On first shift, four operators run the workcell and jointly produce at the rate of 50 pieces per hour. The same cell can be run on second shift with only two operators when a rate of only 30 pieces per hour is required. Significantly, this flexible staffing allows shop-floor management to easily adjust production rates to match variable demands. Furthermore, the proximity of operators to each other enables them to quickly identify problems and solve them through a "cell team" approach.

When designing cellular manufacturing systems, the ideal situation is to reconfigure the entire manufacturing shop floor into a number of distinct cells, each designed to produce a family of parts. Indeed, the greatest benefits occur when a job is completely processed in one cell. However, in most cases it is difficult to organize cells so that each job is processed completely in only one cell because of environmental, safety, machine utilization, and processing constraints. Therefore, it is appropriate, and often necessary, to develop separate yet interlinked workcells to produce a family of parts. Figure 5.2b shows an example of interlinked U-shaped cells that when properly balanced, can yield continuous-flow processing and avoid the disadvantages of batch processing that would otherwise result.

5.1.2 Workstation Capacity and Work loads

The level of capacity available in a workstation is determined by three factors: The number of machines in each workstation, the number of shifts per day of operation, and the number of hours per shift that each workstation (or machine) will be

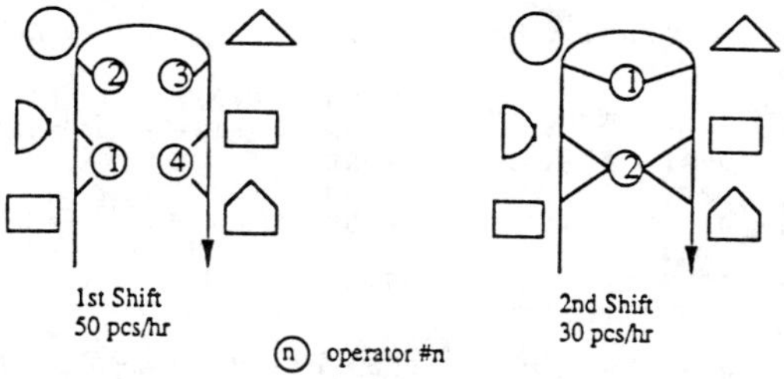

(a) Flexible Staffing and Adjustable Production Rates

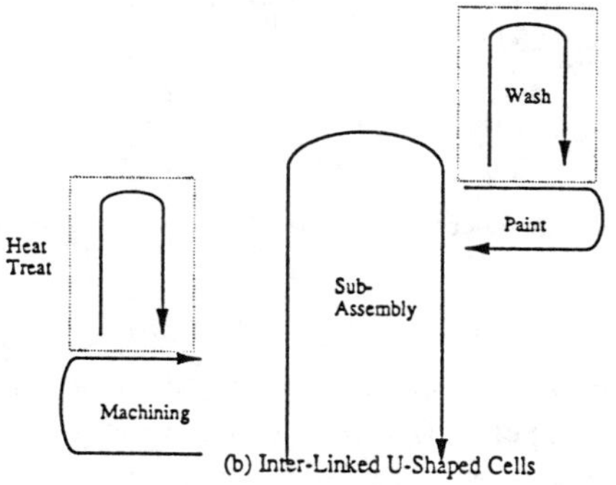

(b) Inter-Linked U-Shaped Cells

Figure 5.2. U-Shaped Layout Pattern for Flow-Line Workcells.

utilized. Selecting appropriate levels of these determinants of workcell capacity is paramount, since these decisions will affect and determine the selection of many subsequent design variables. For example, the number of machines and/or shifts will greatly influence how the workcell will be staffed and operated by the work force. Similarly, the number of machines and shifts will also determine and influence the appropriate levels of interworkstation buffers that are needed to maintain good work-flow balance.

The first step in the workstation-capacity analysis is to collect the following data for each of the parts to be processed in the workcell.

- Job setup times by operation
- Part processing times by operation
- Part handling times by operation
- Expected annual usage
- Production batch size

After this information is obtained, it is possible to calculate the expected annual work load on each workstation in the cell. Table 5.1a shows a workload summary

TABLE 5.1. Work Load Summary for Gear Cell.

W/S	MACHINE SETUP[HRS]	CYCLE[hrs]	TOTAL TIME	ANNUAL DEMANDS PIECES	ORDERS	NO. OF MACHINES	UTILIZATIONS
1	441	2531	2972	18038	413.9	1	148.6%
2	286	391	677	5750	320.7	1	33.8%
3	364	4164	4528	18038	413.9	1	226.4%
4	36	1204	1240	18038	413.9	1	62.0%
5	160	767	927	17921	401.6	1	46.4%
6	289	699	989	8018	339.0	1	49.4%

(a) One Machine per workstation operating one shift per day (2000 hrs/yr per machine)

W/S	MACHINE SETUP[HRS]	CYCLE[hrs]	TOTAL TIME	NO. OF MACHINES	UTILIZATIONS
1	882	2531	3413	2	85.3%
2	286	391	677	1	33.8%
3	1092	4164	5256	3	87.6%
4	36	1204	1240	1	62.0%
5	160	767	927	1	46.4%
6	289	699	989	1	49.4%

(b) Multiple Machines in workstations #1 and #3 (2000 hrs/yr per machine)

for a gear cell being designed to process ninety-eight different parts. This work load summary shows the total number of hours required on each workstation (assuming there is one machine per workstation) to setup and process the expected annual demand of orders and pieces for all of the ninety-eight parts to be processed.

The following equations are used to calculate the work load times shown in Table 5.1:

$$\begin{matrix}\text{SETUP[hrs]}\\ \text{[on WS No. i]}\end{matrix} = \sum_{\text{All jobs}} \frac{\text{Annual Demand}}{\text{Batch Size}} \times \frac{\text{Setup Time (min)}}{\text{60 min/hr}} \times \frac{1}{\text{Efficiency}}$$

The setup hours (setup [hrs]) per job are summed over all the parts routed to workstation number i (WS number i). Note that this formula that assumes only one machine per workstation will be setup on the job.

$$\begin{matrix}\text{CYCLE[hrs]}\\ \text{[on WS No. i]}\end{matrix} = \sum_{\text{All jobs}} \frac{\text{(Annual Demand)(Piece Time)(100\%)}}{\text{(Efficiency \%)}}$$

$$\text{where Piece Time} = \frac{\text{(Load Time + Cycle Time + Unload Time(min))}}{\text{60 min/hr}}$$

Again the cycle hours (cycle [hrs]) per job are summed over all parts routed to workstation number i. (The calculation of workstation work loads is easily accomplished by using PC-based spreadsheet software.)

The workload summary is used to estimate the number of machines required in each workstation. Note that the utilization values for workstations number 1 and number 3 in Table 5.1a indicate that additional capacity is needed. One alternative is to operate these two workstations more than one shift per day (2000 hrs/yr); however, this approach would yield large interworkstation queues to compensate for the imbalance in the workcell. Evidently, this solution is not ideal for this particular case because the advantages associated with workcells would be accordingly lost. A better alternative to consider would be the addition of one machine to workstation number 1 and two machines to workstation number 3. Table 5.1b shows the improved utilizations values of the workstations as a result of these additions of equipment, assuming that all machines in a workstation would be setup to process each job.

It is important to note that the total time shown in Table 5.1 represents only hours of total annual work load required to process the parts. To correctly estimate the number of hours of machine availability required, it is necessary to consider some time allowance for machine breakdowns, machine time lost because of interference, and machine idle time because of "waiting for work." Thus, 2000 hours of total annual work load often cannot be processed in one shift of machine time because of these lost-time factors. However, total machine availability requirements can be estimated quite accurately by using simulation or analytic (rapid) modeling techniques.

5.2 OPERATOR LEVELS AND ASSIGNMENTS

In the past, design techniques for manufacturing systems have often ignored the effects that the number of operators and their assignment to workstations have on system performance. In the case of manually operated machines configured in a job-shop layout, the effects of operators on system performance were often overshadowed by the effects of the machines, and therefore, were commonly ignored or assumed to be constant. However, in the case of manned workcells, an operator can be cross-trained to attend more than one machine, so that the number of operators required is actually less than the number of machines in the workcell. This situation not only increases operator productivity, but also improves flexibility in controlling the production rate of the workcell. If too few operators are assigned to a workcell, however, the workcell may not be capable of producing at the required production rate. On the other hand, too many operators will only increase the cost of operating the cell, and thereby lower its overall productivity. Determining the number and assignment of operators in the workcell is an important design issue that warrants careful consideration.

5.2.1 Effects of Operators

The number of operators in a workcell affects the productivity and production rate of the workcell. The ideal number of operators to have in the cell depends on the type of machines in the cell (manual or automatic) and the rate of production required from the workcell. It is important to determine the ideal number of operators in the design stage so that the operators can be cross-trained to run two or more different types of machines prior to cell implementation. It is also important to start developing the team attitudes and skills necessary for running the cell effectively prior to implementing the machines on the shop floor.

When the machines in a workcell are of the manual operation type, the operator is responsible for performing machine setup between jobs, part processing, and part loading and unloading from the machine. The effects of the number of operators on workcell production rate and productivity, in the case of manual machines, can be illustrated as follows. Consider the simple case of the three machine/three operation workcell shown in figure Figure 5.3a. Operation number 1 requires 80 seconds, operation number 2 requires 130 seconds, and operation number 3 requires 90 seconds. If each machine in the cell is assigned an operator

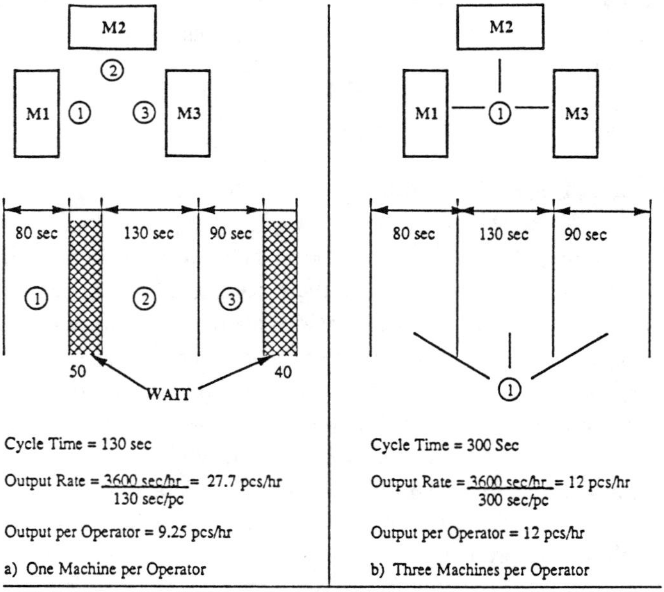

$$\text{Productivity Increase of Operators} = \left(\frac{12 - 9.25}{9.25}\right) \times 100\% = 29.7\%$$

Figure 5.3. Productivity Improvements with Reduced Number of Operators.

to perform these tasks, then the cycle time for the cell will be equal to the longest operation time, in this case operation number 2 at 130 seconds. Accordingly, the workcell output rate for this job would be 27.7 pieces per hour, yielding an output per operator of 9.25 pieces per hour. The case of assigning three machines to one operator is illustrated in Figure 5.3b. The cycle time is now the total of the three elements—300 seconds—which reduces the output rate to only 12 pieces per hour. However, since the output per operator is also 12 pieces per hour, the labor productivity improvement with the reduced number of operators is a significant 29.7 percent. On the other hand, the output rate dropped from 27.7 pieces per hour to 12 pieces per hour. Hence, determining the number of operators to assign to a workcell is a balance between operator productivity and required machine output rate. Nevertheless, it is evident that management has considerable control in planning and manipulating the output rate of the cell to match demand requirements.

When a workcell is made up of machines where the processing cycle of the machine is automatic, operator attention is often not required during machine processing. This results in an even greater opportunity to assign an operator to attend more than one machine. Occasionally, the machine requiring service (such as unloading/loading) will have to wait if all the workcell operators are busy attending other machines. When this situation occurs, machine interference results, consequently slowing down the overall production rate of the workcell. Essentially, the number of operators and their assignments in workstations can drastically affect the workcell throughput rate caused by operator induced machine interference.

To illustrate the effects of the number of operators on the performance of a manufacturing workcell, industrial data for a shaft line cell was modeled using the simulator STARCELL®. The workcell in this case consists of six workstations and eleven machines that perform saw cutting, centering, turning, grinding, milling, and threading operations on a total of twenty different parts. Because each of the machines in the shaft cell has an automatic cycle, the operators' tasks only involve setting up machines between jobs and loading and unloading work pieces.

Figure 5.4 shows the effects of the number of operators on the average throughput rate (pieces per hour) for the shaft workcell. For this simulation-based analysis, it was initially assumed that all the operators assigned to the workcell were fully cross-trained so that any operator could attend any of the machines in the workcell. Also, the capacity of the interworkstation buffers were initially set at high levels in order to exclude any interaction effects with the operator level. Hence, the workcell performance measure of average throughput rate in this analysis only reflects the effect of the number of operators in the workcell. Furthermore, the average throughput rate is based on four replicated runs consisting of eight weeks of processing that were performed following a two-week initialization period. The

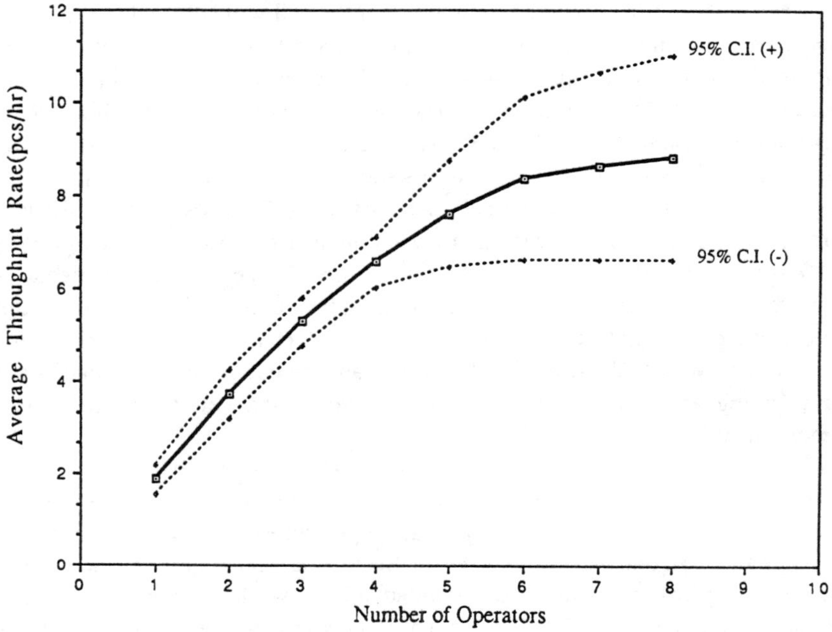

Figure 5.4. The Effects of Number of Fully Cross-Trained Operators on Average Throughput Rate.

center line represents the average throughput rate at different levels of operators. Also shown with dotted lines are the 95 percent confidence interval values for each level of operators.[1]

Several interesting points are illustrated in Figure 5.4. First, the throughput rate of the workcell can be improved by adding operators to the cell. However, as the number of operators increases (above six in this case), there is a diminishing rate of improvement in throughput. This occurs because machine interference (due to limited operator resources) will be reduced at a decreasing rate as the number of operators is increased. The relationship between average machine interference and the number of workcell operators is illustrated in Figure 5.5. Note that as the operator-induced machine interference is eliminated, the throughput rate is no longer affected to any major extent by the number of operators.

Another interesting point shown in Figure 5.4 is the change in confidence interval values above and below (near) the average throughput rate. The size of the

1The confidence interval shows the range of values where the "true average throughput rate" would fall 95 percent of the time.

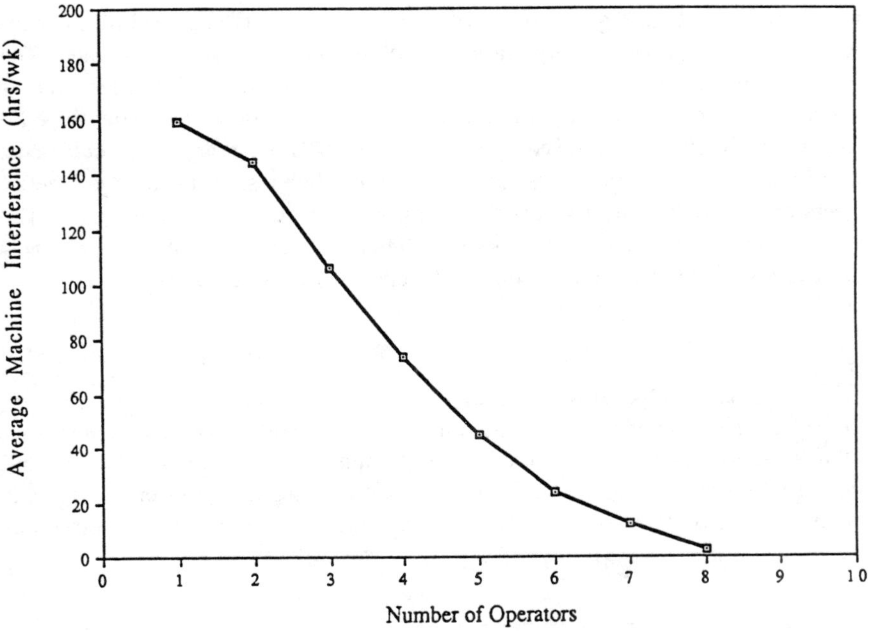

Figure 5.5. Average Machine Interference Versus the Number of Operators.

confidence interval increases as the number of operators increases. This indicates that using a certain number of operators depends on the job mix that is processed through the workcell. For example, if the number of operators is relatively high, then certain job mixes with high demands on the workcell can take advantage of the additional operator capacity available. However, if the job mix requires lower levels of work load demand on the cell, then the higher level of operator capacity will provide little or no improvement in the throughput rate. This is illustrated in Figure 5.4 where the lower limits of the confidence intervals are practically equivalent for the case of five operators or more in the shaft workcell.

The selection of the most appropriate level of operator capacity depends on various factors. A primary factor is reaching a throughput level that satisfies the demand on the workcell, which is dictated by customer requirements. For example, if current demand levels only require an average of 6.5 pieces per hour of throughput, then four operators would be sufficient to fulfill that requirement. This example illustrates one of the advantages of workcells; that is, the workcell output rate can be easily controlled by management.

Since one of management's objectives in implementing workcells is to reduce manufacturing costs, it is important to consider the cost of selecting the appropriate number of operators. Figure 5.6 shows the relationship between machine interference costs and operator costs at various machine and operator expense rates. The intersections of the two sets of curves give the break-even point. These curves show that the most cost-effective level for the number of operators varies depending on the relative cost of the equipment and labor. For example, at a machine cost of $50 per hour and a labor cost of $25 per hour, the most cost-effective level of operators would be four. However, three operators would be appropriate at a $20 per hour machine cost and a $15 per hour labor cost. This type of cost analysis is essential when selecting an appropriate level of workcell operators.

5.2.2 Effects of Operator Assignment

One of the main objectives when assigning a certain number of operators to a set of workstations in a cell is to simultaneously minimize machine interference and obtain a desired throughput rate. However, determining good operator assignments is not a trivial task since the occurrence of operator-induced machine interference

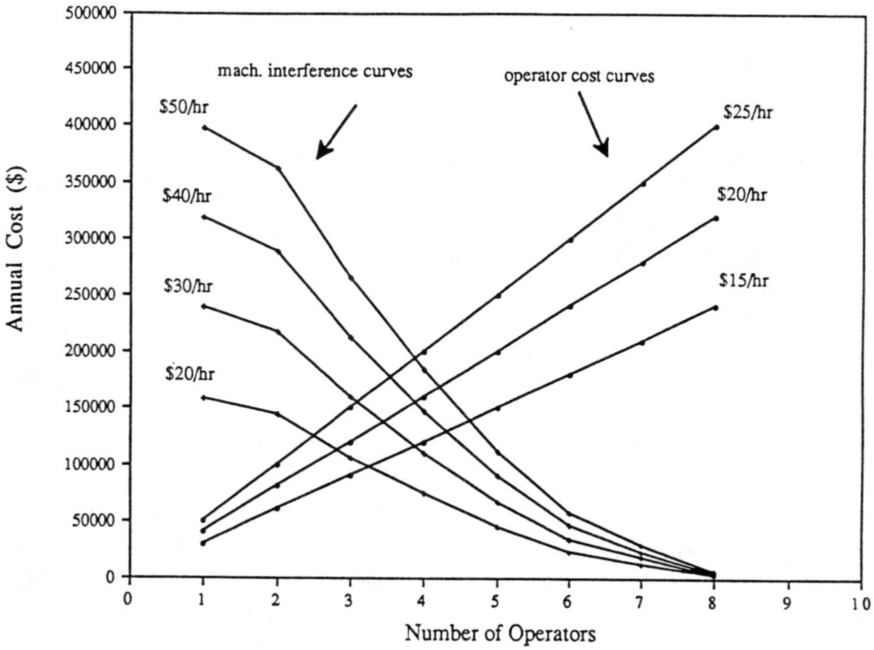

Figure 5.6. Annual Costs Caused by Machine Interference Cost and Operator Cost by Varying the Number of Operators.

is highly combinatorial. In most cases, a solution to this problem is specific to the workcell environment under evaluation.

There are several factors that management needs to consider when assigning operators to workstations. These factors are: (1) operator cross-training (management has to decide on the cross-training to provide for operators assigned to workcell), (2) the physical layout of the workcell (this mainly concerns the traveling distance/time of the operators between workstations and the physical constraints that may limit the movement of the operators), and (3) the type of material transport and handling system available.

To estimate the extent to which operator assignment can affect system performance, four different operator assignment strategies, each involving five operators, were evaluated using STARCELL®. Table 5.2 presents four different operator assignments and the respective average throughput rates (based on four replicated runs of eight weeks of processing, following a two-week initialization period). As expected, the highest throughput is achieved when all the operators are fully cross-trained (assignment strategy number 1). In a workcell where operators are moderately cross-trained (as in the case of assignment strategies 2 and 3), the assignments are based on the objectives of balancing the work load as evenly as possible and minimizing operators' traveling time between workstations. Assignment strategy 4 represents the situation where there is only minimal operator cross-training causing the resulting throughput rate to be relatively low.

These results show that the shaft workcell performance can vary as much as 15 percent with the same number of operators. The problem of operator assignment is usually specific to the manufacturing environment under consideration. (A gear cell showed a 25 percent difference under a similar analysis.) This example does illustrate, however, the importance of using simulation as a tool to evaluate different cell design alternatives. Also, this example shows that additional cross-training may allow for a lower level of operators to obtain a higher level of throughput performance.

5.3 BUFFER LEVELS AND TRANSFER BATCH SIZES

Much has been written and been said about the desirability of keeping work-in-process inventories to a minimum. However, when processing a variety of parts in the same family of parts, the allocation of interworkstation buffer storage is often necessary and practical in order to counteract the work load imbalance that exists between workstations in a workcell. This section addresses the issue of how to determine appropriate sizes of interworkstation buffers so as to minimize the occurrence of machine downtime induced by phenomena called blocking and starving. Also in this section, the effects of transfer batch size will be addressed.

TABLE 5.2. The Effects of Operator Assignment on System Performance.

Assgnmt. Strategy	Opr. no.	Workstation No. (no. of machines) 1 (2)	2 (2)	3 (3)	4 (1)	5 (2)	6 (1)	Throughput (pcs/hr)
1	1	x	x	x	x	x	x	7.76
	2	x	x	x	x	x	x	
	3	x	x	x	x	x	x	
	4	x	x	x	x	x	x	
	5	x	x	x	x	x	x	
2	1	x	x	x				7.34
	2		x	x	x			
	3		x	x	x			
	4	x				x	x	
	5	x				x	x	
3	1	x	x	x				7.21
	2	x	x	x				
	3	x	x	x				
	4				x	x	x	
	5				x	x	x	
4	1	x	x	x				6.74
	2		x	x	x			
	3				x	x		
	4					x	x	
	5	x					x	

(x denotes the assignment of operator to workstation)

CONSIDERATIONS

* Cross-Training Costs?
* Physical Layout of Workcell?
* Desired Production Rate/Flexibility?

5.3.1 Workstation Blocking and Starving

Determining the appropriate level of workstation buffers is an important design consideration since finite buffer capacity can cause blocking and starving within the workcell. Blocking occurs when a work piece, having been processed, is blocked from exiting the machine because there is no space in the downstream buffer. In effect, the machine is prevented from processing more work because the immediate downstream buffer storage is full. Unfortunately, blocking results in a buffer-induced machine interference that can lower the workcell throughput rate. The phenomenon of starving occurs when a machine completes a work piece but is still "starving" for work, because the immediate upstream buffer is empty. Both blocking and starving cause machine interference; that is, induced idle time, which adversely affects the throughput rate of the workcell. It is therefore essential in the design stage to determine appropriate levels of buffer capacity that will minimize the unproductive time on the machine, support high levels of workcell throughput rate, and minimize the amount and cost of in-process inventories.

Figure 5.7 illustrates the phenomena of blocking and starving with the simple case of a two-machine flow-line workcell where the interworkstation buffer has a capacity for two pieces. Blocking occurs on machine number 1 when workpiece 3 of job 1 is completed. Since, at this time, the buffer is already full to capacity with workpieces 1 and 2, workpiece 3 cannot be unloaded from machine 1 until workpiece 1 is loaded into machine 2. Until workpiece 1 is loaded into machine 2,

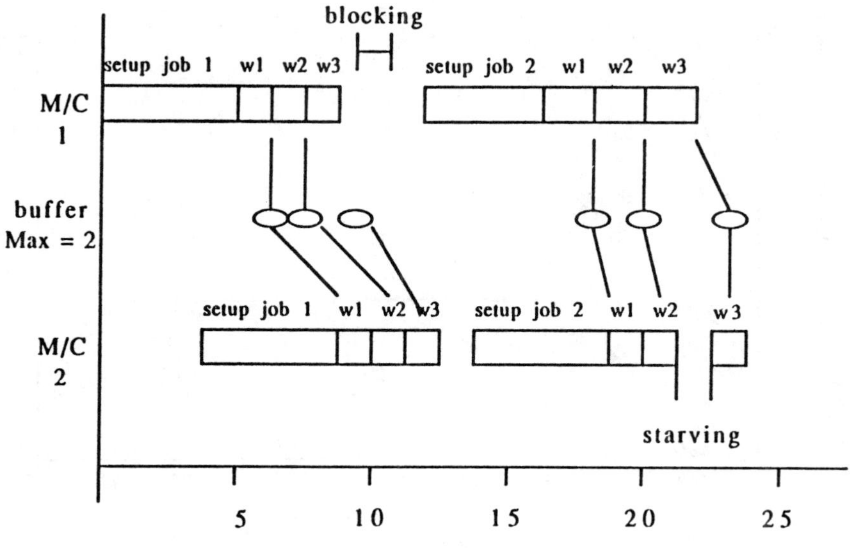

Figure 5.7. FMC Flow Line with One Operator and Maximum Buffer Size of Two.

machine 1 will be blocked from further processing because of the limiting capacity of 2 pieces in the buffer. In addition, the buffer capacity of 2 pieces results in a starvation condition for job number 2 on machine number 2. Since the cycle time of job 2 is considerably less on machine 2 than on machine 1, the buffer is depleted of workpieces after workpiece number 2 is loaded into machine number 2. Consequently, machine 2 is starving for work until workpiece 3 of job 2 is completed on machine number 1. This figure illustrates that buffers of finite capacity can cause blocking and starving, resulting in machine interference or idle time, both of which adversely affect the throughput rate of the workcell.

As illustrated, blocking and starving phenomena can readily occur in a workcell when there is finite buffer storage between the workstations. These phenomena can cause machine interference in addition to that caused by operators (as previously discussed). Since both buffer size and the number of operators can cause machine interference, it is interesting to consider these factors jointly in terms of their effects on the throughput of a workcell.

Figure 5.8 illustrates the effects of buffer size on the workcell throughput rate at the two different operator levels of 3 and 5. Note that with buffer sizes between 1 and 20, the throughput rate in both cases improved significantly as the size of buffer was increased between 1 to 20. Significantly less improvement is observed

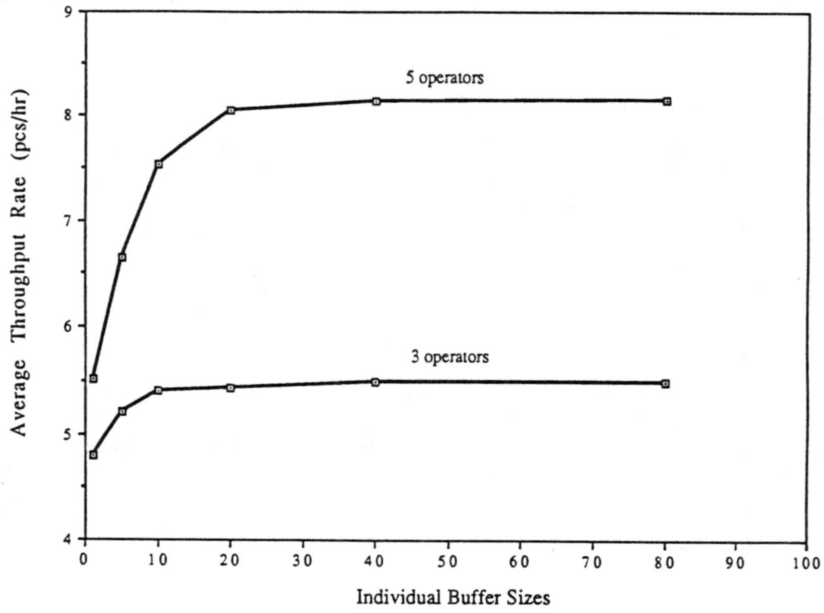

Figure 5.8. The Effects of Interworkstation Buffer Size on Average Throughput Rate.

as the buffer size increases beyond 10 in the case of 3 operators, and beyond 20 in the case of 5 operators. This exhibits the diminishing rate of return of buffer size on throughput rate for the workcell.

In the case of 3 operators, a maximum throughput rate of 5.4 pieces/hour can be achieved at a buffer capacity of 10 items, whereas a throughput rate of 8.1 pieces/hour can be achieved with 5 operators when the buffer capacity is 20 items; indicating that both operator levels and buffer levels have additive interaction effects on throughput rate.

Further illustrating the contributive effects of operator and buffer levels in terms of machine interference, Figure 5.9 shows the machine interference resulting from different levels of buffer capacity for 5 operators (all fully cross-trained). Machine interference, in percentage of the total workcell utilization, is the average of four replicated runs for each run condition. The bold curve represents the sum of machine interference caused by buffer and operator levels. Each dotted curve illustrates the interference independently induced by either buffer size or operator level at different buffer sizes.

An interesting fact illustrated by these curves is that operator-induced interference increases when buffer capacity is increased, however, at a decreasing rate. This behavior of the curve for operator-induced interference is contrary to the buffer-induced interference curve, which shows a decreasing rate of interference

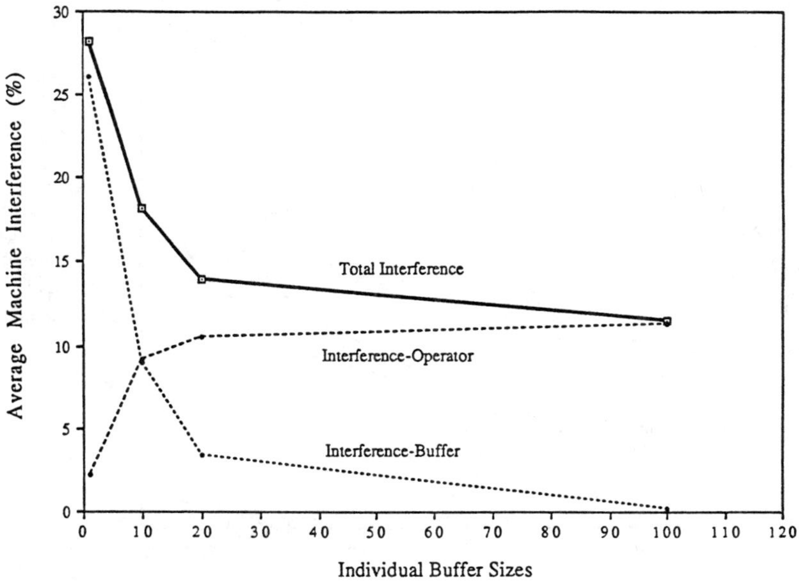

Figure 5.9. Average Machine Interference Versus Buffer Size (5 Operators).

as the buffer capacity increases. In this particular analysis, it is interesting to note that the buffer-induced interference is much more pronounced at all levels of buffer capacity than that of the operator-induced interference. Also, the interference curves of both the factors intersect at a buffer capacity of 10 items, thus indicating the buffer level at which the machine interference, caused by both the factors, is equal—a useful fact in selecting appropriate levels for buffer size and number of operators in a workcell. These results indicate the value of studying the interaction effects of operator level and buffer size.

5.3.2 Transfer Batch Effects

The transfer batch size (TBS) is the number of pieces moved as a group (or entity) between workstations in a flow-line workcell. For example, a batch of 100 pieces could be processed and moved through the workcell piece by piece (TBS = 1), or in a group of 10 pieces (TBS = 10), or any other group size up to the size of the batch size itself (TBS = 100). In designing a workcell, it is often necessary to also design the transport units for moving in-process parts safely and efficiently between the workstations within the workcell. It is therefore important to know the effects that the size of the transport unit will have on the performance of the workcell, which in turn, affects manufacturing costs and the ability to satisfy customer requirements.

Figure 5.10 shows the average effects of transfer batch size on job throughput time and total work-in-process (WIP) (in pieces) for a motor reducer housing

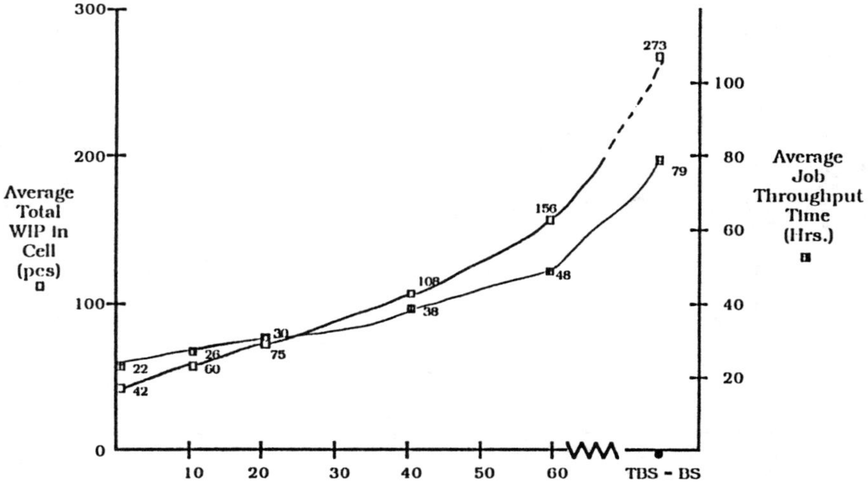

Figure 5.10. Average Effect of Transfer Batch Size on WIP and Job Throughput Time in a Housing Cell.

workcell that processes ten different parts through five workstations encompassing nine machines. The results were obtained from a simulation-based experiment covering fifty jobs completed in five months of production activity. The average batch size was 120 pieces, and it ranged from 60 to 170 pieces. As shown, the effects of increasing the transfer batch size (up to the batch size itself) is quite significant, especially when TBS approaches the batch size. For example, the average job throughput time is approximately four times larger and WIP increases over six times when jobs are run through the cell in a batch mode (TBS = Batch Size) compared to piece-by-piece (continuous-flow) processing (TBS = 1). This simulation analysis supports what many companies learn the hard way through experience: namely, small transfer batch sizes are critical in order to achieve the full benefits of cellular manufacturing.

5.4 IMPLEMENTATION ISSUES

Whereas the previous sections of this chapter focused on the technical issues associated with the design of manufacturing workcells, this section will address those issues that are typical to the successful implementation of manufacturing workcells. Granted, sound adherence to the design methods outlined in this chapter is necessary, but not, in itself, sufficient for the success of a workcell, because obstacles that hinder successful workcell implementation are often organizationally related such as performance measures, incentive pay systems, culture, and insufficient levels of education and training. Although obstacles such as these are considered soft issues, they can make workcell implementation a more difficult and time-consuming process than is the mere design and transformation of workcells; therefore, they need to be strategically addressed beginning with a team approach.

5.4.1 Establishing a Team Approach

Forming an effective group that, together, will function as a team throughout the workcell design and implementation process is a key step to successful workcell implementation. Because significant organizational and behavioral changes are required, a strong team approach is essential to workcell implementation success.

It is recommended that a multidisciplinary core team be developed and have the responsibility for workcell design and implementation. This core team serves as the basis for EI and thus should consist of one or two members from each of the following disciplines:

- Industrial engineering
- Manufacturing engineering
- Production supervision

- Cell team leader
- Operators

The core team will require support from management, maintenance, production planning and control, and other production organizations throughout the design and implementation processes. The core team's relationship to the support groups is illustrated in Figure 5.11.

The selection of the appropriate core team members—especially for the first cell implementation—is critical for establishing a well-functioning team and will increase the likelihood of workcell success. When possible, select employees that, in the past, have demonstrated an ability and willingness to work as a team member to achieve organizational goals. In addition to careful screening of potential core-team members, it may be necessary to provide the core team with formal training in cooperation and in group dynamics.

In some cases, a more efficient and effective workcell design and implementation will occur by working with an outside consultant. The consultant brings to the group a high level of experience and expertise in the area of workcell design and implementation and can also provide project direction for the core team. Acquiring the assistance of a consultant to assist the core team should be openly considered during the planning and budgeting process.

5.4.2 Obsolete Incentive Systems

A common problem facing companies interested in implementing manufacturing workcells is the existence of an incentive-based pay system. The problem is

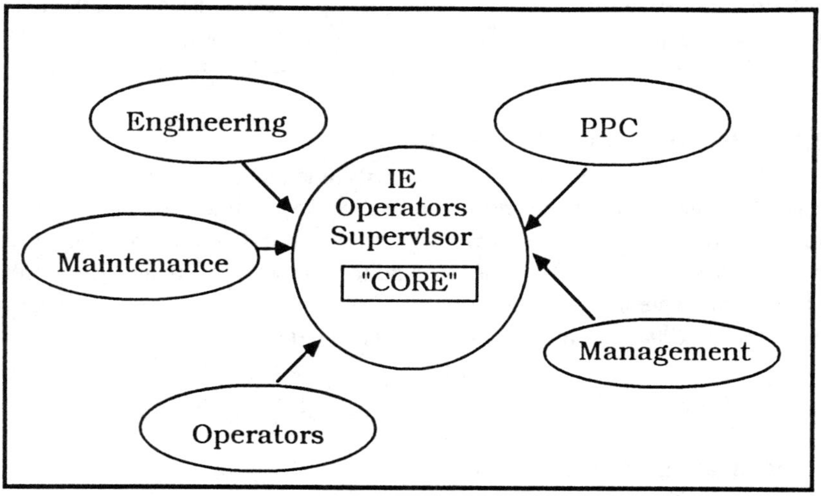

Figure 5.11. The Relationship Between the Core Cell Team and Other Functional Areas.

twofold: First, traditional incentive pay systems promote individual performance and not the teamwork approach that is required for effective workcell operation, and second, when moving to a manufacturing-workcell concept, the methods used to manufacture the product change significantly. These changes, in many cases, render the current production or labor standards obsolete, and therefore the basis for the incentive system is no longer valid. This section discusses the problems associated with existing incentive systems and their impact on manufacturing cell implementation. Alternative approaches to dealing with existing incentive systems are also presented.

Traditional incentive systems are typically developed for an individual operator working on an individual process. Given this pay system structure, there is no incentive for one operator to assist another operator in processing a product. In fact, it is actually self-hindering, an encumbrance for an operator to help others because it takes away time from his or her operation, and thereby reduces his or her output and, consequently, incentive pay. As a matter of course, a protective wall is built around each operator in an effort to maximize their pay.

The behavior generated by traditional incentive systems is not consistent with behavior standards of WCM organizations and can be detrimental to the effective functioning of manufacturing workcells. Effective workcells require the collective, cooperative efforts of all cell operators so that quality and productivity are improved continuously. Having an incentive pay system that promotes individual performance only and, in some cases, discourages teamwork, is a common obstacle that companies must overcome when implementing workcells.

Besides the focus on each individual, the basis typically used to drive incentive systems—namely, production or labor standards—is no longer valid in the workcell environment. New workcell practices, such as mobility of operators from machine to machine, operators working together on a specific task, and the potential for any number of operators to operate a machine, obviously make the existing one-man/one-process standards obsolete.

It is clear that a change in or elimination of the traditional incentive system is required. If the history of work incentives is positive for an organization, it may make sense to retain some form of incentive pay. Also, if the incentive system has resulted in higher paychecks for operators, an alternative or elimination of such a system would meet with significant operator resistance. For these reasons, a modification of the current incentive system to a group (cell) incentive may be the right compensation plan.

A fundamental problem facing organizations interested in developing incentive pay systems for manufacturing workcells lies in the method used for setting standards. How does a company (organization) develop a production or labor standard for a group of operators, operating a group of machines, in a teamwork fashion? Although not a simple task, it can be done effectively with the aid of modeling tools. (These modeling tools are discussed in more detail in the next section.)

The approach used to develop the standards is to treat the cell, rather than the individual workstation, as an entity or production unit. Using the cell output as the standard enables the incentive system to be structured so that the group is rewarded for improved cell performance rather than an operator being rewarded for improved workstation performance.

Another approach that can be used to deal with an obsolete incentive system is to use the workcell implementation as an opportunity to phase out the incentive system altogether. This approach, however, may meet with stiff opposition from employees who are accustomed to receiving paychecks padded with extra pay. Nevertheless, when management considers the administrative costs typically associated with incentive systems that have an obsolete-standards basis, phasing out the incentive system may be an attractive alternative. It should be noted that it is critical that employees do not incur a financial penalty for being part of a workcell. There is no better way to increase the likelihood of failure than to decrease the cell operators' paychecks. In fact, a case can be made that cell operators should receive an increase in pay from their current level because of the higher skill level required in the workcell. For example, operators will need to acquire additional skills such as knowing how to operate additional machines (cross-training), basic problem solving, preventive maintenance, and statistical process control.

Because of the increased skill level that is required of the workcell operator, a pay-for-knowledge system is the answer for some companies. A pay-for-knowledge system increases the cell operator's pay for each new skill learned. This pay system provides an incentive for operators to learn new skills that are important to improving the operation of the cell. As an employee learns a new skill, he or she becomes more valuable to the organization and should be rewarded accordingly.

If a company feels that phasing out the current incentive system is necessary, a pay-for-knowledge system can be phased in simultaneously. When designed and executed properly, cell operators' pay will not decrease during the transition from an incentive system to a pay-for-knowledge system.

Although the issue of dealing with traditional incentive systems is complex, an organization using such a pay system must deal with this obstacle in order to achieve workcell implementation success. Accordingly, the traditional incentive pay system must be either eliminated or adjusted so that the reward for being a member of the workcell team not be in the form of a financial penalty.

5.4.3 Performance and Labor Reporting

Moving from the traditional batch-oriented production methods to workcells typically requires performance and labor reporting systems to be updated or, in some cases, developed from ground zero. These reporting systems were formed around the traditional batch-manufacturing methods of one person operating one or more comparable (similar) machines. An output rate was established for each product and served as the standard. This standard served not only as the basis for performance and labor reporting but also for capacity planning and job scheduling.

As discussed in the previous section, the methods of operation for a flow-line workcell vary significantly from those of batch-manufacturing practices. The foundation of cell operation is flexible workers and team-oriented responsibilities; thus, many of the current performance measures, which are based on standards, are in need of review.

To develop performance (output) and labor standards for flow-line workcells the cell is treated as the production unit instead of the individual workstation being treated as the production unit. The output rate of each product should thus be determined for the workcell, not for the workstation.

To illustrate the calculation of workcell output rate, a flow-line workcell that manufactures gears is considered. The setup and processing times for each operation of part A (which are used in this calculation), are as follows:

Machine	Operation	Setup(min)	Process Time (min/pc)
1	Turning	60	3.00
2	Hob	90	4.50
3	Shave	30	4.00
4	Broach	15	2.00

Assume, for simplicity of illustration, that load and unload times are equal to zero. Figure 5.12 shows the workcell layout with four manual workstations (one machine per workstation) and one operator for each workstation.

The output rate of the cell can be calculated from the slowest operation in the processing sequence. For part A, the hobbing operation has a cycle time of 4.5 min/gear or a rate of 13.3 gears/hour. The output standard for the workcell could therefore be set at 13.3 gears/hour. For every production hour, 4.0 labor hours are required (four operators), and the labor standard can be calculated as follows:

$$1 \text{ Prod. Hr.}/13.3 \text{ gears} \times 4.0 \text{ Labor Hr.}/1 \text{ Prod. Hr.} = .300 \text{ hr/gear}$$

Assume now that the workcell is staffed with two operators, as illustrated in Figure 5.13, where one operator runs the turn and hob machines and the other operator runs the shave and broach machines.

In this case, operator 1 can process a gear every 7.5 minutes (3.0 + 4.5 minutes) or at the rate of 8 gears/hour. Operator 2 is able to process a gear in 6.0 minutes (4.0 + 2.0 minutes) which is a rate of 10 gears/hour. Therefore, the maximum

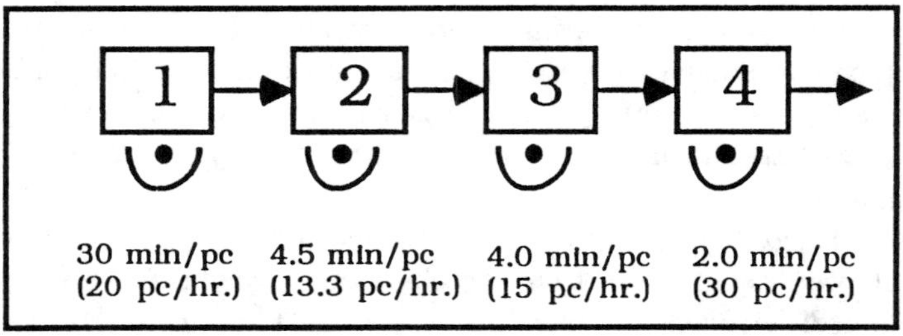

Figure 5.12. Machine Production Rates for Processing Part A in the Gear Cell.

output rate of the cell would drop to 8 gears/hour for a cell manned with two operators. The labor standard is calculated as follows:

$$1 \text{ Prod. Hr.}/8 \text{ gears} \times 2.0 \text{ Labor Hr.}/1 \text{ Prod. Hr.} = .250 \text{ Hrs/gear}$$

It is evident that the output of the workcell is very dependent on the number of workcell operators.

Interestingly, this illustration shows that a 50 percent reduction in the number of cell operators was accompanied by only a 40 percent reduction in output (13.3/hr to 8.0/hr). A better utilization of manpower was achieved when the cell was staffed with two operators.

This example was used to present an approach to developing standards for a flow-line workcell. In such simple cases, standards can be calculated without much trouble. But, typically, workcell designs are more complex and consist of such features as multiple machines per workstation, different workstation setup policies, and operator cross-training overlap (the capability of more than one operator to run

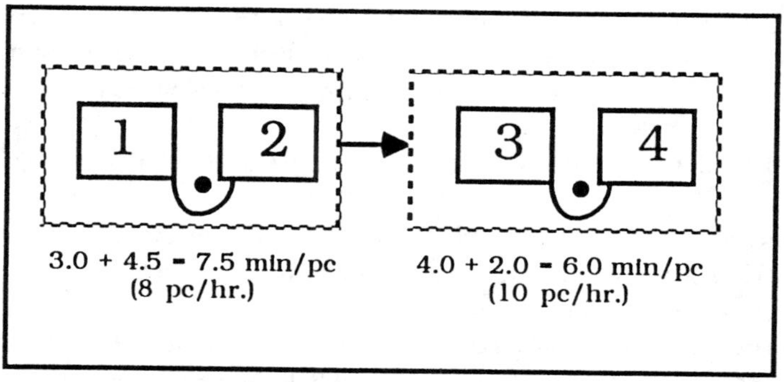

Figure 5.13. Gear Cell with Two Operators Assigned.

a single workstation). Calculating standards of workcells with such complex features can be facilitated with the use of a simulation tool such as STARCELL®.

After establishing the standards for each part, workcell-scheduling, capacity-planning, and product-costing functions can take place.

One approach to measuring workcell performance is to compare recent performance to a baseline performance. Improvements or gains over the baseline can be shared between the cell operators and the company in the form of monetary or other types of reward and is typically referred to as "gainsharing." It is a well recognized fact that people improve their performance when they have the opportunity to share in the gains. Gainsharing lends itself well to the cell concept and is an appropriate incentive system used with manufacturing workcells.

Other manufacturing cell performance measures can be used, such as: defect levels, setup reduction improvement, cell inventory level, employee skill improvement.

It is important that these be tracked and analyzed by those whose performance is being evaluated, namely, the cell operators.

5.4.4 Education and Training

The transition from traditional batch manufacturing to a manufacturing approach based on workcells requires changes in the organizational structure, job responsibilities, reward systems, accounting practices, and performance measurement system. As presented in chapter 2, education and training plays a key role in preparing employees for this level of change and in helping them overcome a natural resistance to change. This section will address the specific education and training issues associated with the implementation of manufacturing workcells.

One of the best ways to get the core team off to a good start is to send the team to visit a company that has already implemented workcells. This visit would also be beneficial to other members of the organization including management. A plant visit offers first-hand, real experience with a cellular manufacturing approach. During the visit, it will be important to discuss the following with people that were involved with the implementation of the cell(s):

- What benefits have they realized?
- What mistakes were made along the way?
- What they would do differently next time?

In addition to the plant visit, early in the life of the project, core team members should receive training in the important workcell fundamentals—the advantages of workcells, how workcells operate, teamwork, and the role of setup reduction and how to achieve it.

A true advantage with manufacturing workcells is the opportunity to have operators run multiple workstations. In order to take advantage of this operator-flexibility potential, appropriate cross-training must take place. To determine the minimum level of cross-training required to achieve the desired workcell performance, the concepts and methods presented in section 5.2 should be used. Further cross-training may provide additional benefits in the way of increased operator flexibility for such cases as absentee coverage and manpower scheduling flexibility.

It is worth noting that different operators will have different cross-training requirements. For example, an operator that can run an automatic screw machine typically has a higher skill level than a heat-treat operator. In most cases, it would be significantly easier to train the screw machine operator to run the heat-treat workstation than to train the heat-treat operator to run the screw machine. In fact, the reason that the heat-treat operator isn't a screw machine operator may have to do with a lack of inherent mechanical aptitude. In this case, it may not make sense to cross-train the heat-treat operator to run a screw machine. If it is necessary for the heat-treat operator to be cross-trained, a different operator may be needed.

If the cell is going to consist of existing machines, cross-training can take place before a cell is formed. Cross-training is best accomplished using a hands-on training approach where the trainee works with an experienced machine operator at the workstation itself. It may take several weeks before the trainee becomes proficient at that workstation. Throughout the training period, the trainee should be allowed to perform an increased number of activities until the experienced operator is no longer needed to supervise the activities.

6

Setup-Time Reduction

In recent years, many manufacturing experts from both Japan and the United States have demonstrated the necessity to reduce lot sizes, thereby also demonstrating the necessity to reduce equipment setup time. Starting in Japanese companies and then in European and American companies, more and more managers have realized that reducing setup time is fundamental to reducing manufacturing lead times, improving quality, reducing work-in-process, reducing inventory costs, and increasing manufacturing process flexibility. In extreme cases, shorter setup time has resulted in reducing lot sizes to one, drastically cutting production lead time and inventory.

Today, in the early nineties, the worldwide manufacturing community is aware that setup-time reduction is a cornerstone of the JIT manufacturing philosophy. However, in many companies, setup-time reduction has not yet become a natural way of thinking, or an integral part of the shopfloor culture. In such companies, the questions that usually arise are:

- How do we achieve setup-time reduction?
- How do we get started?
- Where do we focus our efforts?
- How do we estimate benefits and foster justification?
- How do we ensure the success of a setup–time reduction program?
- How do we make sure that such a program will not generate only a one-time gain, but will lead instead to the continuous improvement of the production process?

This chapter answers these and other questions with a focus on the definition of appropriate objectives. A thorough discussion of the practical aspects and organization of setup-time reduction projects is also presented.

6.1 THE ROLE OF SETUP-TIME REDUCTION IN WCM COMPANIES

In the nineties' highly competitive worldwide market, company survival is conditioned by its ability to offer, at attractive costs, a diversity of high-quality products in order to react quickly to customer needs and variations in demand. Evidently, competitive costs, product diversity, high quality, and flexibility are not compatible with long setup times because long setup times result in wasting resources: Machines are unavailable for long periods, employees are paid for unproductive work, and large quantities of nonneeded products are manufactured.

The JIT production environment combats resource-wasting factors by streamlining manufacturing operations. However, to realize the full potential of streamlined operations setup times have to be reduced. Reducing setup time makes manufacturing cells as flexible as possible, which in turn, increases the company's competitive advantage. For example, in the event of an unexpected low volume order that requires one hour of production, the ability to setup the machines of a cell in fifteen minutes—rather than in three hours—will create satisfied customers and also attract new customers.

Setup-time reduction is considered a cornerstone of JIT because numerous improvements that are the foremost characteristics of companies becoming meaner and leaner in their moving toward WCM are direct effects of setup-time reduction:

- Reduced lot sizes (number of setups can be increased)
- Reduced work-in-process
- Reduced inventory
- Reduced floor space occupancy
- Reduced material handling
- Reduced lead times
- Improved efficiency and flexibility of production systems (production can be adjusted to actual demand)
- Eliminated waste (products, equipment, time, et cetera)
- Reduced manufacturing costs
- Improved quality (through lot-size reduction)
- Increased worker productivity (reduction and elimination of unproductive tasks)
- Increased sense of equipment ownership (operators not only responsible for production, but also for setups)
- Progressive implementation of new attitudes and new culture (workers involved in work condition improvement)

A prerequisite to applying the methodology that brings forth these improvements is to clearly understand the definition of setup time.

6.1.1 Definition of Setup Time

The most comprehensive measure of machine setup time is the time spent between the production of the last part of one lot and the production of the first *good* part of the next lot: The time during which no good part or product comes out of the equipment. It is essential to include the time spent adjusting the equipment until the first good part is obtained because, in many cases, a substantial part of the setup time is spent in machine adjustment during which the products must be scrapped. The machine setup time is usually called "setup time" or even more simply "setup."

Operations performed during setups are easily grouped in a few categories: Equipment preparation and storage, jigs and fixtures exchange, component magazines replacement, machine settings modification (temperature, pressure, intensity, speed, etc.), machine-part adjustment, and trial runs. The chart in Figure 6.1, which is based on the analysis of numerous setups, shows that the area of jigs and fixtures exchange represents a surprisingly low percentage of the total setup time. Observations have shown that most of setup time is actually spent in tool chasing, component preparation, fixture and part adjustment, and trial runs.

All the operations mentioned in Figure 6.1 are commonly performed when the machine is stopped, thereby wasting operator time and reducing machine utiliza-

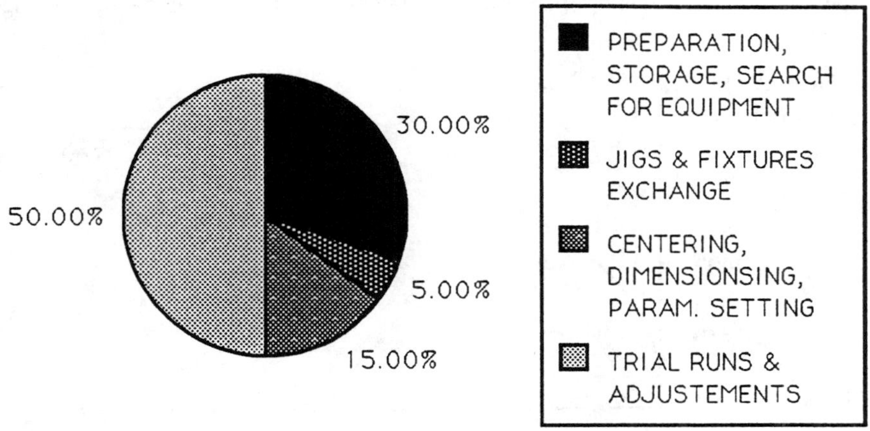

Figure 6.1. Main Categories Setup Breakdown.

tion. However, this need not be the case. For example, there is no reason why a machine should be stopped during fixture preparation and storage.

An immediate and most interesting conclusion of this analysis is that a significant percentage of setup time (at least 30 percent) can be eliminated at almost no cost at all. Actions as inexpensive as cleaning the work area, organizing the storage racks, and storing the most commonly used tools close to the corresponding machine can significantly reduce setup time. This result is illustrated by the typical time reduction versus investment curve shown in Figure 6.2. This curve shape is characteristic of any setup study on any machine. Usually the curve origin is significantly above 0 percent. (Note that more than a 10 percent setup time reduction is possible at no expense.)

6.1.2 Conventional Wisdom on Setup

Setup time is unproductive time. Traditionally, certain strategies have been used to reduce the impact of long setup times on equipment utilization. Some of the most popular are:

- Train machine operators and setup people to improve their skills.
- Use experienced people.
- Increase lot sizes.
- Balance setup cost versus inventory cost by determining "Economic Lot Sizes."
- Combine and sequence back-to-back jobs with similar setup conditions.

Unfortunately these strategies are based on the false assumption that setup operations cannot be simplified. Another drawback is that they aim at optimizing

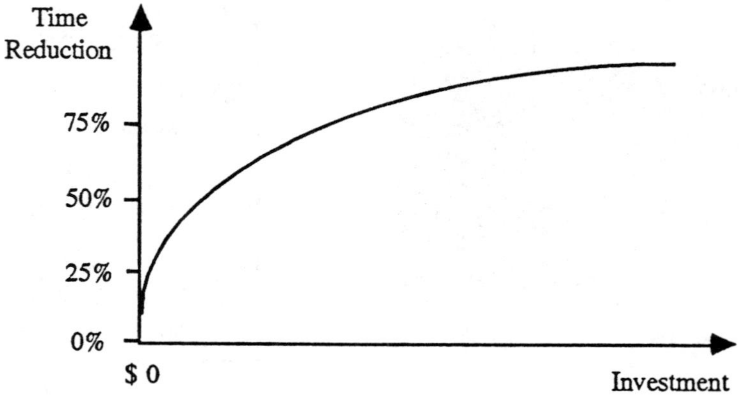

Figure 6.2. A Typical Time Reduction Versus Investment Curve.

unproductive operations rather than eliminating them. Significantly, using any of the above strategies actually results in *increasing* production costs—the latter three even add complexity to production and inventory management. For a long time the credo of manufacturing managers, these strategies have now become prohibited tools in WCM companies. The reasons for the gradual demise of these strategies is discussed in the remainder of this section.

Traditional setup operations require more skills than are necessary to run the machines. To reduce setup time, conventional wisdom has placed emphasis on raising the skill level of the people in charge of setups. However, why spend money to raise the skill level of people to perform unproductive operations? It is indeed more cost-effective to use the current skills and knowledge of machine operators or setup men to simplify and eliminate setup operations that in turn will lead to lowering the required skill level.

From a mathematical perspective, increasing lot sizes appears to be the most appropriate solution to the problem of long setup times. As a result, products are manufactured in large quantities and stored for future shipments. Usually more products than are immediately needed are manufactured to prevent any future shortages caused by unexpected scrap levels or demand variation. However, this strategy does not consider the additional costs that are actually incurred by high inventory levels:

- Monetary value of stored products (usually the only cost considered when estimating inventory costs on a day-to-day basis).
- Material handling and storage equipment.
- Storage facilities.
- Inventory management and production scheduling systems.
- Man-hours for inventory management, production scheduling, materials handling and facilities housekeeping.
- Product loss, obsolescence, and damage.

Actually there are so many drawbacks associated with large lot sizes that long setup times appear to be their only justification.

The economic order quantity (EOQ) model balances holding costs versus job-order costs. Although the EOQ model is correct in theory, it assumes that job-order costs (setup costs) are fixed; therefore, it is not applicable to modern manufacturing conditions. Indeed, as will be discussed in the next section, drastic reduction in setup times—and consequently in setup costs and inventory-holding costs—leads to the obsolescence of the model.

Combining jobs with similar setup conditions certainly reduces setup time; yet, it also increases production costs. Usually, the combination sequence is the result of a foreman reasoning or perhaps of the application of an optimization algorithm.

It is easy to conclude that applying this strategy results, again, in producing parts when they are not needed. Job combination and large-lot production present similar drawbacks.

When applying these conventional solutions, companies lose sight of the real problems (for example, long setup times), and utilize resources to solve the wrong problems (for example, determine "economic" lot size).

6.1.3 Fallacy of Economic Lot Sizing

Traditionally, the EOQ model (also known as the economic lot size model) has been used to determine lot sizes. As shown on the graph in Figure 6.3, the model determines the so-called optimal lot sizes that will minimize total costs given job order costs, annual demand, and inventory-holding costs.

The effect on the inventory level when using the EOQ model is shown in Figure 6.4. In most cases, the lot size is several times larger than the shipping quantities (size of shipments to customers). Note that the average inventory level is equal to nearly half the lot size and is several times higher than the shipping quantities.

Using this method not only results in high inventory levels, but can also lead to making wrong decisions because of model assumptions and approximations. For example, job order costs (setup costs) are assumed to be fixed, and inventory-holding costs are approximated. (It is very tedious and practically impossible to precisely estimate—for every product on hand—the cost of material handling, storage facilities, inventory management, and lost or damaged items.)

One must keep in mind that the EOQ model is used with the objective of minimizing inventory costs. However, is not the best way to minimize inventory cost to *eliminate inventory?* Is the goal of a manufacturing company to produce at minimum cost parts and products that will be sold to customers, or is it to calculate

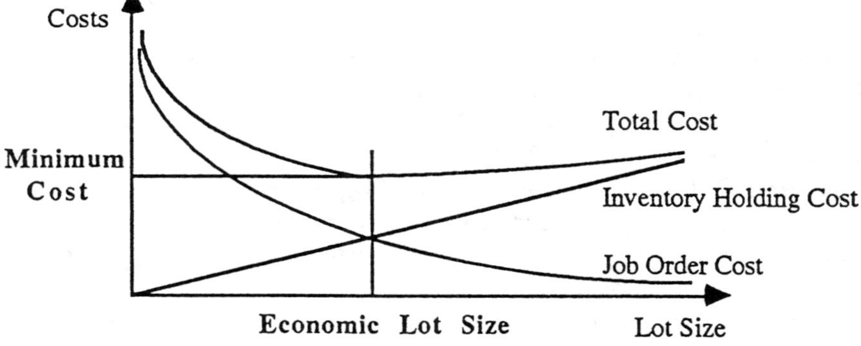

Figure 6.3. The Economic Lot Size Model.

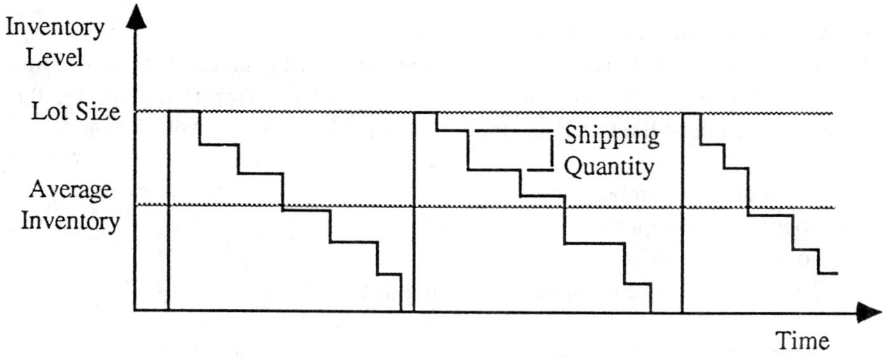

Figure 6.4. Inventory Level Variation when using the Economic Lot Size Model.

lot sizes that will balance estimated job order costs against estimated holding costs as a function of a forecasted customer demand? The EOQ model is clearly obsolete when setup times are reduced and allow drastic setup costs reduction: When the cost of individual setups approaches zero, lot sizes can be directly determined as a function of the customer demand in order to keep balanced production operations and low inventory levels.

In a JIT environment, depending on lead times and shipment frequency, lot sizes can be set equal to shipping quantities. The corresponding variation in inventory level is shown on the graph in Figure 6.5. Note that the average inventory level does not exceed the average shipping quantity.

Conventional solutions, such as the application of the EOQ model, are aimed at adapting production variables to long setup times. The scope of this chapter is to present strategies that lead to the elimination of these long setup times.

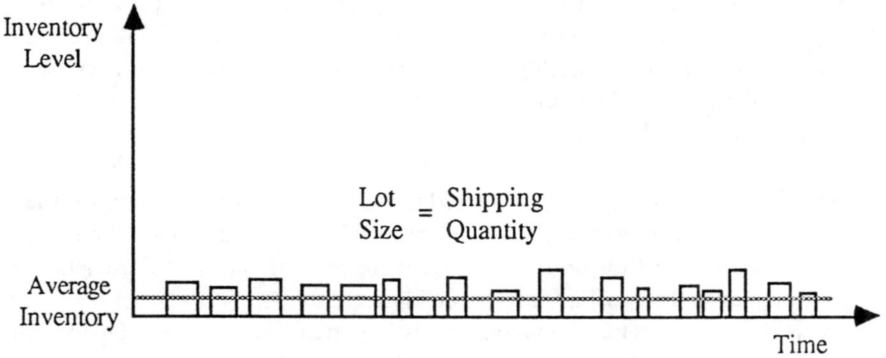

Figure 6.5. Inventory Level Variation in JIT Environment.

6.1.4 The Importance of Setup-Time Reduction

Why is it important to reduce setup time? If manufacturing managers are asked this question, depending on the company and its specific shop-floor culture, the responses will most likely be one or several of the following:

- To reduce the number of setup people.
- To increase equipment capacity.
- To reduce lot sizes.
- To increase flexibility in manufacturing operations.

The first two responses are not objectives aimed at by WCM, because the objectives lack foresight and do not justify the investment in a setup-reduction program. A reduction in individual setup times does not necessarily lead to an overall setup-time reduction. The goal of reducing individual setup times is to be able to do more setups without reducing the available machine capacity. The overall capacity required for setups is likely to remain the same, as is the overall time spent for setup by machine operators.

The second two responses are objectives aimed at by world-class companies. Lot size reduction is the immediate result of setup time reduction, while the main benefit from reducing setup time is increased flexibility in manufacturing operations. Shorter setups and runs reduce production lead times and enable companies to adjust production to customer-order quantities and frequency. Gradually reducing machine setup time will decrease lot sizes with the following desirable results:

- Work-in-process is reduced (parts waiting in front of equipment).
- Finished product inventory is reduced (parts or products waiting for shipment).
- Lead times are reduced.
- Quality is improved (detected defects impact fewer parts, and less material handling and less storage reduce transport risks and time deterioration).
- Production-facility flexibility is improved and allows for a quicker response to customer demand variation.
- Operating capital requirement is reduced.

This is a clear consequence of the reduction in work-in-process and finished products inventories. Although not necessarily the most important long-term benefit of setup-time reduction, this decrease in operating capital requirement has a great impact on the company's financial big picture.

In the long-term, setup-time reduction is important because—combined with workcell implementation—it allows traditional manufacturing facilities to be transformed into flexible production tools.

6.1.5 Illustrative Example of the Importance of Setup-Time Reduction

This short example illustrates the drastic effect that reducing setup time has on work-in-process and lead time.

A production process consisting of two machines and which produces five different parts is considered. Figure 6.6 illustrates the before-setup-reduction-production schedule, and Table 6.1 shows the before-setup-reduction-key-performance measures. The two illustrations together indicate that each machine is utilized ten hours per week for setup and thirty hours per week for production.

Figure 6.7 and Table 6.2 show the production schedule and performance measures achieved after a setup time reduction of 80 percent (the percentage commonly attained).

Reducing the individual setup times from two hours down to twenty-four minutes enables the setup frequency to be multiplied by five and the lot sizes to be reduced by 80 percent. Both machines are still utilized ten hours per week for setup and thirty hours per week for production. Figure 6.8 gives an additional comparison of lead times and work-in-process before and after setup-time reduction.

After setup-time reduction, lead times and work-in-process are both reduced by 80 percent. Although this reduction level might seem satisfactory, further lead time and work-in-process reduction can be attained by moving machine A next to machine B in the same workcell. With the same lot size, but changing the transfer batch size to one part (as soon as one part is finished on A it is made available to B), work-in-process between machines drops to one part and the lead time drops to one hour and twelve minutes added to the time of processing one part. This comparison clearly illustrates the drastic effect that reducing setup time has on work-in-process and lead time.

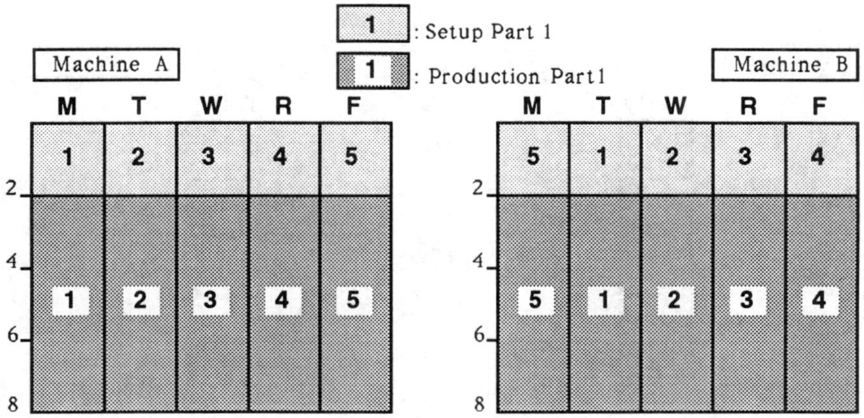

Figure 6.6. Weekly Production Schedules on Machines A & B Before Setup-Time Reductions.

TABLE 6.1. Key Performance Measures Before Setup-Time Reduction.

Setup Time (per machine)	2 Hours
Work-in-process (between machines)	6 Hour Production
Lead Time	16 Hours

6.2 ORGANIZING FOR SETUP TIME REDUCTION PROJECTS

Is a significant setup-time reduction achievable at low cost, in a short time frame, even if only internal resources are employed? With proper project organization the answer is yes. Using a well-planned and systematic approach will ensure success as early as with the first setup analysis. Because the project is highly visible on the shop floor and shop-floor workers are involved in the project, an immediate success is important. Early success will gain the people's support for the entire project and will foster EI. In addition, setup-time reduction is usually a primary phase of JIT implementation; if this initial phase is successful, shop-floor people will be more likely to accept the project and be more actively involved in the reorganization.

A typical setup project involves the phases shown in Figure 6.9.

To ensure the success of the project, it is crucial that the plant's or company's top managers devote sufficient time and attention to planning the first two phases, which appear in the diagram on Figure 6.9: Organization of the project and

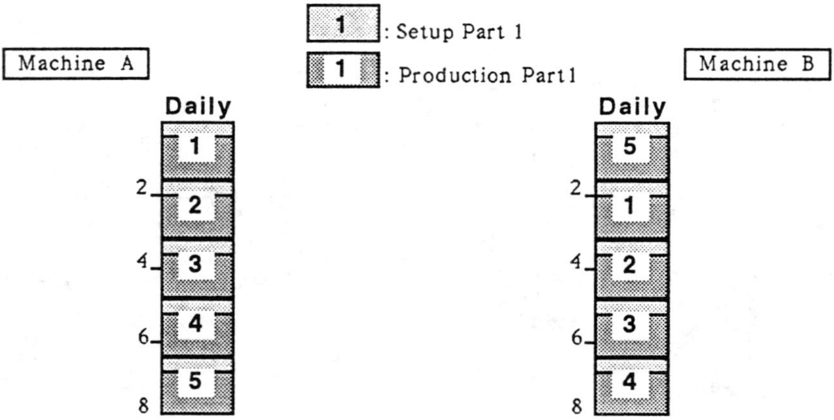

Figure 6.7. Daily Production Schedules on Machines A & B After Setup-Time Reduction.

TABLE 6.2. Key Performance Measures After Setup-Time Reduction.

Setup Time (per machine)	24 MN
Work-in-process (between machines)	1 Hour 12 MN Production
Lead Time	3 Hours 12 MN

education of the personnel. The following key points should be addressed before starting to analyze the first setup: (1) a clear definition of the objectives, (2) constitution and training of setup teams, and (3) information of the personnel affected by the setup project. In addition, top managers should be committed to the success of the project and provide active support to the entire project. These key points will be detailed in the following sections.

6.2.1 Defining the Objectives of the Setup Project

Reducing setup time is not a primary objective in itself. Shorter setup times are a means to achieve broader objectives. It is unrealistic to attempt to justify setup-time reduction as a stand-alone project, because the broader objective of setup-time

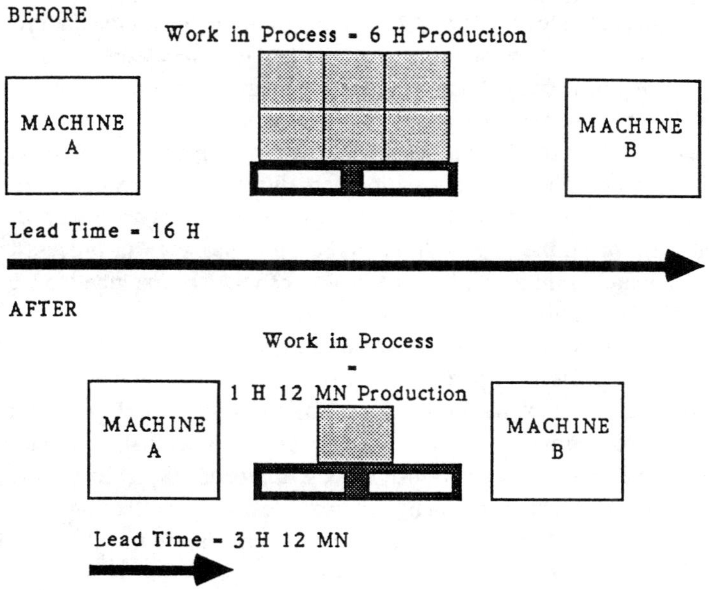

Figure 6.8. Performance Measure Comparison.

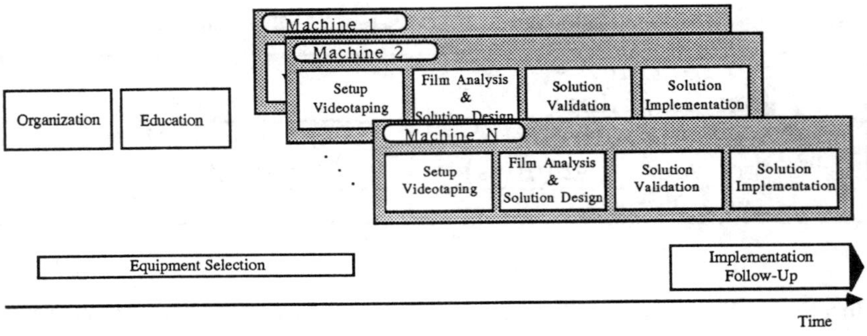

Figure 6.9. Typical Setup Project Schedule.

reduction is to eliminate some of the obstacles on the way to achieving a WCM system. The "setup project" is only one of the first stages of a global reorganization and evolution from traditional manufacturing methods toward WCM.

General objectives for the reorganization might be expressed in terms of lead time reduction, increases in productivity and quality levels, and the like. These objectives need, of course, to be communicated to the work force when the reorganization program is presented to them. However, in order to motivate the people who participate and are involved in the setup project, more precise objectives need to be assigned. These objectives will be specifically expressed in such terms as the expected level of setup-time reduction, the plant area and number of machines concerned by the setup project, the budget allocated to each machine, and the project's time span. Answering the following questions will be helpful in assessing the project objectives and defining its scope:

What area of the plant is concerned with the project? The setup project should start in the pilot area chosen for the entire reorganization—for example, in the first manufacturing cell to be implemented. If the setup project is not a part of a larger improvement project, the area with the highest product work-in-process levels is the best candidate for initial studies; the project can then be extended to a group of cells, a department in the plant, or the entire plant.

What is the estimated number of machines to be analyzed? What is the budget allocated to the project? The number of machines on which to reduce setup depends on the extent of the area covered by the project and on the proposed budget. The setup team's assignment is to design solutions that are as simple and as inexpensive as possible: A low-cost, no-cost approach. Typically, an average of \$3,000-\$5,000 is allocated for each machine, depending on the machine's complexity, age, and manufacturer. The number of machines and the allocated budget can be estimated on the basis of these factors.

What level of time reduction is expected? Using simulation and modeling tools can help in evaluating optimal levels of setup time reduction to meet target lead times, work-in-process levels, and output rates. A reduction of 75 percent, the current industry standard, represents a minimum value that is easily attainable for a first setup reduction on any type of equipment, when the study is conducted properly. All numerical information given in the rest of this chapter assumes that a reduction of 75 percent can be attained.

What is the project time span? The project's time span depends on the number of machines to be studied and the number of people that can be assigned to the setup teams. Usually a study will deal with one machine or a group of similar machines. The estimated work load is eight to twelve person-days per study for a team of two to three people working full time. This time frame does not include solution implementation—the final phase of each machine study—which does not require full-time commitment.

Once the project objectives have been defined, setup project teams can be constituted.

6.2.2 Forming Setup Project Teams

Selecting the setup team members is a critical step of the project organization. Choosing the right people will ensure that generated solutions will work and, more importantly, will be used. (Henceforth, please note that "setup team" refers to a team that is formed specifically for the setup reduction project, whereas "setup people" designates the people who traditionally perform machine setups.)

What is the mission of the setup team? Once the equipment has been selected, the team members must record machine setups on videotapes, analyze the videotapes, generate ideas, and find and implement solutions. This mission requires full-time commitment until the first solutions are implemented.

Who are the best qualified people? More explicitly,

- Who has the best knowledge of the machines on the shopfloor?
- Who is continually confronting problems with machine operation and setups?
- Who already has many ideas on how to improve the working conditions, but has never been asked for them?
- On whom will the proposed improvements have the most impact?
- Who has the most incentive to propose practical solutions?

The answer is neither the shop supervisor, who is permanently fighting several fires at the same time, nor the people in the offices upstairs who rarely go down to the shop floor; The machine operators and the setup people are the experts. From working on these machines for years, the machine operators and setup people know every little detail about them. Unfortunately, in many companies, no one has ever had the idea—or taken the time—to ask these people how machine operations and working conditions could be simplified and improved.

Another advantage of involving machine operators in the setup teams is that the project automatically becomes a shop-floor project: It is not just another study made by a team of engineers whose solutions are imposed on the shop-floor people. Shop-floor ownership ensures that the proposed solutions will be applied effectively. In addition, ideas coming from machine operators and setup people will be more simple and more practical, and often less expensive than so-called engineered solutions.

Several teams can be formed depending on the number of machines to analyze and on the project time span. The selected people should be familiar with the equipment that will be analyzed. Setup teams typically consist of two setup people or machine operators who are supported by a technician: The machine operators and setup people will generate most of the ideas, and the technician will help the other team members to formalize ideas (technical writing, drawings, justifications, et cetera) and to communicate with external peripheral participants (design engineers, manufacturing engineers, equipment vendors, et cetera). Occasional support will be necessary from design engineering, maintenance, purchasing, and other departments related to the manufacturing process. Company personnel need to be informed about the setup project and should be ready to participate when required. The setup team(s) should be located as close as possible to the machines that are being analyzed.

The project leader (or team leader)—usually from manufacturing engineering—guides the different teams. The project leader's position also requires 100 percent commitment throughout the solution implementation. It is important to choose a project leader who already knows the setup-reduction methodology or one who is open to new challenges. The project leader has to communicate his or her enthusiasm to the team members by being actively involved in machine selection, team member's training, relations with other departments and vendors, conflict resolution, and solution implementation and follow-up. In addition, the project leader monitors and is responsible for the expenses so that the setup teams can purchase equipment and request vendor estimates without waiting for a dozen signatures.

The team leader's first task is to purchase the tools for setup analysis. The following equipment should be purchased for every team so that it is available the day of the team member's training: A video camera with tripod (for filming long setups), a TV monitor, and (recommended) a desktop computer with spreadsheet software and printer. Consider buying a video camera with an incorporated VCR that runs standard tapes and has the date and time (in seconds) display option directly on the image—the display option will greatly facilitate the detailed analysis of the setup. In addition, the video camera should also give a clear image in normal shop-floor lighting conditions.

Two prerequisites to sending the setup teams on to the shop floor, are (1) to train team members in setup-reduction methodology, and (2) to present the setup project to the plant or company work force.

6.2.3 Informing the Work Force and Training Setup Teams

An initial informational meeting should be held with the plant or company work force involved in the project or affected by the project. The objectives of this meeting are to present the project and to demonstrate management's commitment to the project. The meeting date must be as close as possible to the date of the first setup videotaping, ideally immediately before team members' training. The underlying objectives are to drive out any fear of the project, to arouse interest, and to prepare the atmosphere for EI.

Because this meeting is often a part of the overall JIT project presentation, one or two hours should be designated for the presentation of the setup project. Management and the setup project leader should conduct the presentation.

To start the meeting, explain the project objectives and the expected results. It is essential that the employees understand that the project goals are (1) to reduce setup time—not reduce setup people (explain that it will allow more setups to be done economically), and (2) to make people work more efficiently by eliminating wasteful operations—not make people work faster. It is also important to explain the project methodology to the workers because people must understand why they will be videotaped. (For example, play a tape from an actual setup reduction study that shows excerpts from the different study phases: The setup videotaping, the tape analysis, the different meetings, and the resulting machine improvements.)

Setup team members should also attend this general informational meeting, thereby becoming aware of the objectives and importance of the project. Formal training will then familiarize team members with the project's tools (video equipment and computer) and methodology. The training is conducted by the project leader and supported by people who have previously been involved in setup projects. Training typically lasts two days which are broken down as follows:

Day 1:

- Recall and detail project objectives.
- Introduce the setup teams.
- Explain the project methodology in detail.
- Present the tools (assume no previous experience in videotaping or with personal computers).
- Show a videotape that illustrates an actual setup-reduction study, which details the following steps:

1. Setup videotaping;
2. Analysis and idea discussion (insist on the low-cost/no-cost approach);
3. External consultation;

4. Solution validation meeting; and
5. Improvements implementation.

- Introduce the next day's exercise by showing a videotaped setup (fifteen - twenty minutes).

Day 2:

- Breakdown the group according to the different setup teams.
- Let each team analyze the tape (using their own equipment) shown at the end of the previous day (each team generates a list of ideas).
- Bring the teams back together and discuss, classify, and modify ideas.
- Show potential problems with generated ideas (complexity, cost, time to implement, et cetera).
- Emphasize the need for outside support (shop-floor people, other departments, and vendors).
- Complete the analysis and formalize solutions (compare the different team's solutions).
- Announce the project's schedule—or at least the first setups to be videotaped.

Education plays a major role in the implementation of WCM culture in a company. The first actions in the implementation of this new culture are usually taken by correctly informing the work force about the setup project and thoroughly training team members. In many companies there is no direct communication between top management and shop-floor people. The setup project is an opportunity for creating these communication channels.

6.2.4 Showing Management Support

The role of top management is to support the setup project from start to finish. The proposed improvements will impact most of the plant or company departments, and certain departments may be reluctant to modify their way of working because machine operators, and not management, will propose the new procedures. Unless the people in these departments feel strong management support for the project, they will not be receptive to the improvements proposed by the team members.

Team members will also be more interested in the project and more productive if they understand the project objectives and feel management's commitment to its success. However, if they believe that they are just working on another "project of the year," poor results can be expected.

It should be remembered also that the setup project is usually the first visible step of JIT implementation. Reorganization will dramatically modify the way every individual works in the plant. Managers should use the setup project as a warming-up period for the implementation of cultural change (including their own cultural change). Managers should invest time and energy in the project by closely following the setup studies, talking with the setup team members, visiting the shop floor to talk with the operators and supervisors, and showing interest in the implementation of the solutions designed by the setup teams.

After the setup project has been carefully organized and presented to the work force, and after the setup teams have been trained, the actual hands-on phase can begin.

6.3 HOW TO ACHIEVE SETUP-TIME REDUCTION

This section will present and discuss in detail how to achieve setup-time reduction by addressing the following topics:

- How to select which setups to study.
- How to record and analyze setup activities and data.
- How to convert internal setup activities to external setup activities.
- How to simplify and eliminate setup operations.
- How to validate and implement solutions.
- How to document setup studies.

6.3.1 Selecting Machines and Setups

A setup-time reduction project will improve the overall efficiency of a production system only if the machines to be studied and the specific part setups to be analyzed are properly selected. Such factors as machine utilization, assigned product mix, and current setup lengths have a significant impact on how the time reduction of a particular setup on a particular machine will translate into an overall system-performance improvement. For example, setup-time reduction on a machine that is utilized at 50 percent and dedicated to two or three parts yields a different gain than if the machine is utilized at 80 percent and manufactures a dozen part types. Similarly, reducing the time of a setup that represents 10 percent of a machine's setups, again, yields a different gain than if the setup represents 70 percent of the machine's setups.

Obviously, efforts should be focused on the setups that will yield the most gain—keeping in mind that the number of studies will be limited by the budgeted resources (personnel, time, and money). Similarities in parts, fixtures, machines, and production environment will allow the generalization of many improvements from analyzed setups to nonanalyzed setups; thus, it is not necessary to analyze every part setup on every machine. The purpose of the equipment-selection phase is to discern (1) which machines should be studied in order to reach the greatest impact, (2) which part setups of the machine selected should be analyzed, and (3) in what sequence these setups should be analyzed. Thus, the results of the equipment-selection phase are a list of machines with an indication of the specific setup(s) per machine and a tentative analysis schedule.

Equipment selection should start as soon as management has defined the scope of the setup project (cell, group of cells, department, or entire plant) and fixed either the number of machines to study or the budget allocated to the project. The first machines and setups should be selected prior to training the team members so that the first setup studies can start immediately after the training. Subsequent setups can be selected and scheduled as the setup project progresses.

A combination of two different approaches can be used to identify the machine and setups to be analyzed: First, an analytical approach using a modeling tool (computer simulation of manufacturing operations); and second, a somewhat more subjective approach based on the shop floor people's knowledge. PC-based modeling packages are becoming more and more popular with industrial engineers; however, it is essential when selecting the machines and setups to keep in mind that there are knowledgeable people on the shop floor that are an invaluable source of information. If the modeling approach is selected, the list of machines and setups identified through computer simulation should be discussed with the shop-floor people. Indeed, using a simulation tool to select machines and setups and not submitting the resulting list to shop supervisors will get the cultural-change implementation off to a poor start. (Modeling approaches and their appropriateness are discussed in chapter 7 on "Lot Size Reduction.")

Whichever approach is taken, the guideline that will help identify the machines and setups that will yield the greatest benefits in setup-time reduction is to select the machines that exhibit high utilization (bottlenecks), long setup times, large product mix, and/or large lot sizes. Additionally, when deciding which setup(s) to analyze on a selected machine, the longest setups or setups for recurrent parts (parts that represent a high percentage of the machine utilization) should be chosen. The most methodical approach to use to identify recurrent parts is to draw Pareto charts that will show, in decreasing order, the percentages of the volumes of the different parts assigned to the machine. Figure 6.10 shows two examples of such diagrams. In the first chart, part C is the only candidate; therefore, the analysis of only one setup is appropriate on machine 1. In the second case, parts A and C are both candidates; if these two parts have significantly different setup operations, then analyzing both setups in the study of machine 2 should be considered.

Identifying the appropriate setup(s) on a selected machine should be fast and simple. The point is to not spend weeks analyzing route sheets and production reports; in fact, no more than a day should be spent gathering the information and selecting the parts or products. If the data cannot be obtained easily, then seek out people with knowledge and experience in order to identify parts causing known and repeated setup problems. Accordingly, valuable information can be obtained

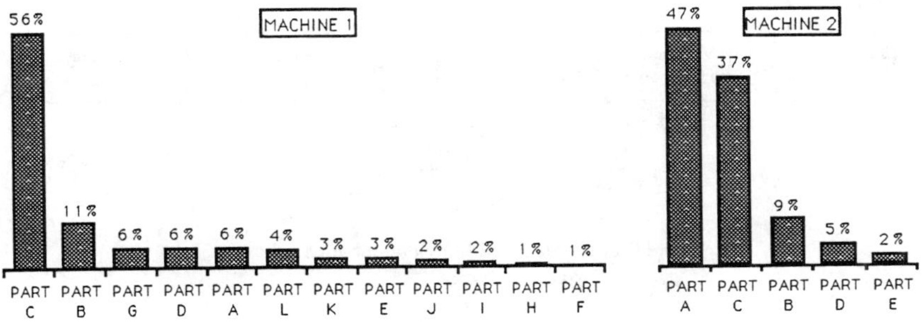

Figure 6.10. Examples of Pareto Diagrams for Setup Selection (Volume Percentages of Parts Assigned to the Machines.)

from people such as shop-floor supervisors, setup people, process planners, production schedulers, and manufacturing engineers.

The last step—and in many production environments the most challenging—is to create a detailed project schedule from the list of machines and selected setups. The schedule is initiated by the project leader with the help of the production planning department. In order to keep the analyzed setups natural and as representative as possible of normal setups, the production schedule should *never* be modified in order to provide a setup for videotaping; rather, videotapings must be planned as a function of the actual production schedule. The schedule should be kept flexible since many last minute changes can be expected—never count on establishing a definite detailed schedule. In addition, enough days should be allowed between each videotaping to keep the teams work-in-process at a minimum level. Accordingly, a study should be, at least, at the solution-validation step before the next study starts. If two teams share the same video equipment—although this is not recommended—then two videotapings should not be scheduled for the same day. If a machine requires more than one videotaping, then the videotapings should be scheduled as closely together as possible for the same team, so that both analyses can be made in the same study. The actual setup starting time should always be checked with the production supervisor the day before the scheduled setup time in case any changes in production are to take place.

A final recommendation is to start the project with the setup that shows the most potential for success. Starting the project with a success will incite enthusiasm and interest and facilitate involvement in the project.

6.3.2 Adopting a Systematic Approach

Conducting the individual setup studies requires method. The difference between marginal and substantial gains lies in whether the proper approach is adopted. The main steps involved in a setup study are shown in Figure 6.11. These steps are detailed in the following paragraphs.

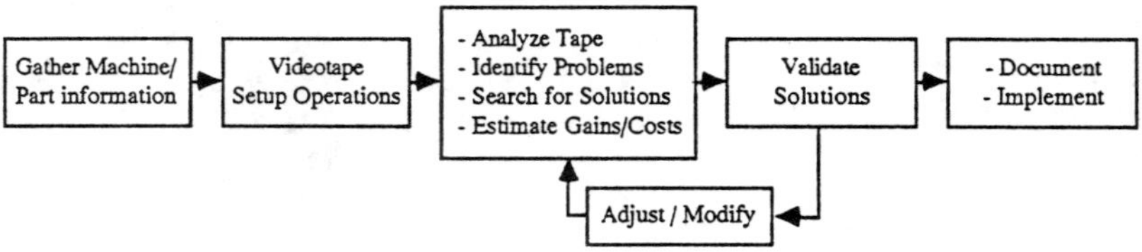

Figure 6.11. Detail of Individual Setup Study.

Gather machine and part information.

The objective of this initial step in the machine study is to gather both specific information relevant to the videotaped setup and documents that will be used to support equipment modifications. The documents consist mainly of part blueprints, fixture and pallet blueprints, machine technical descriptions, and machine blueprints.

These documents will not be used until during the setup videotape analysis. However, they should be gathered as early as possible to avoid interrupting the analysis process when the information is needed. In addition, having the information at hand will ensure a quick response to questions that team members will be required to answer as the study progresses.

Videotape setup operations.

After scheduling a setup videotaping, a short informal meeting (ten–fifteen minutes) should be conducted with the machine operator responsible for setting up the machine: One or two of the team members must explain to the operators the reason for the videotaping and how they should react to the presence of the camera. Since the setup has to be kept as natural as possible, the operator or setup people should perform the work as usual and not pay attention to the camera or the (project) setup team. This meeting should be conducted near the machine to be studied on the day before the actual videotaping. In addition, this is the occasion for the team members to consider where to install the camera during the setup.

The setup recording starts when the last part of the preceding lot is processed, and the recording stops when the first good part of the next lot is processed. The entire setup must be recorded. As some setups last several hours, enough blank tape should be prepared for the entire setup. In addition, the camera must not be stopped when the operator leaves the machine for any setup related operation: Remember that among the objectives of the study is the reduction of the time spent searching for tools, parts, and documents. If several people work on the setup, focus the camera on the main setup person, and make sure that his or her hands are in the camera field.

Take notes! Taking notes enables a precise and fast film analysis to be produced.

One team member takes notes while another operates the camera. The current time should be reported regularly on the notes and then coordinated with the time displayed on the videotape. If not familiar with the setup being recorded, the person taking notes should be assisted by the operator supervisor or someone knowledgeable about the setup operations because, to keep the setup natural, the operator or setup people must not be interrupted with repeated questions about what they are doing.

Analyze the videotape.
This step of the study consists of rigorously decomposing the setup into elementary operations (whose lengths vary from seconds to minutes) and of identifying main classes in which the operations are grouped. The analysis can be done manually; however, since the computations of the different gains are rather tedious, it is more efficient to use spreadsheet software on a desktop computer. Another advantage of using a computer is that high-quality graphs can be quickly generated from the spreadsheet data. Any commercial package can be used as long as it can display the data in the form of graphs. If necessary, team members should be familiarized with the computers and the software during the training phase: Experience shows that machine operators can learn basic computer and software functions in a few hours. (Occasional assistance by someone familiar with the software is useful for building the first worksheets and graphs.)

In this section, it is assumed that the tape is to be analyzed with the help of a computerized spreadsheet. The chart presented in Figure 6.12 is an example of a list of operations from the tape analysis done with a computerized spreadsheet.

The information shown in Figure 6.12 can be obtained by (1) defining each operation, (2) determining the operation's beginning and ending times, (3) calculating the operation length, and (4) grouping the operation in a specific class. (Every operation depicted on the videotape is decomposed and charted in this manner.)

Defining each operation. An operation can be defined as the smallest identifi-

	A	B	C	D	E	F	G	
1	OPERATION DESCRIPTION	Hr	Mn	Sec	Lgth	Class		
2					(Sec)	#		
.	.	.	.	.	.	.		
.	.	.	.	.	.	.		
56	Opens Machine Jaws	0	59	52	71	1		
57	Releases Mold	1	01	03	10	1		
58	Lifts Mold with Crane	1	01	13	62	1		
59	Places Mold on Pallet	1	02	15	21	1		
60	Gets New Mold from Rack	1	02	36	201	2		
61	.	1	05	57	.	.		
.	.	.	.	.	.	.		

Figure 6.12. Example of List of Operations.

able action performed by the operator: Secure bolts, get wrench from tool bench, get cutting tools from tool room. Why identify extremely detailed operations? It is much easier to efficiently apply the reasoning "eliminate, simplify, or transfer to external time" to a small well-defined operation than to a long part of a setup. For example, if all operations involved in fixture installation on one machine where grouped under a macro-operation called "fixture installation," the setup team members would have difficulty reducing the time spent doing such jobs as part preparation, tool searching, and actual fixture securing on the machine table. On the other hand, if each of these jobs is broken down into the smallest identifiable operations, eliminating or simplifying the operations becomes less troublesome.

Determining beginning and ending times. Each operation has a well-defined starting and ending point, such as the moment when the operator takes a wrench from the workbench, when he or she starts turning the bolts, or when the part is secured. The operation beginning time is read from the videotape at the operation starting point. The operation ends when the next operation begins. Operation lengths can vary from several seconds to several minutes. Each operation is entered in a single row on the spreadsheet so that, at this point in the analysis, the entry consists of the operation description and its beginning time (hr, min, sec).

Calculating operation length. After the tape has been decomposed and beginning times determined, then each operation length is computed. The operation length (in seconds) is obtained by subtracting the current operation's beginning time from the next operation's beginning time.

Grouping operations in classes. The next step is to identify main classes in which to group the operations. (No more than ten classes should be defined.) Classes of operations are selected by breaking down the entire setup into sequential categories. Former fixture removing, new fixture installation, and trial runs are a few examples of classes.

As shown in Figure 6.13, a list of the determined classes is setup on the worksheet by entering in a row the class number (1, 2, 3, and so on) and the class

	AA	AB	AC	AD	
1	CLASS	CLASS DESCRIPTION	TIME	TIME	
2	#		(SEC)	(%)	
3	1	Former Fixture Removing	389	11.60%	
4	2	New Fixture Preparation	202	6.02%	
5	3	Fixture Install. on Machine	653	19.47%	
6	4	Adjustments	1010	30.11%	
7	5	Trial Runs	887	26.44%	
8	6	Paper Work	213	6.35%	
9					
10		TOTAL TIME	3354	100.00%	
11					

Figure 6.13. Example of List of Classes

description. As each operation is designated to a particular class, the corresponding class number is entered on the operation row (see Figure 6.12).

The objective of defining classes is to build a basis for discussion with people external to the setup project, such as shop supervisors or managers. When presenting a setup analysis to these people, it can be explained that the observed setup time is broken down in fixture installation (12 percent of the total time), trial runs (30 percent), material handling (24 percent), and so forth. From the setup team perspective, the class level is the communication level, and the operation level is the improvement level.

The last important step in the videotape analysis consists of computing and charting class-time percentages *before time reduction.* This synthesis recaps the videotape analysis. The time spent on each class of operations and the percentage of the initial setup time represented by each class are computed. Each class's total time is computed by adding together the individual times of each operation in that class. The total time per class is noted on each row of the list of classes. The total setup time is then computed, and relative class percentages are marked down in the class rows, as shown in Figure 6.13.

Figure 6.14 shows a pie chart that can be constructed to present a graphical synthesis of the setup before time reduction.

At this point in the setup study, the videotaped setup has been decomposed, an initial synthesis before time reduction has been established, and setup team members have some ideas about how to reduce, eliminate or simplify some of the operations. The next steps will consist of identifying potential problems in the way setup operations are currently performed and searching for possible solutions.

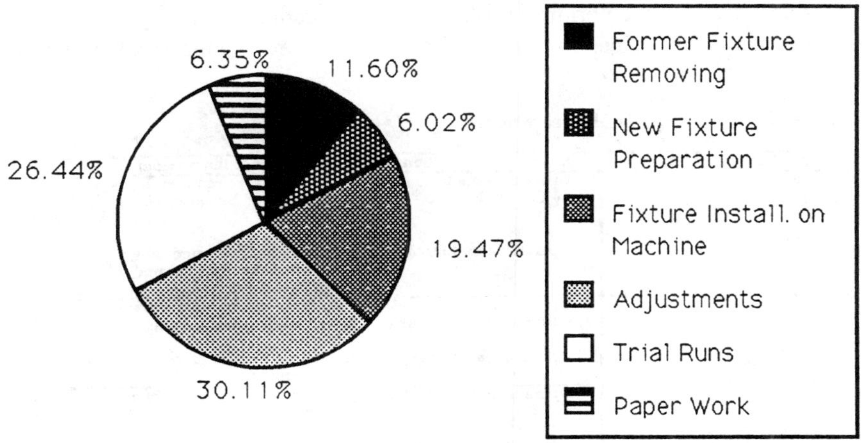

Figure 6.14. Initial Setup Time Breakdown in Classes.

Identify problems and Search for solutions

An efficient approach to setup efficiency is found in the philosophy "eliminate, simplify, or transfer to external time." Thus, when identifying problems, each setup operation should be considered as a candidate for elimination, simplification, or transfer to an external operation. A few examples are given in Table 6.3.

Note that there exists no precise limit between elimination, reduction, and transfer to external operation. An operation such as getting a fixture from the storage room, could first be transferred to external time and then further simplified by storing the dedicated fixtures a few feet away from the workstation. (Fixture dedication to workstations is a direct result of part dedication to workstations when manufacturing cells are formed.) In addition, dedicated material handling equipment such as jib cranes, conveyors, and carts could eliminate the wait for forklift trucks or bridge cranes. The appropriate strategies and techniques to use for operation elimination, simplification, and transfer to external time are reviewed in sections 6.3.3 to 6.3.6.

The search for improvements is a continuous process. During the setup study, setup team members can start developing ideas as soon as they analyze the tape. The machine operator who was filmed is involved in the analysis as early as possible, in part to criticize the team's ideas, but mostly to express his or her own ideas. Advice should also be sought from shop supervisors, process planners, tool designers, manufacturing engineers, design engineers, and equipment vendors. To effectively lead the teams, the project leader should review, at least on a daily basis, the state of completion of each ongoing study.

Informal discussions are a better ground for developing ideas and formulating solutions than are formal meetings; however, formal meetings might be necessary to solve problems that involve a larger number of people. It is the responsibility of the project leader to determine when a formal meeting is appropriate—these meetings should be exceptions and not the rule. A few examples of proposed

TABLE 6.3. Examples of Improvements.

Type of Improvement	Example of Operation	Example of Improvement
Elimination	Operator waits for forklift truck	Lot sizes are reduced, parts are moved by operaors on carts
	Operator goes to toolcrib	Tools are dedicated to workstations and stored at workstations
Simplification	Operator adjusts fixture on machine table	Marks on the table allow one-touch positioning
	Operator clamps part	Toggle clamps allow one-touch clamping
Transfer to external operation	Operator pre-heats mold	Mold is pre-heated while previous job ends
	Operator gets fixture at storage room	Fixtures are prepared and brought to workstation before end of previous job

improvements that might require a call for a formal meeting are: Improving the quality of purchased castings, simplifying procedures for NC programs update, and improving procedures to set design tolerances. To give the meetings structure, easy-to-use problem solving techniques (such as cause-and-effect diagrams or Pareto charts) can be used.

Every idea that is generated during the analysis must be discussed. It is clearly better to spend time evaluating an idea that might not be implemented, than to reject an idea that at first seems unrealistic but eventually would have resulted in great improvements. Ideas should be written down with names of people to contact for discussion. Sketching the proposed improvements makes explanations easier.

After discussion, as ideas become potential solutions, the solution's impact on individual operations should be estimated in terms of:

- Elimination: resulting in a 100 percent time reduction.
- Simplification: resulting in a partial time reduction (50 percent for instance).
- Transfer to external time.

Estimated time reduction can be entered directly on the spreadsheet and the new operation times computed (copies of the spreadsheet can provide a means for solution comparison).

After the screening, discussion, testing (in some cases), and modification and adjustment of initial ideas, the setup team selects a set of solutions for evaluation. The list of solutions is entered on the worksheet as shown on Figure 6.15.

Evaluate the improvements: Estimate the gains.

Estimating the gains (for example, time reduction) that each solution can provide is an important step in evaluating the improvements. Each proposed solution leads to the elimination, simplification, or transfer to external time of one or several setup operations. As shown in Figure 6.16, individual operation improvements are estimated by entering, on the operation rows in the list of operations, solution numbers and estimated time reductions.

Each solution number is followed by the estimated time reduction, and by an indication (yes or no) of whether the solution can be transferred to external time.

1	SOLUTION	SOLUTION DESCRIPTION	
2	#		
3	1	External Prep./Report	
4	2	Workstation Organization	
5	3	Fixtures Standardization	
6	4	Marks on Machine Table	
7	5	Modified Crane	
8	6	Preventive Maintenance	
9			
10		TOTAL TIME	
11			

Figure 6.15. Example of List of Solutions.

	A	.	E	F	G	H	I	J	K
1	OPERATION DESCRIPTION	.	Lgth	Class	Solution	% Time	Ext.	Int.	Ext
2			(Sec)	#	#	Reduc.	Oper.	Time	Time
.	.	.	.	.	.	.	.	.	.
.	.	.	.	.	.	.	.	.	.
56	Opens Machine Jaws	.	71	1				71	
57	Releases Mold	.	10	1				10	
58	Lifts Mold with Crane	.	62	1	5	50%		31	
59	Places Mold on Pallet	.	21	1	5	75%		5	
60	Gets New Mold from Rack	.	201	2	1	75%	Yes		50
61	.	.	.	.	.	.	.	.	.
.	.	.	.	.	.	.	.	.	.

Figure 6.16. Example of List of Operations after Computation of New Times.

Then the new times per operation are computed: First, the internal time, which refers to the setup time when the machine is not running; second, the external time, which refers to setup time while the machine is running.

Then, to allow for comparison of class time after improvement to class time before improvement, the list of classes is updated. New times and percentages, decomposed in internal times and external times, are computed (as shown on Figure 6.17).

In addition the total setup time after improvement is computed at the bottom of the list of classes by summing individual class times. In this example, remaining operations performed in internal time represent 16.99 percent of the original setup time, and remaining operations performed in external time represent 5.02 percent of the original setup time. The internal setup-time reduction (the reduction in the time that the machine is stopped for setup) is then [1 - 16.99%] = 85.01%.

The list of the solutions can now be updated by computing the internal and total setup-time reductions per solution (as shown on Figure 6.18). The total-time reduction is the internal-time reduction subtracted from the setup time to be performed externally. The internal-time reduction refers to the reduction in the time that the machine is stopped for setup, and the total-time reduction refers to the reduction in time spent on setup operations by the machine operator or the setup people.

	AA	AB	.	AE	AF	AG	AH	
1	CLASS	CLASS DESCRIPTION	.	INT. TIME		EXT. TIME		
2	#		.	(SEC)	(%)	(SEC)	(%)	
3	1	Former Fixture Removing	.	150	4.47%	20	0.6%	
4	2	New Fixture Preparation	.	0	0%	80	2.39%	
5	3	Fixture Install. on Machine	.	280	8.35%	0	0%	
6	4	Adjustments	.	140	4.17%	0	0%	
7	5	Trial Runs	.	0	0%	0	0%	
8	6	Paper Work	.	0	0%	68	2.03%	
9								
10		TOTAL TIME	.	570	16.99%	168	5.02%	
11								

Figure 6.17. Example of List of Classes after Computation of New Times.

	BA	BB	BC	BD	BE	BF	
1	SOLUTION	SOLUTION DESCRIPTION	INT. TIME RED.		TOTAL TIME RED.		
2	#		(SEC)	(%)	(SEC)	(%)	
3	1	External Prep./Report.	268	7.99%	248	7.39%	
4	2	Workstation Organization	162	4.83%	82	2.44%	
5	3	Fixtures Standardization	1080	32.20%	1080	32.20%	
6	4	Marks on Machine Table	850	25.34%	850	25.34%	
7	5	Modified Crane	70	2.09%	70	2.09%	
8	6	Preventive Maintenance	354	10.56%	286	8.53%	
9							
10		TOTAL TIME	2784	83.01%	2616	77.99%	
11							

Figure 6.18. Example of List of Solutions after Computation of the Time Reductions.

Figure 6.19 shows a pie chart that represents a graphical synthesis of the setup after time reduction. The different wedges represent the total time reduction percentages for each type of solution. Note that the remaining setup time is also shown.

The pie chart is a simple tool that allows the setup team to evaluate the impact of the different solutions. In addition, simple charts enhance communications with people outside the setup team, and particularly with management.

At this stage of the analysis the setup team should have identified the solutions that will be recommended for management approval.

Evaluate the improvements:Estimate the costs.

Cost estimation is also an important step in the analysis because it will finalize the list of solutions which will be then proposed to management for approval. The cost of each solution is estimated, and then priorities are given to the solutions according to their cost/gain ratio. Depending on the solution type, the cost can be estimated either by consulting external vendors or, if the modification can be implemented by internal resources, as a function of internal hourly rate and cost of materials.

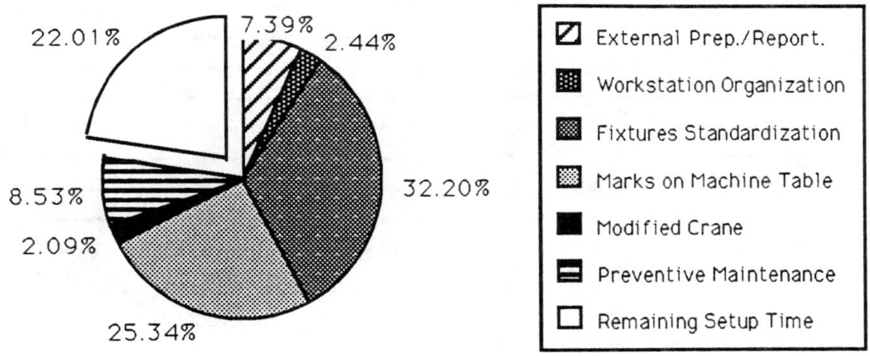

Figure 6.19. Setup Time Reduction Breakdown.

Estimated solution costs are entered on the list of solutions, then cost/internal time reduction ratios are computed, and the solutions are finally sorted in increasing order according to their cost/gain ratio, as shown on Figure 6.20.

Internal-and total-time reductions versus solution cost can then be presented graphically as shown in Figure 6.21.

Note that this graph shows that a 50 percent time reduction can be obtained by implementing a limited set of inexpensive solutions (namely external preparation and report, marks on machine table, preventive maintenance and workstation organization), and that the remaining 33 percent require a much larger investment. This is typical of many setup studies.

This graph is a simple, yet reliable aid for investment decision making, and should be used by the setup team to finalize the list of solutions that will be proposed to management validation.

	BA	BB	BC	BD	BE	BF	BG	BH
1	SOL	SOLUTION DESCRIPTION	INT. TIME RED.		TOTAL TIME RED.		COST	$/GAIN
2	#		(SEC)	(%)	(SEC)	(%)	$	RATIO
3	1	External Prep./Report.	268	7.99%	248	7.39%	50	6
4	4	Marks on Machine Table	850	25.34%	850	25.34%	500	19
5	6	Preventive Maintenance	354	10.56%	286	8.53%	250	24
6	2	Workstation Organization	162	4.83%	82	2.44%	250	52
7	3	Fixtures Standardization	1080	32.20%	1080	32.20%	5000	155
8	5	Modified Crane	70	2.09%	70	2.09%	500	239
9								
10		TOTAL TIME	2784	83.01%	2616	77.99%		
11								

Figure 6.20. Example of Prioritized List of Solutions.

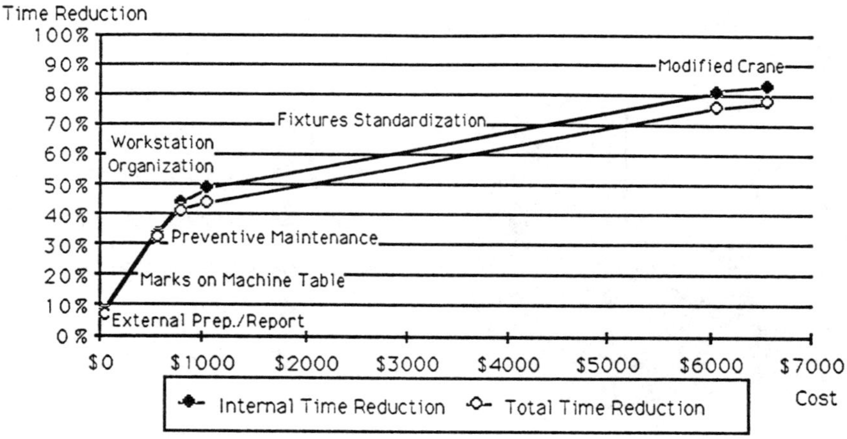

Figure 6.21. Setup Time Reduction Versus Cost.

Validate the solutions.
Solution validation is the last step before solution implementation. The objectives are (1) to inform management about the proposed solutions and the expected costs and gains, and (2) to obtain approval for implementation. A prerequisite to final validation is that the solutions must already have been discussed and accepted by the persons that they will impact: Machine operators, shop supervisors, production schedulers, process planners, manufacturing engineers, and design engineers. These people are the main participants during the implementation phase and will have to carry out the validated solutions.

Depending on a company's style, a setup study validation can either be a simple review of the proposed solution costs and gains, or a more thorough discussion of the solutions. However, regardless of the type of validation, the following information should be presented:

- Brief machine description.
- List of classes and percentage-breakdown pie chart.
- List of solutions and time-reduction-breakdown pie chart.
- Time reduction versus cost diagram.
- Brief description of each of the proposed solutions.

Many companies have found that it is a good idea to let the team members themselves present the solutions to management, because this can enhance team member motivation and allows direct communication between the decision makers and the people with the most knowledge about the setup problem. Team members can be supported by the project leader during this presentation.

If the project scope exceeds two or three different machines, individual studies should be proposed to management for validation as soon as they are completed; this has the advantage of keeping full project momentum. If solution validation and implementation closely follow the study phase, problems that will occur are more likely to be solved than if there is a time lag of several weeks between the different project phases. Another advantage is that early identification of false assumptions or errors made by the team will improve the quality of subsequent studies.

Implement the solutions.
Solution implementation immediately follows solution validation. A final study report consisting of the information that has been validated by management is compiled to document individual setup studies. During solution implementation, equipment (such as tools and workbenches) is purchased, and machine modifications are studied in detail with vendors or the manufacturing department. Team members are involved on a part-time basis (a few hours per week), while the project leader supervises the implementation and monitors expenses. Some solu-

tions will generate more than the forecasted improvements-other will not; therefore, a small part of the project budget (2–5 percent) should be reserved for solution modification. Hopefully, the shop floor will gradually take possession of the solutions, and—as part of the new continuous improvement philosophy—will feel committed to perfecting these solutions.

6.3.3 Identifying Internal and External Activities

Many setups involve operations that could be performed while the machine is running, although it is common that the setup preparations (getting the next job order, preparing tools, preparing fixtures, et cetera) are done entirely when the machine is shut down. There is, therefore, a substantial time-reduction potential to be gained by transferring these internal operations to external time.

Remember that an internal operation is one performed when the machine is shut down, and an external operation is one performed when the machine is running. Simplifying or eliminating internal operations reduces the internal setup time (also called internal setup)—time when the machine is down between two consecutive lots of parts. On the other hand, simplifying or eliminating external operations is aimed at reducing the external setup time (also called external setup)—time spent before and after internal setup (for preparation, tool storage, time report, et cetera).

To determine whether internal operations can be transferred to external time, setup team members must be able to identify to which of the following categories individual setup operations belong:

- *Intrinsic external operations.* These are internal operations that can be transferred to external time under current setup conditions.
- *Potential external operations.* These are internal operations that require modifications of the current setup conditions to be transferred to external time.
- *Intrinsic internal operations.* These are internal operations that cannot be transferred to external time under any reasonable modification of setup conditions.

The strategy to use to maximize the number of internal setup operations transferred to external time has three steps: First, identify intrinsic external operations and potential external operations; second, transfer the intrinsic external operations to external time; and, third, transfer the maximum number of potential external operations to external time. Figure 6.22 illustrates the three steps involved in this approach. The shaded wedges indicate the external setup time after transferring operations previously done in internal time.

To identify intrinsic external operations and potential external operations, one must differentiate between the two. Intrinsic external operations can be identified by determining which operations are performed while the machine is shut down

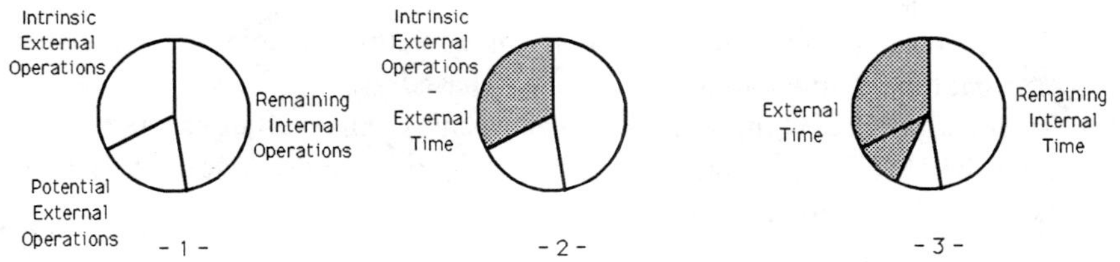

Figure 6.22. Three Steps Involved in Transfering Operations to External Time.

but could easily be performed while the machine is running. Examples are: hand tools and fixtures preparation, fixtures repair, time report, et cetera. These operations can be transferred to external time at almost no cost. Potential external operations can be identified by determining which operations are performed while the machine is shut down but could be performed while the machine is running—if current conditions were modified. Examples are: fixture adjustment, die preheating, part adjustment on the fixture, et cetera. Transferring these potential external operations usually requires purchasing or modifying equipment.

The second step, transferring intrinsic external operations to external time, is completed on every operation identified as intrinsically external. However, the transfer feasibility depends mostly on the production environment. For example, if the setups are performed by setup crews, there is no obstacle to transferring the operations to external time, since the setup crews can operate when machines run as well as when they are shut down. On the contrary, in the case of a typical shop-floor environment, with one operator per machine and where the machines are usually distant from each other, if the setups are performed by machine operators, then unless the machines are automatic the operators will not be available to perform the external operations. However, if the machines are grouped into cells and the setup operations are assigned to the cell production team, external operations can be performed by any available team member. More efficiency can even be achieved if parallel operations are performed by two or three cell production team members.

The final step, transferring potential external operations to external time, involves an in-depth analysis of the remaining operations. Typically these operations are the ones that could be transferred to external time if the current equipment was modified. The objective is to find the simplest and least expensive ways to allow the transfer. The following are a few examples of possible ways to transfer operations:

- Mold preheating on an injection molding machine can be done in external time before installing the mold on the machine if some equipment to preheat the dies is purchased.

- Fixture installation and adjustment on a metal-cutting machine can be done in external time if an additional pallet is purchased.
- Reels and tubes of components on an electronic component insertion machine could be installed on a rack in external time, and the racks could be simply exchanged in internal time.

Note that all internal operations are actually potential external operations. An extreme situation would be an entirely duplicated machine; one machine would be setup in external time while the other machine runs, and production would be switched from one machine to the other with each new lot. (To a lesser extent this is what is done when parts and fixtures are installed on pallets in external time, and the pallets are simply exchanged on the machines.) However, the setup teams are asked to design low-cost solutions. When examining potential external operations, the role of the team members is to determine what level of equipment duplication (preheating oven, tool offset measuring gauge, additional dye tank, second machine table, et cetera) is appropriate given the allocated budget and the estimated time reduction. The time-reduction-versus-cost diagram (refer to Figure 6.21) is a valuable tool to use when comparing different solutions.

To evaluate the feasibility of transferring to external time, team members should examine each operation with a critical attitude and use common sense. When evaluating the feasibility of a transfer, the first question to ask is, could the machine be running during this operation? For example, "Could the machine be running while the operator is getting cutting tools from the tool crib?" If the answer is yes, then the operation is an intrinsic external operation and should be transferred to external time. If getting the tools from the tool crib takes six minutes, and if the machine is automatic and can be left unattended during several minutes, then the operator can go to the tool crib and get the tools for the next job while the previous job is running. If the answer is no, then ask why must the operation be performed when the machine is stopped. For example, "Why can't the operator leave the machine unattended for six minutes?" The answer might be that expensive parts are manufactured on the machine, and the risk of breaking tools is high. Also, the cost of scrapping a part is too important compared to the marginal savings in setup time. For either a yes or no answer, team members should analyze all the potential solutions, evaluate these solutions, and then classify the operations as either a potential external operation or an intrinsic internal operation. Next, they should either transfer the operation to external time or leave it in internal time (at least for the moment). For example, a potential solution to this problem might be to dedicate the tools to the machine (or the cell) and to store them a few feet away from the machine; thus, the operator can prepare the tools in external time and yet stay close to the machine.

Identifying external operations and transferring them to external time usually reduces the internal setup time by 50 percent. Yet this transfer has, theoretically, no effect on the total setup time (internal plus external time). The next section will review strategies and techniques to eliminate or simplify both internal and external operations.

6.3.4 Strategies and Techniques

Although setups vary from each other, most of the problems encountered are similar; therefore, the strategies and techniques discussed in this section can be successfully applied to different setup reduction projects. The strategies and techniques listed in Table 6.4 are employed to eliminate and simplify setup operations, and also to transfer operations to external time. Each of the listed strategies or techniques are developed in this section.

Schedule and plan the setups. The first step in organizing a setup is to schedule at what moment the machines should be setup, and to plan what operations should be accomplished during that setup. In most companies setup schedules are accurate and followed. However, setup operations planning is generally poorly done and results in a costly waste of time: Material handling equipment is not available, additional operators are not free for parallel operations, special tools are not ready, et cetera. Thus setups must not only be scheduled, but also setup operations must

TABLE 6.4. Strategies and Techniques for Setup Time Reduction.

ORGANIZATION OF SETUP OPERATIONS:

1) Schedule and Plan the Setups
2) Use Checklists
3) Use Kits
4) Repair Tooling and Perform Function Checks
5) Use Preset Tooling
6) Review the Operation Sequence
7) Implement Parallel Operations
8) Dedicate Tools to Machines
9) Dedicate Fixtures to Workstations
10) Organize Fixture and Tool Storage
11) Organize the Workstations
12) Eliminate Waiting Periods
13) Preheat in External Time

MODIFICATION OF THE EQUIPMENT:

14) Use Dedicated Material Handling Systems
15) Standardize Fixtures and Tooling
16) Eliminate Adjustments
17) Use Single-Motion Clamping Devices
18) Decrease the Number of Bolts, or Eliminate Them.

be planned so that waiting periods are eliminated during internal setup. After the production environment is organized in autonomous cells, setup scheduling and planning become easy tasks that should be accomplished by the cell team leader.

Use checklists. Checklists provide verification that all tools and fixtures are present at the time of the setup; checklists can also list the operation sequence and mention information such as "Second operator needed for die insertion (2 min.)." Checklists are specific to a setup and a machine; thus they are stored at the machine workstation. They can be maintained by operators and by process planners.

Use kits. All components specific to a setup must be stored together: There is no reason why all the dies should be on one side of the storage racks, all the clamps on the other side, and all bolts and nuts in some cabinet drawer. Dedicate specific components to each setup and then store each set in kits so that the operators will find everything they need for a particular setup at a single location.

Repair tooling and perform function checks. These are the two rules to follow when tools and fixtures are transferred between machines and storage locations:

1. Because stored tools and fixtures are assumed to be in working condition, no tool or fixture should be returned to storage without being checked, and then repaired if necessary (the checks must be performed in external time).
2. Before internal setup, perform a quick function check on the tools and fixtures that are being prepared in external time. If a problem is then detected, it can either be remedied before the internal setup, or a different setup can be substituted.

Use preset tooling. Tooling adjustments can be more efficient by performing the adjustments in external time, when possible. Obviously, machines should not be stopped for the sole purpose of using the machine table as a workbench.

Review the operation sequence. Operation sequences can be improved and become more efficient by eliminating repeated tasks. For example, grouping together operations that are performed with the same tool will eliminate the need to continually retrieve this tool, and thereby will significantly reduce the time spent getting the tool.

Implement parallel operations. Some cases call for implementing parallel operations that are performed simultaneously by two or more operators. For example, in a setup operation that requires an operator to move repeatedly from one side of the machine to the other, a second operator can alleviate the performance of the operation on one side, thereby eliminating the first operator's displacements. Parallel operations should be implemented only if the resulting time is less than half of the initial time.

Dedicate tools to machines. Because operators waste a significant amount of time searching for tools, cutting tools and hand tools should be dedicated to machines or workstations, in order to be readily located. Enough hand tools should be purchased so that each one can be assigned to a particular machine and stored at the machine. Operators should become responsible for the tools. When a tool is broken a new one is obtained from the tool crib—not from the next machine. Hand tools should not move from machine to machine.

Since no more than a dozen different tools usually represent 80 percent of the total number of cutting tools used on a machine, the most commonly used cutting tools could be assigned to the specific machines and should also be stored at the machines. Special tools can still be stored at the tool crib and then be brought to the machine before internal setup. Note that when machines are grouped into cells and when part families are assigned to the cells, the number of cutting tools used on each machine is reduced, thereby simplifying the problem of tool dedication to the machines.

Dedicate fixtures to workstations. As tools should be dedicated to machines, fixtures should be dedicated to workstations (in this context a workstation is defined as a group of two or three identical machines adjacently placed; a cell is composed of several workstations). Commonly used fixtures should be stored next to the workstations, occasionally used (bimonthly for example) fixtures should be stored outside the cell. In both cases, fixtures should be prepared in external time.

Organize fixture and tool storage. Using the techniques of dedicating tools to machines and fixtures to workstations, tool and fixture storage should be organized so that the most commonly used tools and fixtures are the most easily accessible. In addition, storage should be organized so that dedicated material handling equipment such as manual carts and jib cranes can be used to transport the fixtures. Great storage improvements are achieved by transporting the fixtures on roller equipped carts with the same height as the machine table, and also by storing the fixtures on roller equipped tables that are again the same height as the machine table. Fixtures can then be moved by a single operator without any other material handling equipment. Remember also, that to reduce searching time, storage should be organized per setup (all elements for a given setup at the same location) rather than per element type (all clamps together, all bolts together, all fixtures together, et cetera).

Organize the workstations. Workstations should be organized in such a way that there is one place for every item, and every item is in its place when not used. A substantial amount of time can be gained if operators know exactly what is stored at their workstation, and where they can find hand tools, checklists, working papers, and other essential.

Eliminate waiting periods. Every time an operator has to stay idle waiting either for something to be brought to the machine, or for someone to come and help, there is opportunity for improvements in efficiency. The two main solutions to the waiting problems are (1) to improve setup planning to achieve better coordination between, for example, material handling and setup operations (for example, schedule to bring specific tools to the machine before internal setup), and (2) to eliminate the reason for the wait (for example, by using dedicated material handling equipment to reduce component moving by forklift truck.)

Preheat in external time. If applicable, preheating should be done in external time. After time reduction, internal setups are so short that dies and molds can be preheated in external time and remain warm enough to start producing correct pieces immediately.

Use dedicated material handling systems. Dedicated material handling equipment should be used whenever possible. Forklift trucks should not be allowed next to the machines. For example when machines are grouped into cells, the area between the machines should be a forbidden zone to forklifts. For the most part, autonomous carts are the best solution for moving tools and fixtures, although depending on the weight of the fixtures motorized carts can be used. The ability to move dies, fixtures, and tools without having to wait for material handling equipment availability is the goal. Jib cranes, used for loading or unloading machines, can also help mount or remove fixtures. A better solution, if feasible, is to use carts the same height as the machine table, and to roll or slide dies and fixtures from the cart to the machine table, and vice versa. Bridge cranes should be used only for moving heavy parts.

Standardize fixtures and tooling. When dedicated to a machine, fixtures and tooling should be standardized to reduce or eliminate the need for adjustments. For example, standardizing the die height eliminates any machine height adjustment when changing dies. Hand tools should also be standardized. For example, ideally, all bolts and nuts that could not be eliminated should be tightened or untightened with the same wrench, yet at times this is not feasible, and then no more than two or three wrenches should be required.

Eliminate adjustments. Most machines are designed to adapt to a wide range of part sizes and working conditions. As a result, every time the type of job to be performed on a machine changes, there is a need to adjust the machine to the new working conditions, for example to adjust material feed pitch, fixture position on the table, machine opening height, et cetera. However, on the shop floor a limited number of part sizes or working conditions are usually encountered, and out of the quasi-unlimited machine adjustment capabilities (in a given range), only a few are commonly used. To alleviate the operators from searching through numerous settings and positions, machine settings should be marked for each part assigned to that machine so that the machine can be set correctly without adjustment. (Designing simple dedicated gauges, placing marks on the machine, recording positions on digital readouts are a few examples, more solutions will be given in section 6.3.5.)

Use single-motion clamping devices. Bolts and nuts are the most commonly used fasteners for clamping fixtures or parts. Providing high clamping force, they are inexpensive and can be ordered in every possible size. Unfortunately, they have two major drawbacks: (1) They can only be tightened or untightened with a wrench, and (2) the operations of tightening and untightening are not efficient operations. The need for using a wrench poses the problem that the wrench has to be found before starting the work; moreover, several wrenches are usually needed because of different bolt sizes. Additionally, a bolt or a nut has to be turned at least ten times before the surfaces come into contact, and only then can the part or fixture be secured by tightening the bolt. The same is true for unsecuring. In effect, less than 10 percent of the work involved in using bolts and nuts to clamp fixtures or parts is directly productive. To modify the attachment components so that single-motion clamping becomes the rule, use toggle clamps (they usually provide more than the required clamping force), pear-shaped holes, U-slots, U-shaped washers, et cetera. (these modifications as well as other clamping techniques will be discussed in more detail in section 6.3.6).

Decrease the number of bolts or eliminate them. Because nuts and bolts are inefficient clamping devices, they should be eliminated, or at least the number of them should be decreased. Remember that each bolt or nut that is turned more than 360 degrees during a setup indicates a potential for improvement. Different types of improvements can be considered:

- Standardize the size of bolts and nuts so that only one wrench is necessary.
- Reduce the number of bolts. (Where six or eight bolts are used to secure a part, it is most likely that four can suffice.)
- Modify equipment so that only one turn is necessary (see how in section 6.3.6).
- Replace the remaining bolts by single-motion securing devices (see how in Section 6.3.6).

Design parts on a family basis. Standardization, when applied to fixtures and toolings, results in substantial time and cost-savings. When applied to the products themselves and to their individual components, impressive results are obtained. Grouping parts into families, assigning these families to specific workcells (refer to chapter 4 on GT), can result in never having to completely "re-setup" some of the machines. For example, a common tooling basis can be used for an entire family of parts to be punch-pressed, and die pins can be added or removed according to part variations. Grouping parts into families extends beyond the scope of setup-time reductions and can only be the consequence of decisions made at the company's top management level. It is a long-term approach. Its success depends mostly on how effectively manufacturing, design engineering, and marketing, supported by top management, work together to standardize part designs.

Strategies and techniques listed in this section are useful guidelines for reducing setup time. In addition, they are applicable to many of the problems that are confronted during machine setups. Designing solutions requires a little imagination and a lot of common sense, therefore setup team members are encouraged to design original solutions adapted to their environment.

6.3.5 Adjustments

Because machines have been designed to be able to adapt to many different conditions, they obviously will have to be adjusted for specific jobs. A significant part of most setups is spent on adjustments such as adjusting tooling on a machine, adjusting machine settings, and adjusting a part on tooling. It is therefore worthwhile to make these tasks more efficient. For example, if ten parts are dedicated to a machine, then at most ten different machine speeds and openings will be used. The same settings will be used again and again each time the same job is rescheduled on the machine. A way to reduce the time spent adjusting the machine is, in a first time to record for each job the values of the machine speed and opening, and then to directly set these parameters to the recorded values when the jobs are rescheduled on the machine. Adjustments would be eliminated without reducing machine flexibility, and the setup time would be significantly reduced.

There are two strategies to achieving efficiency in adjustments: (1) Record manually, mechanically, or by any other means the different settings on the machine so that single-motion positioning is achieved (for example, using dedicated gauges), and (2) reduce the number of positions used (ideally to one) so that the settings do not have to be changed for every job (for example, standardizing die heights).

Adjustments can be classified in three types according to their functions:

Type 1: Mounting tooling and fixtures into the correct position (mostly aligning and centering).

Type 2: Setting an infinitely variable linear, circular, or helicoidal displacement (for example, along a slide or a screw) to the correct position for the next job.

Type 3: Zeroing on the correct position of tools, temperatures, and pressures to get a good piece.

Notably tooling and fixture standardization will eliminate most type 1 adjustments. By using stops, dogs, and notches to make the tooling position itself in the correct location, single-motion positioning can be achieved. Figure 6.23 shows how fixture locating on a machine table is improved by using quickly installed dedicated locating plates. This method can be further improved by standardizing the fixture bases so that the plates can remain on the machine table.

Recording the different job-specific positions is the first step toward eliminating type 2 adjustments. Then, the machine is set according to the recorded positions. "Hard" or "soft" methods can be used to record the positions and ensure accuracy in repositioning. For example, a soft method would be to affix a graduated scale along a slide and to keep track of the relevant position per job, another example for a soft method is to use a digital read-out to keep track of positions. On the other hand, a hard method would be to use a job-dedicated gauge to set the distance between the moving part and a fixed reference surface. Soft methods are more flexible than hard methods. However soft methods are more subject to errors and modifications of initial calibration, and they also require that additional information be recorded on the setup sheets. Hard methods are more reliable. Once a gauge has been fabricated for a specific job it can be used again and again every time the machine is setup for that job. It is important that gauges are stored in immediate proximity of their point of use or with the job fixture kit brought to the machine for the setup. To avoid errors, large reference numbers should be painted or printed

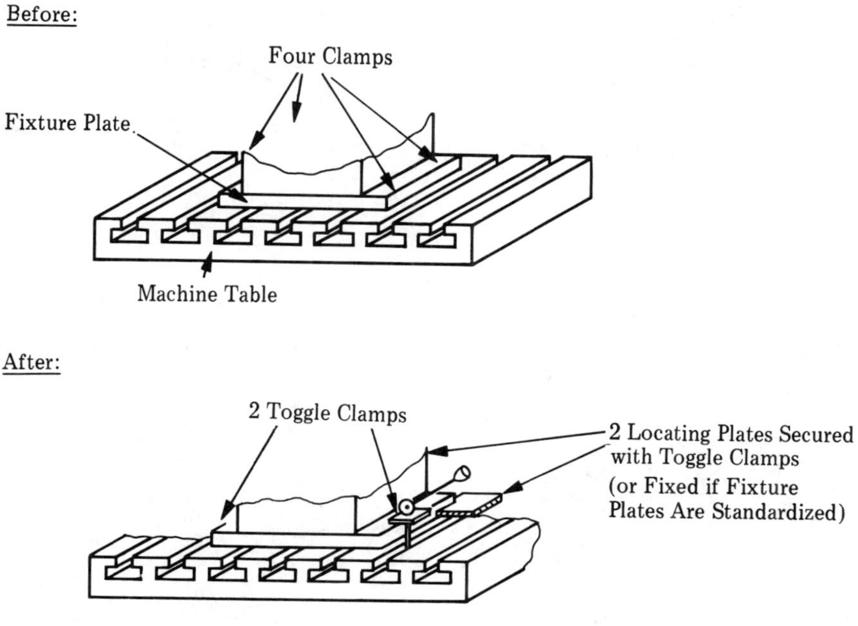

Figure 6.23. Fixture Locating Improvement.

on the gauges, and the gauge number should be indicated on the checklist, on the setup sheet, or on a specific table placed where the gauges are stored. These two methods are illustrated in Figure 6.24.

Type 3 adjustments are the most challenging to eliminate. The objective is to get the machine properly set so that the first produced part is a good part. Experimentation and adjustments might remain necessary when a new part is put into production; however, once the correct settings have been found they should be recorded, as described for type 2 adjustments, to ensure fast setup every time the part is rescheduled on the machine.

Many people familiar with shop-floor operations have difficulty believing that adjustments can be eliminated or reduced to seconds. The mission of the setup

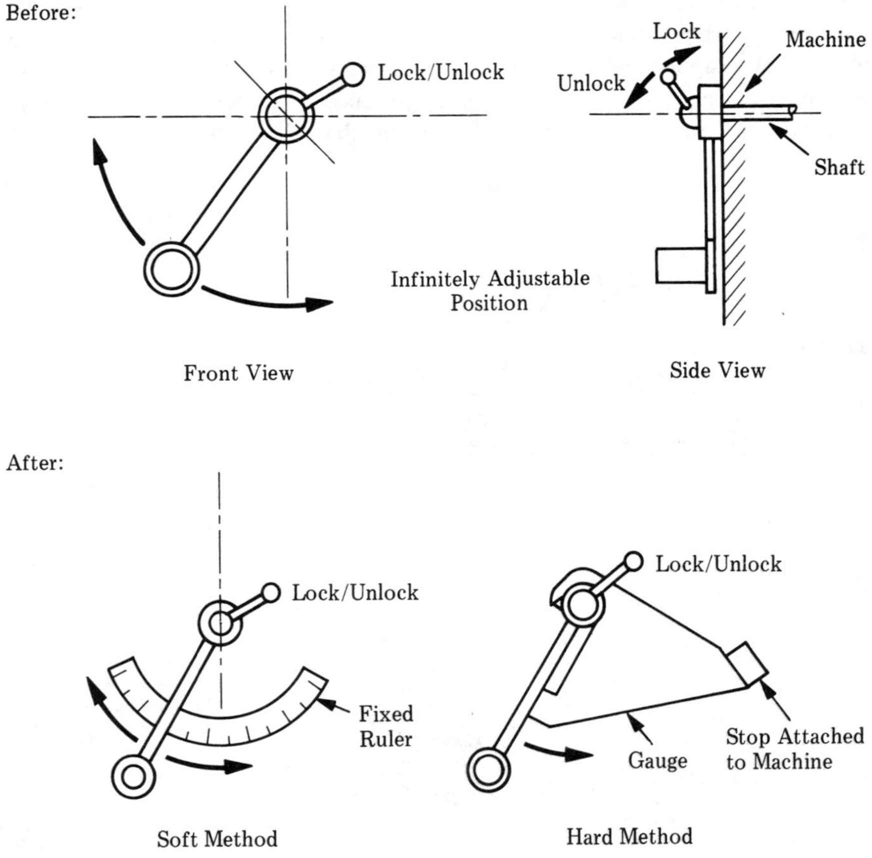

Figure 6.24. Two Methods for Repositioning a Movable Part along a Circular Slide.

team is to change this aspect of the shop-floor culture. The team members must explain that traditional adjustments waste the operator's time, and show that simple techniques such as the ones discussed in this section (use self-positioning tools, use dedicated gauges, record positions, et cetera) can provide one-touch-setting of machine and tooling parameters and thereby simplify the operator's jobs.

6.3.6 Clamping

Bolts and threaded fasteners are among the least efficient means of clamping parts because of the many turns they require when only one turn is sufficient to tighten or loosen an attachment; in addition they have the disadvantage of requiring the use of tools that in turn have to be located and brought to the work site. When fastening and unfastening operations are performed in internal time (fixture clamping on a machine table for example), internal setup time can be significantly reduced by modifying fastening devices, from bolts and threaded fasteners to single-motion clamping.

The attachments that should be considered for modification are the ones that (1) require more than one turn, (2) cannot be easily performed by one operator, or (3) require different types of wrenches. In addition, when two parts are secured by more than two bolts, eliminating unnecessary bolts should be considered.

When reducing the number of bolts, eliminating bolts, or implementing single-motion clamping, the direction and magnitude of the forces acting on the attachment should be examined. If the main direction of the force acting on the attachment is lateral (perpendicular to the bolt axis), then the bolts can be replaced by unthreaded pins or wedged parts that will counteract the effect of the lateral force, and simple loaded-spring devices can be added to prevent the parts from moving along the attachment axis. If the main direction of force is axial (along the bolt axis) and if the force applied on the bolts is less than approximately 500 kg (1,100 pounds), then bolts can be replaced by single-motion spring-type devices such as toggle clamps. If again, the main direction of force is axial, but if the force applied on the bolts is more than 500 kg, then the parts should be modified to allow single-turn bolt fastening and unfastening such as U-shaped slots and pear-shaped holes.

However, because bolts cannot always be reduced in number, eliminated, or modified to fast clamping devices, it is necessary to reduce the fastening and unfastening time of the remaining bolts. Ideally a single wrench should be used to turn every bolt on a machine, because then the operator will not waste time looking for different tools. Practically, depending on the size of the parts to be assembled, a maximum of three different bolt head and nut sizes can be considered as still being reasonable, and accordingly a maximum of three different wrenches would be used. For maximum efficiency, tools should be stored in the immediate proximity of their area of use (less than one arm length). Examples of fastening improvement are given in Figure 6.25.

The examples in Figure 6.25 represent only a small sample of the various solutions that can be designed to reduce fastening and unfastening time. All the solutions described in this section are simple and inexpensive, thereby demonstrating that simple clamping means should be attempted before expensive automated solutions, such as hydraulic clamping systems, are considered. The range of solutions is more limited by the team members imagination, than by the allocated budget.

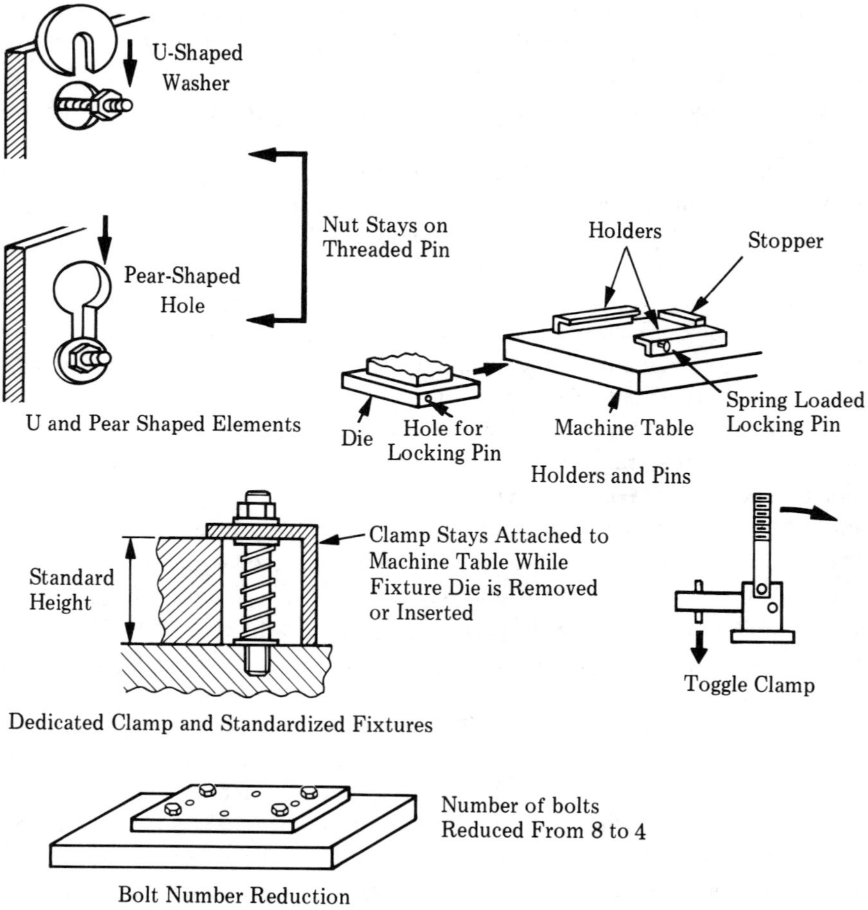

Figure 6.25. Examples of Fastening Improvements.

6.3.7 Problems and Solutions

During the entire setup project, team members must deal continually with problems. A technique to help minimize problems and initiate problem solving is to:

1. Identify and confront existing problems.
2. Eliminate as many problems as possible.
3. Solve remaining problems.
4. Prevent or prepare for future problems.

Confronting, eliminating, solving, and preventing problems is an important aspect of the new culture that management has committed to implement in the company. It is crucial that setup team members understand what is expected from them in terms of how encountered problems should be addressed, that they become enthusiastic about their mission, and that they communicate their enthusiasm to the other people on the shop floor. The remainder of this section will discuss the four steps that team members should use for problem solving during a setup project.

Identify and confront existing problems. Problem identification is the objective of videotaping and analyzing the setup operations. Team members will discover that many of the operations they used to consider appropriate or acceptable (such as waiting periods, displacements, and lengthy adjustments) are not compatible with quick setups. Problem confrontation means that team members must be committed to finding a solution to the identified problems or, better, to eliminate the identified problems. When proposing solutions, the team members will often hear negative comments such as "It can't work," "We have tried that already," "Why change? I have been doing this operation for years that way." These comments will be heard especially at the beginning of the setup project because shop-floor people have lost critical sense over the years and do not believe that it is possible to change the way the work is performed. It is essential that the setup team members do not let problems go unsolved because of such comments. For example, when methods are proposed to eliminate adjustments, the best thing to do is to go ahead in spite of the negative comments, and test the improvements on a pilot machine. If the solution works, then it is extended to other machines and the negative comments will discontinue. If the new method does not work as well as forecasted, then it should be modified until satisfactory performance is achieved on the pilot machine.

Eliminate as many problems as possible. Eliminating a problem is always better than solving it, and usually faster. For example, standardizing die heights is more effective than finding a way to set the machine opening more quickly. Similarly, replacing bolts by single-motion devices is more effective than using power tools to turn the bolts.

Solve remaining problems. Unfortunately not all problems can be eliminated. Tooling will always have to be placed on the machines and parts on the fixtures, some sort of tool will always be needed to turn remaining bolts or nuts. Common sense will usually help resolve most of the problems. In more difficult cases, simple techniques can help in finding the appropriate solution. A few examples of these techniques are histograms, Pareto diagrams, and cause-and-effect diagrams.

- Histograms can help to evaluate quantitative data. Figure 6.26 shows how an histogram is used to classify parts assigned to a machine per weight in order to identify the dedicated material handling equipment required capacity.
- Pareto diagrams provide a ranking of quantitative data according to the number of occurrences of the data. Figure 6.27 shows how a Pareto diagram can be used to identify the most frequently used cutting tools on a CNC machine in order to determine which tools should stay in the tool changer.
- Cause-and-effect diagrams (also called Ishikawa diagrams) can help to identify the multiple causes that have an effect on a specific problem. Figure 6.28 shows how a cause-and-effect diagram can help formalize the different causes that result in long CNC machine on-line program debugging.

Prevent or prepare for future problems. A truth about manufacturing operations is that numerous problems are likely to occur every day. Setup problems, such as difficulties caused by new parts, or by production shifts to different machines because of machine breakdowns, have to be expected. However, some problems—

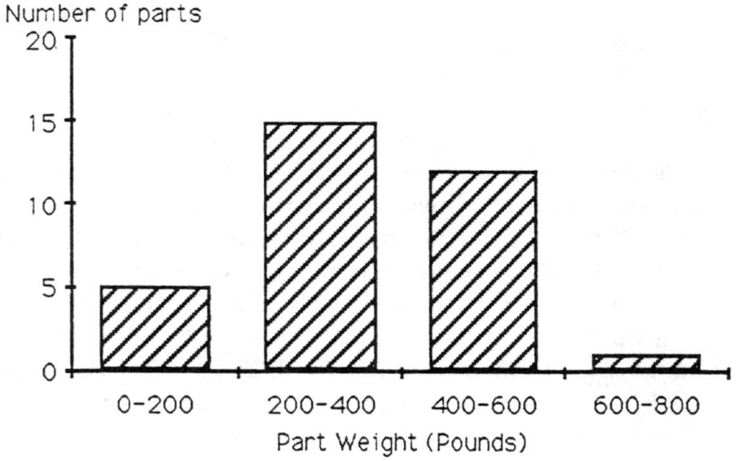

Figure 6.26. Example of Histogram.

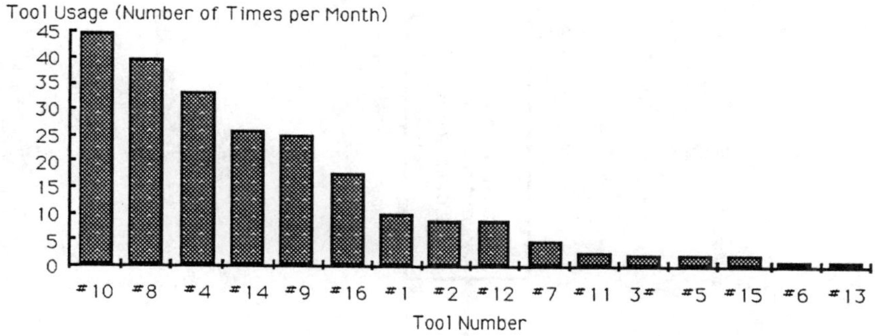

Figure 6.27. Example of Pareto Diagram.

related or not to setup operations—can be prevented. For example, by the following steps:

- Checking the tooling systematically after its use will reduce the number of tooling breakdowns.
- Reinforcing a sense of ownership of equipment by the operators will enhance the detection of equipment failure by the operators and help prevent machine breakdowns.
- Keeping the work area clean and in order will allow the easy detection of missing parts and tools.

Nonetheless, not all problems can be eliminated or prevented. People must understand that they will have to continuously solve new problems, and that they should report the problems they cannot solve. Temporary task forces involving operators, supervisors, and engineers can be formed to tackle these problems. Before the onset of the setup project most of the problems are unidentified or are circumvented. With the setup project, a temporary structure (composed of the setup teams) is created to confront problems that have been existing for a long time, but either have not been recognized or more simply have been considered as normal constraints in a manufacturing environment. The setup project initiates a momentum aimed at confronting problems and finding solutions. This momentum is a characteristic of world-class companies and it is crucial that it is not lost after the formal termination of the setup project.

6.5 SUMMARY AND STRATEGIES FOR GETTING STARTED

Setup-time reduction is an essential step for companies that need to become leaner and meaner in their evolution toward world-class competition. Many potential

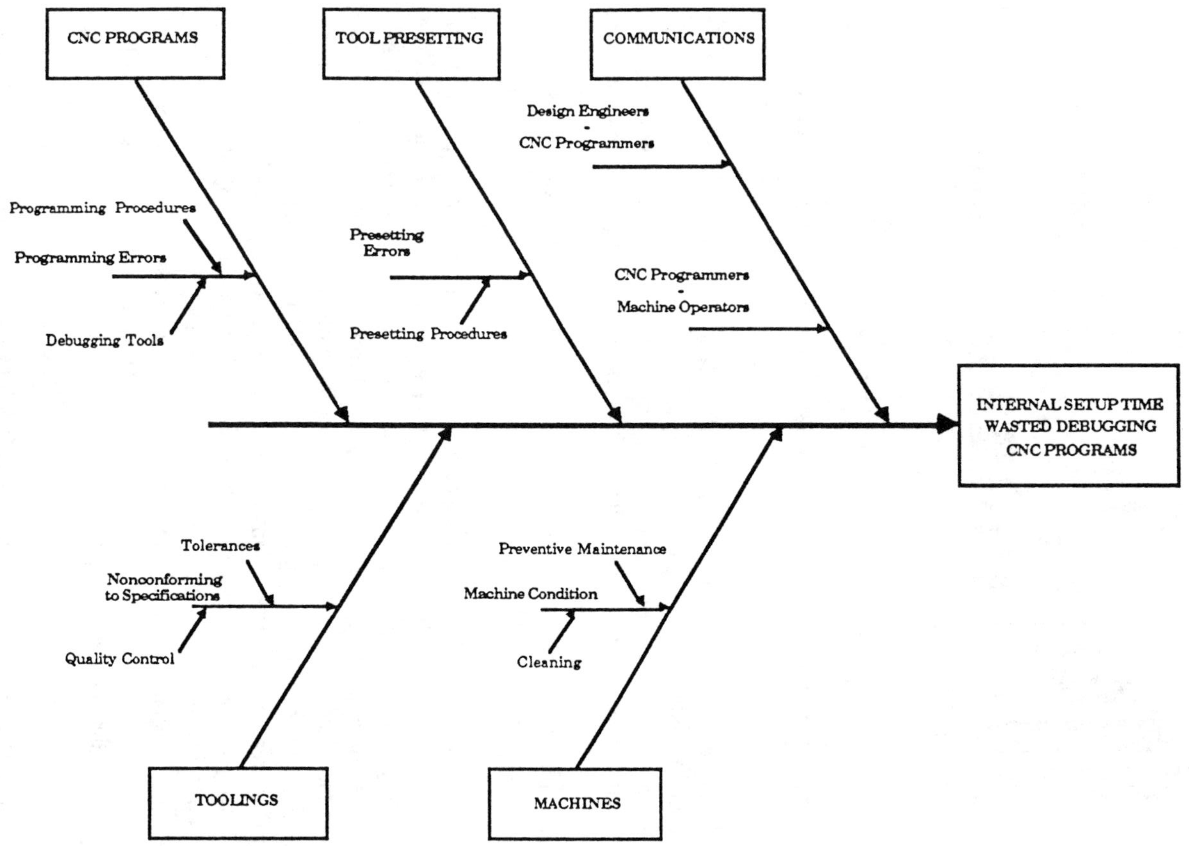

Figure 6.28. Example of Cause-and-Effect Diagram.

benefits of cellular manufacturing organization—shorter lead times, reduced inventories, reduced work-in-process, and reduced floor-space occupancy—cannot be realized if setup times are not reduced.

The machine setup time, which is the time between the previous lot's last part production and the next lot's first good part production, can easily be reduced, at low cost, by at least 75 percent, provided that the proper approach is applied. Applying the proper approach starts by clearly defining the project objectives, carefully organizing the project, and informing the entire company or plant work force about the project. To be successful, the project should be strongly supported by management.

The setup project belongs to the shop floor. Setup project teams are composed of machines operators or setup people supported by technicians. Team members are trained on the basis of earlier setup-time reduction studies, and, if necessary, they learn how to use the video equipment and computer tools during the training.

The project leader has a key role. He or she is in charge from the time of the initial work force information meeting until solutions have been implemented and adopted by shop-floor personnel. The project leader is actively involved in the machine and setup selection, team member training, and relations with other participants, including those with other company departments, with vendors, and with management—in effect, he or she is an orchestra conductor. The orchestra is composed of many different musicians who are not accustomed to playing together, they are the people directly related to the company's products and production systems. From vendors and design engineers to shop-floor supervisors and machine operators, each person has to give their best whether it is during tape analysis, informal discussions, meetings, or brain storming sessions. The project leader knows the completion stage of the ongoing studies, is aware of and helps solve the problems faced by the team members, and shows the direction to follow in order to find better solutions. In addition, the project leader is in charge of monitoring the expenses.

A systematic approach is used to record and analyze setup data. The entire setup is videotaped and then decomposed in individual operations, after that, every operation is analyzed. Solutions are searched for with the aim of eliminating or simplifying operations, or transferring them to external time. Meanwhile, ideas coming from setup team members, machine operators, machine shop supervisors, and other participants are discussed, criticized, evaluated, tested, modified, and approved or rejected. Also, solution costs are weighted against the estimated time reduction. A simple PC-based tool can be used to estimate these time reductions and to present results. Using these results, each setup study is summarized in a brief report for management follow-up and solution validation. Next, the validated solutions are implemented and the resulting improvements are measured, with any solution being modified if necessary. Meanwhile, shop-floor personnel will progressively take ownership of these solutions.

This chapter has explained why setup times need to be reduced and has presented a simple and effective approach to obtain drastic setup-time reduction. Using the proper approach is mandatory for obtaining impressive results—but cannot guarantee them. Remembering and continuously enforcing the application of a few key principles will make the difference between mediocre and world-class results. These principles are: objective definition, organization, information and education, management support and commitment, and shop-floor involvement and ownership. Setup time reduction opens the way to lot-size reduction.

7

Lot-Size Reduction

Lot production is commonplace among low-, medium-, and even large-volume manufacturers, when the dedication of machines to the production of a single part cannot be justified. In an effort to minimize setup costs, the size of production lots is usually set as large as possible. Similarly, when components or raw materials are purchased from outside suppliers, the size of the purchase orders are also often set as large as possible in order to minimize the impact of the fixed costs incurred by processes such as ordering, shipping, and receiving. Consequently, large amounts of purchased materials and components, work-in-process (WIP), and semifinished or finished products have to be stored for long periods.

However, because in the nineties more and more companies compete in markets where delivery speed and the adaptability to consumer demand are determinant factors for remaining in business, the less time the products spend inside the company's plant walls, the greater is the company's competitive advantage. Therefore, those companies that are able to manufacture products in small lots that flow quickly through the production process benefit from a definite competitive edge over companies where the products spend most of their time waiting to be processed.

In addition, as already mentioned in chapter 6, WIP reduction that results from cutting lot sizes can dramatically reduce the operating capital, if properly implemented. Figure 7.1 shows the reduction in average inventory (WIP) that occurs when lot-size computation is based on the EOQ formula (a) or is based on shipping-quantities (b).

The preceding chapters have presented ways to streamline production flows (cell formation) and to reduce the equipment changeover time from one lot to the next (setup-time reduction). In this chapter, new methods for determining lot sizes are presented as well as practical ways to reduce the costs associated with decreas-

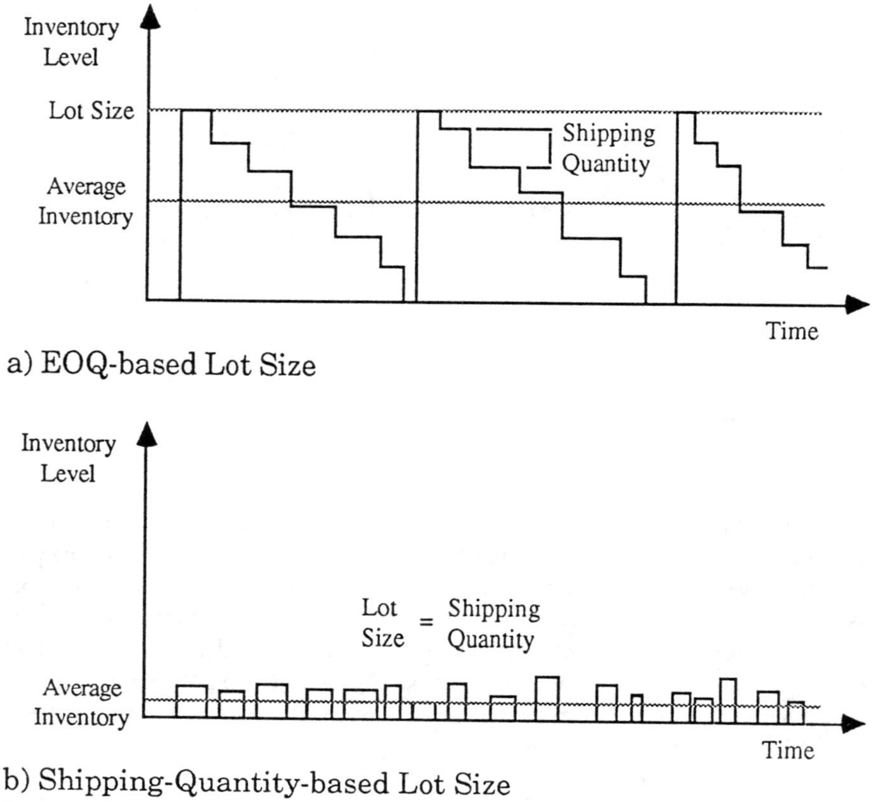

Figure 7.1. Effect of Lot-Size Reduction on Average Inventory.

ing these lot sizes. Also discussed are some simple approaches, which evaluate the effect of lot-size reduction on the performance of manufacturing systems; for example on flow times and inventory levels.

7.1 WHY, WHEN, AND HOW TO REDUCE LOT SIZES

The objective for reducing lot sizes must be clearly identified and understood. It could be, for example, to reduce the amount of capital immobilized in inventory, or to reduce product flow time, or to increase the quality of the products. A precise goal should be established by the management team, then the parts or products for which to reduce lot sizes should be determined, and, finally, practical ways of reducing these lot sizes should be identified.

After a definition of the different types of lot sizes, the remainder of the section will address these issues for production orders and for purchase orders.

7.1.1 Defining the Different Types of Lot Sizes

Every company uses its own vocabulary and, without proper definition, the use of terms such as lot sizes and order quantity can lead to misunderstanding. Therefore, the different types of lot sizes that are reviewed in this section are each assigned a name that will be used throughout this chapter. The names for these lot sizes are found in Table 7.1, which also summarizes their functions.

Note that the *shipping quantity* is the quantity that must be shipped to a customer on or by a specific date. The *order quantity* is the quantity of products that are ordered by a customer through a single production order. Also, when an order generates several shipments, the *order quantity* is not equal to the *shipping quantity,* which is determined as a function of the shipping frequency negotiated with the customers.

TABLE 7.1. Types of Lot Sizes.

	Type of Lot Size	Action	From	To	Environment
Shipping	Order Quantity	Sell		Customers	
	Shipping Quantity	Ship		Customers	
Producing	Setup Quantity, Batch Size, Lot Size,	Produce			
	Move Lot Size	Move	Machine	Machine	Job Shop
	Move Lot Size	Move	Cell	Cell	Cellular Manufacturing
	Transfer Batch Size	Move	Machine	Machine	Cellular Manufacturing
Purchasing	Purchase Quantity	Purchase	Suppliers		
	Receiving Quantity	Receive	Suppliers		

The *setup quantity* (also called batch size or, simply, lot size) is the quantity of parts produced by a machine or process between two setups.

The *move lot size* is the size of lots that are moved from one machine to the next in a traditionally organized machine shop, or from one cell to the next in the context of cellular manufacturing. The *move lot size* can be equal to the setup quantity, or be only a fraction of the setup quantity.

The *transfer batch size* is—in a cellular manufacturing environment—the size of the batches that are transferred from one machine to the next within a cell. Short distances between machines justify small transfer batch sizes; therefore, the transfer batch size is often equal to one.

For purchase orders, the *purchase quantity* must be distinguished from the *receiving quantity*. The *purchase quantity* is the quantity that a company has contractually agreed to buy from a supplier according to a specific purchase order. The *receiving quantity* is the quantity of parts that are received on a given date. When a purchase order covers several shipments, the *receiving quantity* is then only a fraction of the *purchase quantity*.

The remainder of this chapter will focus on why, when to, and how to reduce the size of the setup quantity (hereafter simply referred to as lot size), and the size of the receiving quantity. Although reducing the shipping quantity has a definite impact on the production lot size and on the smoothness of production, it is not addressed in this chapter. However, because customers might require more frequent shipments of smaller lots the shipping quantity is an input used to determine production lot sizes and purchase order sizes (as explained later in the chapter).

7.1.2 Reducing the Size of Production Orders

Reducing the size of production orders (lot size) not only reduces production lead time—which, in turn, increases manufacturing flexibility—but also reduces the costs associated with large lot sizes. Typically, the costs associated with large lot sizes are (1) the cost of holding and managing large amounts of inventory and work in process, (2) the cost associated with damages to or obsolescence of stored inventory, and (3) the cost associated with lost production when parts fail inspection tests. Reducing the lot size undoubtedly will reduce these costs. On the other hand, other costs might be significantly increased when the lot sizes are reduced, for instance costs associated with setup and materials handling.

However, the onset of manufacturing cell design (described in chapter 5) and setup-time reduction (described in chapter 6) has generated new techniques for reducing the cost of handling materials between machines and for reducing the cost of individual setups. Material-handling requirements can be reduced by reorganizing the equipment layout to streamline production flows, and individual setup costs can be reduced by significantly decreasing both internal and external setup times.

Thus, reducing lot sizes in parallel with applying these techniques will significantly decrease the costs associated with large lot sizes without generating additional costs due to small lot sizes.

The reason for reducing lot sizes is clear. When it is appropriate to reduce lot sizes can be shown by a few easily identifiable indicators: (1) High inventory costs, (2) a need to expand or modernize storage facilities, (3) large scrap rates because of large lots (QC), and (4) the need to increase the company's ability to respond to varying customer demand. A discussion of each of these indicators follows.

High inventory cost. Various costs result from high inventory. Some costs are the direct effect of large inventories, such as the cost of the storage facilities, equipment, and personnel, as well as the cost of operating and supporting inventory management information systems. Some other costs are secondary effects, such as the rework or scrap costs of products that are time sensitive (for example, metallic parts that can corrode, or chemical materials that can desiccate), or the cost of lost sales that results from products that have become obsolete. Following these different costs for each product will show the need to reduce inventory and, consequently, lot sizes for that product.

Storage facility expansion or modernization. Because of the inevitable high cost, managers in manufacturing firms should always cautiously consider the proposed projects for expanding warehouses or modernizing existing storage facilities. An example of modernization is the purchase of an automatic storage and retrieval system (AS/RS), which could help reorganize existing storage facilities, and therefore seems very appealing to plant managers buried in inventory-management problems. However, even if an AS/RS is the appropriate solution for distribution companies, it should not be used in the plant to store WIP. Plant managers can control WIP management much more efficiently and economically by reducing the inventory—through lot-size reduction.

Quality control. Large lot sizes cause high levels of scrap and rework because parts are normally inspected after completing an entire lot, and if a defect is found at this stage, the entire lot has to be reworked or scrapped. In addition, even though informal inspections occur when parts are delivered to subsequent processes, the long delays between the processes prevent fast feedback to the initial process and prevent a speedy correction of the equipment settings—which could limit the amount of parts to be reworked or scrapped. Note that in a cellular manufacturing environment, with a transfer batch size from machine to machine equal to one part, the feedback can be almost immediate. Reducing the lot size leads then to *free* quality improvement. In addition, rework and scrap costs are lowered since the number of parts to be reworked or scrapped is smaller.

Response to customer demand. A part production lead time that is tenfold or more of the total processing time indicates that the part spends too much time either waiting in storage areas and buffers between machines or being moved from process to process. This waste of time is a major drawback of large-lot production—especially for companies with unstable customer demand and short delivery times. Reducing the lot size will result in reducing the time that each part must wait for the other parts of the same lot and for the preceding lots to be processed.

Lot sizes must not necessarily be reduced for every product. By monitoring these four lot-size reduction indicators, production managers can identify which products pose the most serious problems and thus select which lot sizes should be reduced.

The minimal size of lots must be determined by considering (1) machine capacity constraints, (2) the constraints of external setup time (the time in which external setup operations are performed), and (3) the adjustments to shipping quantities.

Machine capacity constraints. Machine capacity constraints clearly limit the number of setups that can be performed on a machine, and therefore dictate the minimal lot-size value. Consider a machine or a process running two shifts per day for an average of twenty-one days per month. The machine is utilized at 80 percent of its capacity for processing parts, and at 5 percent for maintenance. Thus, 15 percent of the capacity can be used for setups, for 50.4 hours per month (21 days x 2 shifts x 8 hours x 15% = 50.4 hours). If a setup takes on the average one hour, a maximum of fifty setups per month can be performed. As a result, the minimal lot size for each part processed by the machine is given by the relation:

$$\text{Minimal lot size} = \frac{\text{Average monthly demand/Part}}{\text{50 setups/Number of different parts}}$$

For instance, if the machine processes ten different parts, the minimal lot size for a part with an average monthly demand of 10,000 units is given by:

$$\text{Minimal lot size} = \frac{\text{10,000 units/month}}{\dfrac{\text{50 setups/month}}{\text{10 parts}}} = \text{2,000 units}$$

Machine capacity constraint is the major factor that prevent companies from producing small-lot sizes.

External setup time constraints. External setup times can also limit the size of the lots. Consider again the machine in the previous example, with the same demand characteristics. The internal setup time is still one hour (time during which the machine must be stopped for setup); however, now there is an additional external setup time of seven hours. The external setup time is the time during which setup operations can be performed while the machine is running or being maintained (for example, clamping a part on the second pallet of a vertical lathe). The total time required for setup is equal to the internal setup time added to the external setup time, or eight hours. Thus, only one setup per shift, or two per day can be performed. Therefore, the maximum number of setups per month is reduced to 42 setups (21 days x 2 setups = 42 setups).

In this way, the minimum lot size is constrained by the external setup time rather than the machine capacity. Again, by looking at the same part as in the previous example the minimal lot size is determined as follows:

$$\text{Minimal lot size} = \frac{\dfrac{10{,}000 \text{ units/month}}{42 \text{ setups/month}}}{10 \text{ parts}} = 2{,}381 \text{ units}$$

However, the constraints imposed by external setup are less often encountered than constraints imposed by internal setups.

Adjustments to shipping quantity. To reduce the costs associated with inventory it is reasonable to deliver—and thus produce—the quantity needed at the time it is needed. Producing larger quantities than what is needed results in storing WIP for subsequent usage, and producing smaller quantities requires the production of parts in advance and, again, the storage of WIP. Therefore, starting with the finished product shipping quantity and moving up to raw material processing, the lot sizes should be set as close as possible to the needed quantities during the desired time interval.

For example, assume that the part considered in the two previous examples must be shipped weekly. The shipping quantity is then equal to 10,000 units per month/4 weeks per month = 2,500 units per week. The lot size should therefore be set as close as possible to 2,500 units because this lot size would result in the smallest level of finished product inventory if the products are shipped JIT after the completion of every lot. Note, however, that this rule must be applied flexibly. Producing lot sizes that are equal to shipping quantities is not always desirable if shipping quantities are large and if other parts are assigned to the same machines. The shipping quantity must then be fractioned into several lots so that the different part lot sizes match required order quantities and shipment frequencies. Retaining especially large lot sizes for some of the parts can result in long delays of the subsequent lots waiting to be processed, which leads to unnecessary inventory and eventually to missed shipment dates. Modeling approaches (described in section 7.2) are particularly helpful in identifying appropriate lot sizes when the interaction between parts prevents the use of intuitive or arithmetic approaches.

Note that the limit imposed by machine capacity and external setup time can only be considered a constraint as long as setup times are considered fixed. The application of techniques and strategies detailed in chapter 6 leads to drastic reductions in setup time. As a result, there is more flexibility in adjusting lot sizes to order quantities so that the overall WIP throughout the process is minimal. For example, with a reduced limit on minimal lot sizes, companies that use materials requirement planning (MRP) can produce the exact quantity needed per MRP time bucket rather than producing in advance to store for later use.

The adjustment of production lot sizes to order quantities aims at reducing WIP and finished-products inventory throughout the production process. The upstream progression of this strategy leads reviewing purchased materials and components, inventories, and possibly to reducing the size of purchase orders.

7.1.3 Reducing the Size of Purchase Orders

The size of purchase orders (receiving quantities) should be reduced for exactly the same reasons that the size of production orders should be reduced: First, to reduce lead times, and second, to eliminate the problems caused by large amounts of purchased-materials inventory.

The same indicators as those used to determine the necessity of lot-size reduction (for WIP inventory) can also be used to determine if purchased-items inventory reduction is needed. The following situations call for inventory reduction: Long inventory rotation periods, damages to products due to aging, unmanageable inventories (when parts are lost or ordered twice, parts in inventory are no longer used in the new products—or nobody knows in which product they should be used), discrepancies between the actual inventory and the information in the inventory-management information system, or a need to expand the storage facilities.

The first step in reducing the size of purchase orders is to determine target levels for receiving quantities and frequency, in relation to rate of consumption by production processes, in order to minimize the average level of purchased material and component inventories. Because reducing the size of purchase orders will increase the number of purchase orders, the second step is then to identify the costs associated with individual purchase orders. The third step is to take actions to reduce these costs; thereby making economical the reduction of the purchase orders size.

Determining target levels for receiving quantities and frequency. This initial step is a simple extension of the strategy (described in the previous section) for determining the size of production orders in relation to shipping quantities. This is done by closely matching the receiving frequencies with the rate of consumption. Once the appropriate target levels for receiving frequencies are obtained, the resulting receiving quantities are computed. This adjustment often results in increasing the receiving frequency and usually increases operating costs—unless the cost drivers are identified and proper action is taken to reduce those costs.

Identify the costs associated with individual purchase orders. Although it does not seem appropriate to speak of "setup costs" in regard to purchase orders, there obviously are costs associated with preparing each purchase order. As in production, where the setup costs are independent of the size of the produced lot, the costs associated with individual purchase orders are basically independent of the quantity of purchased items. These costs are mainly generated by the ordering, transporting, and receiving processes. However, actions can be taken so that the increase in receiving frequency does not result in inflating operating expenses.

Take actions to reduce cost. A major difficulty in establishing and implementing a cost reduction program is coordinating communications among the many participants who are involved in ordering, shipping, transporting, and receiving the products—both those outside the company (vendors and transportation companies) and those within the company.

A prerequisite to take action that reduces costs is to establish durable relationships with vendors that are based on trust and effectiveness. To decrease the costs associated with large inventories, delivery schedules that have small quantities and high frequency must be negotiated. For vendors to accept this increase in number of deliveries they must be kept informed of monthly, weekly, or daily production schedules so that they can adapt their own production to the demand.

Transportation costs can be decreased by using smaller trucks for short distances or by grouping transports for long distances. For example, in a case where several vendors from the Chicago area regularly ship components and materials to a large manufacturing company in Cincinnati, the vendors could instead frequently ship to a common hub in the Chicago area with smaller trucks, and then semitrucks could transport the products from the Chicago hub to the Cincinnati plant. This strategy not only reduces costs, but also decreases transportation time.

Within the company, actions to reduce costs can have either short-term results or long-term results. Among the actions that have short-term results are the following:

- Use more blanket orders to reduce the paperwork associated with each order.
- Transfer the responsibility for taking orders from the purchasing department to the production department.
- Move receiving docks closer to the production areas, and allocate small and temporary storage areas to the production areas.
- Work with the vendors to eliminate or reduce the need for incoming material inspection (see in chapter 9 how to improve relationships with vendors).
- Have the production quality inspectors (or production personnel) perform quality checks on all received materials, if incoming material inspection is still required (see additional recommendations in chapter 3).
- Agree with suppliers on using standard containers to eliminate depackaging, repackaging, and material handling. The goal is to use the same containers between the end of the vendor's production lines and the company's production lines.

All these actions aim at decreasing delays and increasing efficiency from order placing to components and materials delivery to the line, and thereby reduce costs.

Long-term, cost-reduction results can be achieved by reinforcing the autonomy of the production department in ordering, receiving, and controlling products by reorganizing the company structure around the products. In addition, the information systems that support activities such as purchasing and cost accounting need to be adapted to the reorganization of the company.

Reducing lot sizes and receiving quantities is a necessity for companies that want to reduce manufacturing delays and be able to react quickly to customer demand. Although managers faced with the challenge of reducing lot sizes usually are knowledgeable about the problems inherent at the source of large lot sizes and large inventory levels, they face difficulty in justifying the appropriate actions that need to be implemented to reduce these lot sizes and levels of inventories. In particular, they often do not have the tools to predict the effects of these actions.

7.2 PREDICTING THE EFFECT OF LOT-SIZE REDUCTION

It is usually agreed that reducing lot sizes leads to smaller inventories and decreased production lead times. However, it is often difficult to quantify the effects of a specific percentage of lot-size reduction, thereby making it difficult to determine target levels for lot-size reduction that are needed for the objectives for lead times and inventory levels to be achieved. The dynamics involved in production systems can hardly be captured in simple inventory-theory mathematical models; thus, more sophisticated approaches, such as computer modeling, are being used more frequently.

The remainder of this section will focus on how to predict the quantitative effects of lot-size reductions and will explain how to determine target values for setup-time reduction.

7.2.1 Evaluating the Effect of Lot-Size Reduction on Inventory Reduction

Reducing inventory levels has always been a concern of production managers mainly because reducing inventory levels decreases the burden of financing the inventories. The way inventory reduction has been traditionally planned is to first establish target percentages for reduction based on financial constraints and then to effectively decrease the on-hand inventory by this percentage. With this approach, even minimal levels of reduction (5 percent, for example) have often resulted in stock-outs, missed shipping dates, rush orders, products shipped with missing parts, bad quality, and, hence, the loss of customers. Since then managers have learned that their inventories alone are not the problem, but instead are the effect of more profound problems, such as unreliable suppliers, long setups, and machine breakdowns.

The preceding section listed these problems, and showed that attacking (and solving) them is a prerequisite to commencing the reduction of lot sizes and inventories.

The effects of lot-size reduction are illustrated in this section by considering two examples, one in an MRP environment, and the other in a Kanban system environment.

Table 7.2 lists, for a part maintained in an MRP environment, the net requirements during the periods numbered 1 to 11, the quantity received at the beginning of the period to cover the net requirements, the actual quantity on hand at the beginning of the period, the remaining quantity on hand at the end of the period, and the resulting average inventory—which is the performance measure of interest.

In this first example (Table 7.2), the item lot size is 100. This lot size could be the receiving quantity for purchased parts or the setup quantity for produced parts. The received quantity is then a multiple of 100 because a lot consists of 100 items. For example, for period 5 the net requirement is 60 items. However, at the end of period 4 the inventory on hand is 20 items, indicating that at least 40 additional items are needed. Thus, an order is placed for reception at the beginning of period 5 in the quantity of 100 (one lot). The inventory on hand at the beginning of period 5 is then 120 (100 + 20 = 120), and the inventory on hand at the end of period 5 is 60 (120 - 60 = 60). Hence, the average inventory in period 5 is 90 items [(120 + 60)/2 = 90]. The inventory level is plotted in Figure 7.2 (it is assumed that the demand rate is constant for a given period).

According to Table 7.2 and to Figure 7.2, the average inventory level for periods 1 through 11 is 67 items when the lot size is equal to 100 items. However, when the lot size is reduced to 50 items, the average inventory level can be reduced to 40 items. Both Table 7.3 and Figure 7.3 show the variations in inventory level when the lot size is reduced to 50 items.

With a lot size of 50, the received quantity is a multiple of 50. For example, in

TABLE 7.2. Inventory with Lot Size = 100.

Period	1	2	3	4	5	6	7	8	9	10	11
Net Reqt.	70	80	70	60	60	50	35	25	20	30	40
Received Qty.	100	100	100	0	100	0	100	0	0	0	100
Parts On hand Beg. of period	100	130	150	80	120	60	110	75	50	30	100
Parts On hand End of period	30	50	80	20	60	10	75	50	30	0	60
Average Inv.	65	90	115	50	90	35	92.5	62.5	40	15	80

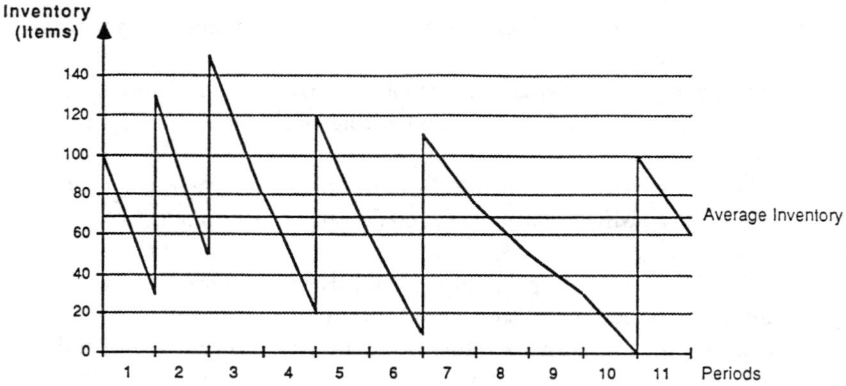

Figure 7.2. Variations in Inventory Levels with a Lot Size of 100 Items.

periods 1 and 3, two lots of 50 items each are received; in the other periods only one lot is received, except for in periods 8 and 10 where no additional items are required. The graph in figure 7.3 shows the inventory-level variation based on the reduced lot size.

The average inventory level for periods 1 through 11 is now equal to 40 items. Therefore, a reduction in lot size of 50 percent results in an average inventory reduction of 40 percent. Note that this reduction of 40 percent is not a maximum reduction but a minimum reduction. If the lot sizes for every item being produced (or purchased) can be reduced by 50 percent, not only will the WIP inventory be reduced, but also every item will flow faster through the plant because of the decreased waiting time in queue behind the other batches. The global inventory level will thus be further reduced proportionally to the reduction in waiting time. Because this reduction cannot be shown with a simple static model, section 7.2.3 of this chapter describes how to use dynamic models to predict this additional reduction in inventory.

TABLE 7.3. Inventory with Lot Size = 50.

Period	1	2	3	4	5	6	7	8	9	10	11
Net Reqt.	70	80	70	60	60	50	35	25	20	30	40
Received Qty.	100	50	100	50	50	50	50	0	50	0	30
Parts On hand Beg. of period	100	80	100	80	70	60	60	25	50	30	50
Parts On hand End of period	30	0	30	20	10	10	25	0	30	0	10
Average Inv.	65	40	65	50	60	35	42.5	12.5	40	15	30

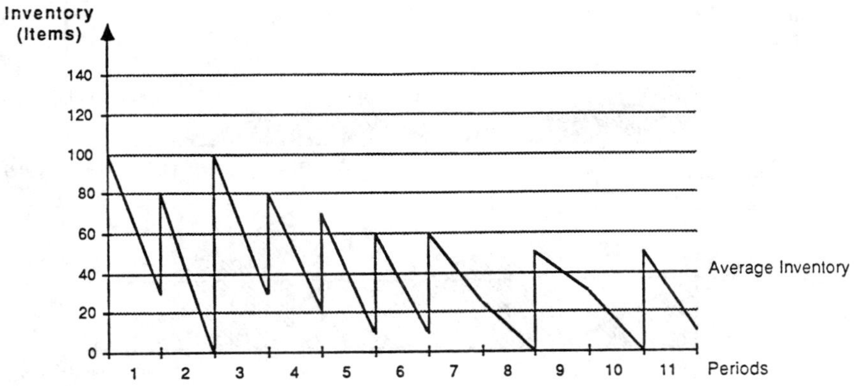

Figure 7.3. Variations in Inventory Levels with a Lot Size of 50 Items.

To illustrate the Kanban system environment, the case of a pull inventory system will be considered by looking at two items (A & B) being processed by the same machine. Assume that the demand rates are constant and equal to 10 parts per hour for A and 20 parts per hour for B. Initially, the lot sizes are 100 for item A and 200 for item B. The graph in Figure 7.4 shows the different production phases of the machine as well as the WIP of the finished items waiting for transfer.

Figure 7.5 shows the resulting inventory of finished products in the buffer following the machine where lots are transferred on completion.

The average inventory is 50 items for A and 100 items for B, or half the lot size. It is clear that if the lot sizes are reduced by any percentage an identical reduction in inventory follows. Furthermore, the WIP waiting after the machine will be reduced by the same percentage (see Figure 7.7). Figure 7.6 shows the level of WIP

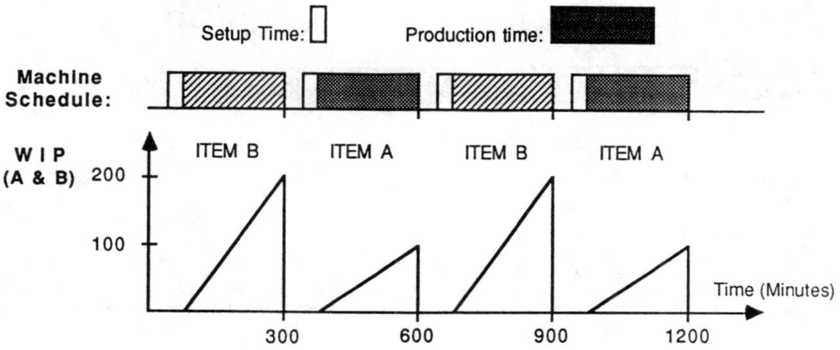

Figure 7.4. Production Phases and WIP with Initial Lot Sizes.

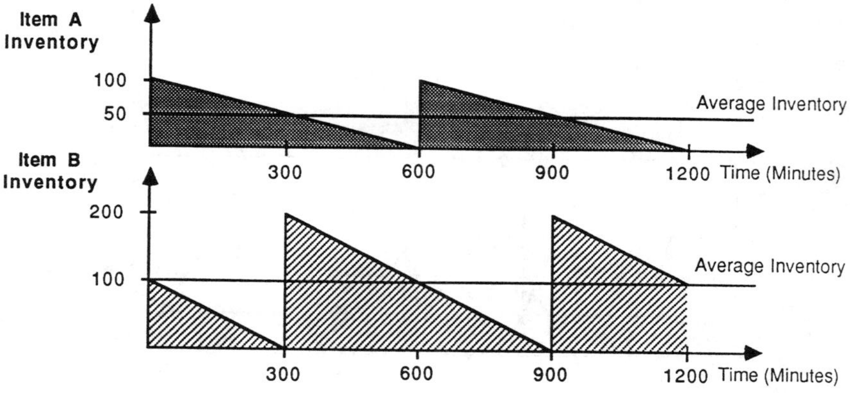

Figure 7.5. Inventory with Initial Lot Sizes.

with the maximum lot size reduction (50 percent) that is permitted by the current setup times.

Figure 7.7 shows the resulting inventory levels for items A and B in the downstream buffer. Note that the production rates and demand rates for parts A and B have not changed.

The positive result of a reduction in WIP and in inventory can be attributed to the reduction of lot sizes. However, the negative effect is that more machine capacity is used for setup. With a machine schedule like the one shown in Figure 7.6, further lot-size reductions cannot be achieved unless the setup time is reduced for both items A and B.

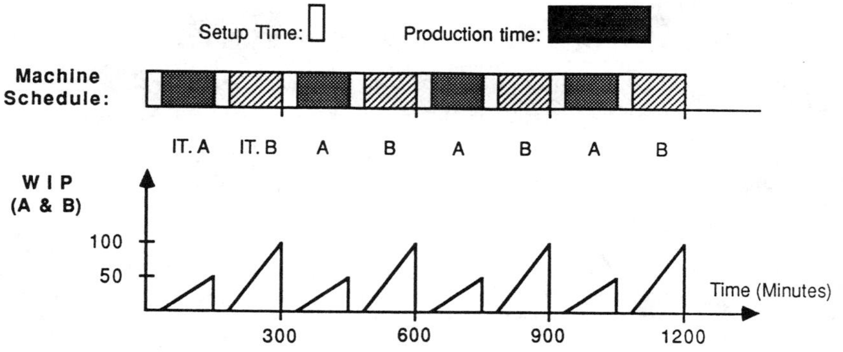

Figure 7.6. Production Phases and WIP with Reduced Lot Sizes.

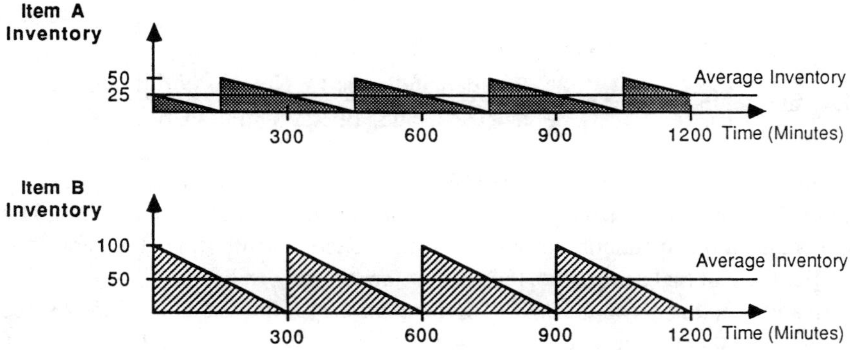

Figure 7.7. Inventory with Reduced Lot Sizes.

7.2.2 Relating Lot-size Reduction to Setup-Time Reduction

Because the minimum value of lot size is limited by available machine capacity and setup time, it is necessary to consider how they are related. The relationship between the internal setup time per part, the machine (or cell) capacity available for that setup, and the minimum lot size is depicted in the following equations:

$$\text{Minimum lot size} = \frac{\text{Part demand/period}}{\dfrac{\text{Machine capacity for setup/period}}{\text{Internal setup time per part}}} \times \text{Number of different parts}$$

or,

$$\text{Minimal lot size} = \text{Internal setup time per part} \times \frac{\text{Part demand/period} \times \text{Number of different parts}}{\text{Machine capacity for setup/period}}$$

When assuming that the terms on the right-hand side of the second equation are constant, except for the internal setup time per part, it is obvious that the minimal lot size is a linear function of the internal setup time per part. Therefore, the smaller the internal setup time, the smaller the lot size can be.

Similarly, the minimal value of the lot size is bounded by the time available to perform external setups. The time to perform external setups cannot exceed the shop's operating time. The relation between external setup time per part, shop operating time, and minimal lot size is the following:

$$\text{Minimum lot size} = \frac{\text{Part demand/period}}{\dfrac{\text{Shop operating time/period}}{\text{External setup time per part}}} \times \text{Number of different parts}$$

or,

$$\text{Minimal lot size} = \text{External setup time per part} \times \frac{\text{Part demand/period} \times \text{Number of different parts}}{\text{Shop operating time/period}}$$

In this relation, all terms on the right-hand side of the equal sign, except for the external setup time per part, can also be considered constant; thus, the minimal lot size is also a linear function of the individual external setup time. Therefore, the smaller the external setup time, the smaller the lot size can be.

In addition, the observation of both the relations for internal and external setup times shows that the larger the demand, or the larger the number of different parts, or the smaller the available time for setup, the larger the resulting lot size—unless the setup time is reduced.

To illustrate this observation, consider a machine (or a cell) that processes six different parts and to which is assigned a seventh part. Inevitably, assigning this new part will decrease the machine capacity available for setup. If the average demand for each of the six parts is 1,000 per month, the time available for setup represents 30% of the total machine capacity or 100.8 hours (21 days x 2 shifts x 8 hours x 30% = 100.8 hours per month) and the average time for each individual setup is 30 minutes, then the average lot size for each of the initial six parts is:

$$\text{Minimal lot size (parts 1 to 6)} = 0.5 \text{ hour} \times \frac{1{,}000 \text{ parts/month} \times 6 \text{ parts}}{100.8 \text{ hours}} = 30 \text{ units}$$

This lot size corresponds to 33.3 setups per part and per month (1,000/30 = 33.3), or a total of 200 setups per month for the 6 parts (6 x 33.3).

Now consider that an additional 10 percent of the machine capacity is required for processing the seventh part. The machine capacity available for setup becomes 67.2 hours (21 days x 2 shifts x 8 hours x 20% = 67.2 hours), and the minimal lot size for parts 1 to 6 becomes:

$$\text{Minimal lot size (parts 1 to 6)} = 0.5 \text{ hour} \times \frac{1{,}000 \text{ parts/month} \times 7 \text{ parts}}{67.2 \text{ hours}} = 53 \text{ units}$$

Assuming that part 7 has a demand of 1,500 units per month and the setup time is still 30 minutes, its minimal lot size is then 79 units based on the following equation:

$$\text{Minimal lot size (part 7)} = 0.5 \text{ hour} \times \frac{1{,}500 \text{ parts/month} \times 7 \text{ parts}}{67.2 \text{ hours}} = 79 \text{ units}$$

This equation corresponds to a number of setups per part equal to 19 per month (1,500 units per month/79 parts per setup = 19 setups per month), or a total of 133 setups for the 7 parts (19 x 7). The maximum number of setups has been decreased from 200 to 133 by the addition of the seventh part. At this point, the example shows the relationship between demand increase and lot-size increase with constant setup times.

To further illustrate how to coordinate lot-size reduction with setup-time reduction, consider that these parts are to be shipped twice daily, and therefore lots are produced that represent no more than half the average daily demand. Hence, a lot of each of the seven parts must be produced during each shift, and, as a result, the number of setups per shift must be increased to seven.

For parts 1 to 6 the average daily demand is 47.61 (1,000/21 = 47.61), and for part 7 the average daily demand is 71.41 (1,500/21 = 71.42). Therefore, the maximum lot sizes for parts 1 through 6 is 24 (47.61 daily demand/2 lots = 24), and the maximum lot size for part 7 is 36 (71.42 daily demand/2 lots = 36).

According to the linearity of predefined relations, since the lot sizes are decreased by 55 percent (from 53 to 24, and from 79 to 36) the setup times for all parts must be decreased by the same percentage, or in this case from 30 minutes to 13.5 minutes.

This simple example clearly illustrates the relation between lot-size reduction and setup-time reduction, as well as the relation between demand and lot size. Note that it has been assumed that the minimum lot size. Note that it has been assumed that the minimum lot size is constrained by machine capacity, which is often the case, and not by the span of external setups. The equations described in this section can help to roughly estimate lot sizes and the impact reduced lot sizes have on allowable setup times. Coordinating lot-size reduction with setup-time reduction has the advantage of being simple and easy to apply, but implies that assumptions are made that are often not verified in real systems: stability of demand, consistency of setup times, scrap rates, and breakdown rates. Moreover, these equations do not integrate the interactions often present in real systems, such as machines waiting for operators and operators waiting for parts.

To fine tune the results of these first-cut static analyses, modeling production systems through computer simulation is rapidly gaining popularity among industrial and manufacturing managers.

7.2.3 Modeling the Effect Of Lot-Size and Setup-Time Reductions

In this age of decreasing computational-power costs, high performance—yet easy to use tools—are arriving on everyone's desk. This invasion by the PC and the recent explosion in the variety of modeling softwares explain why more and more manufacturing systems (workcells, flow lines, job shops, etcetera) are designed and evaluated with the assistance of PC-based modeling tools. These tools enable manufacturing systems behavior to be analyzed under diverse conditions and can be used to predict the impact of variables such as setup times and lot sizes on system performance.

When and how modeling tools should be used to achieve improved system performance is explained in the following section. Two powerful modeling approaches—simulation and rapid modeling techniques (RMT)—are presented, with practical examples using STARCELL® and MANUPLAN II®.

Determining When to Model the Operations

This section will discuss (1) the information provided by the modeling tools, and (2) when modeling tools should be used depending on different company situations.

1 Information provided by modeling tools. What information of interest can a computer-based modeling tool provide about a manufacturing system that is being developed or modified? Modeling tools enable a mathematical representation of a real system to be built and its behavior under various operating conditions to be observed. Time reduction is one of the most interesting characteristics of modeling tools because months of production can be modeled in minutes. Traditional manufacturing performance indicators can be monitored to assess the performance of the system being studied and operating strategies being tested. For example, part-flow times, WIP levels, and machine utilization can easily be measured for various lot-size and setup-time reductions. Thus, optimal values of lot sizes and setup times can be determined, which, in turn, allow the objectives concerning flow-time, WIP, and throughput rate to be achieved.

As illustrated in the preceding sections of this chapter, lot sizes usually cannot be reduced unless setup times are reduced on the bottleneck machines. Model analysis will help in identifying setup-related bottlenecks and also allow the investigation of the effects of various levels of setup-time reduction. (This analysis can be used as an input to the setup project described in chapter 6.) Note that in the context of existing-facilities improvement projects, it is strongly recommended to also discuss the list of identified machines with shop-floor supervisors and production schedulers. Since these persons are permanently confronted with scheduling problems that have constraints such as job priority, lot size, and setup time, they know particularly well which machines are inflexible because of their long setup time, and their advice can prove invaluable in validating the list of machines that have been identified for setup-time reduction.

2 When modeling tools should be used. Whether a modeling tool should be used to obtain information depends mainly on the answers to two questions:

- Does a computer model of the manufacturing system already exist?
- If not, can a model of the manufacturing system be easily built?

When designing a new manufacturing system based on existing equipment (for example, grouping machines into manufacturing workcells as described in chapter 5), a computer model—if it has been created by the design team—should be used to identify the appropriate lot sizes and, if necessary, recommend levels of setup-time reduction at specific machines.

When a computer model has not yet been created, it is wise to first consider whether the benefits gained from using a computer model will outweigh the costs of developing the model. When planning operations are forecasted to be drastically different from the existing operations, and particularly if the user has some familiarity with computers and simple modeling tools, then a model of future operations could be created and efficiently used. When choosing to develop a model it is important to keep in mind that enough time should be reserved to explore, analyze, and compare alternatives after the model has been completely built and debugged. A potential pitfall is to spend too much time developing a model and then not having enough time left to evaluate the results. Too often the computer model becomes a "black box," which produces results that are accepted with blind faith; yet, computers should be used to enhance the decision-making process, not dictate decisions. System behavior (characterized by job throughput time, amount of WIP, and queue levels) under different sets of input conditions (such as lot sizes, setup times, and scheduling sequences) should be understood in order to identify problems and adopt appropriate solutions.

Modeling Approaches

Because there are a variety of modeling tools on the market today, it can be difficult to choose one that may actually be helpful. In the remainder of this section, different types of computer-based tools for system modeling (those tools that can be helpful for setup-time and lot-size reduction analysis) will be reviewed, and examples of information that can be obtained from two popular modeling tools will be discussed.

There are four main types of approaches that are used for manufacturing system modeling: (1) Spreadsheet models, (2) simulation packages based on a programming language, (3) menu-driven simulation packages, and (4) tools based on RMT.

Spreadsheet models are best suited to evaluate static work load summaries and perform rough-cut estimations of the number of machines required per workstation. For each part, the typical inputs consist of the annual demand and the setup and operation times at each workstation. Times are then summed over the different parts to estimate the total work load per workstation, which in turn, determines the required number of similar machines per workstation. The model can then compute each machine utilization for production and for setup.

The main advantages of such a model are that it is simple to use and it can be easily built using any commercial spreadsheet software. However, such a model cannot be used to estimate the effects of setup-time reduction on manufacturing system performance for the following reasons:

- Machine and operator interactions, part routings, machine breakdowns, and machine idle times are not considered in the model, and therefore provides inaccurate utilization results.
- The model cannot compute any dynamic data, such as the average job throughput time, average job throughput rate, or the average WIP, which is often critical information for making the appropriate decision.

Simulation tools are becoming increasingly popular in manufacturing systems modeling. However, a major drawback of such tools is that the majority of them require extensive programming for building the model and simulating the manufacturing operations. In addition, every simulation package uses its own programming language. Therefore, unless the project leader or a setup team member is extremely familiar with a simulation language, building a model of the manufacturing system through programming for the sole purpose of estimating the effect of lot-size and setup-time reduction could be a waste of time and money.

Simpler and easier-to-use packages exist on the market that do not require any programming, and using them instead of the extensive-programming type can cut model development from days to hours. These tools are the PC-based, menu-driven simulation packages and the packages that are based on RMT. The typical inputs to any of these tools consist of the following variable parameters:

- Annual demand
- Lot sizes
- System configuration and part routings
- Setup and fabrication times per part and per workstation
- Number of machines per workstation
- Workstation breakdown and scrap rates

The typical information returned to the user after run execution consists of:

- Machine utilizations (for production, setup, breakdown, and idle time)
- Job throughput time
- Job throughput rate
- Level of WIP (at the workstation and waiting in front of the workstation)

These PC-based, menu-driven tools are easy to master and to use. For example, the model of a manufacturing cell consisting of approximately ten workstations with an assignment of approximately forty jobs can be built in half a day to a day. The basic differences between simulation tools and RMT are the length of the runs, the characteristics of the results and the manner in which to interpret them, and their animation capabilities. RMT and simulation tools are based on two different families of mathematic algorithms, but both achieve the same goals. It is not the purpose of this discussion to enter into the details of the differences between the two types of algorithms; yet, from the user's point of view the effect of these dissimilarities can be summarized as follows:

- RMT runs are shorter than simulation runs (minutes with RMT, tens of minutes with simulation).
- Simulation requires the replication of runs, thereby requiring the results to be interpreted with the application of some basic statistics; on the other hand, RMT runs give deterministic results and need not be replicated.
- Simulation can provide the animation of the manufacturing system on the computer screen, thereby allowing the observation of the system's behavior in real-time. RMT does not provide this direct animation.

Two popular PC-based, menu-driven modeling tools are STARCELL®[1] and MANUPLAN II®[2]. STARCELL® is a simulator particularly adapted to the design and evaluation of flow-line manufacturing workcells. The following example is taken from the design of a five-workstation workcell for fabricating motor-reducer housings. The assigned objective was to produce at least 2,900 pieces per period with a maximum throughput time of seven hours for each piece.

Table 7.4 and the graph in Figure 7.8 show the results of the system analysis. They particularly show the results of a step-by-step approach which was applied to reach the design objective:

- Run No. 1: Lot size reduced by 75 percent on all jobs.
- Run No. 2: Setup time reduced by 50 percent on all workstations.
- Run No. 3: Setup time further reduced to 75 percent on workstation No. 4.
- Run No. 4: Setup time further reduced to 75 percent on workstation No. 3.

In Run No. 1, the lot sizes were reduced by 75 percent, which not only had the effect of reducing the parts flow time, but also drastically decreased the system's throughput rate. The design team used the model to test the effect of various levels

---

[1]STARCELL: a product of H.J. STEUDEL & ASSOCIATES, INC.
[2]MANUPLAN II: a product of NETWORK DYNAMICS, INC.

TABLE 7.4. Results of Simulation Runs.

Run	Red. Factor for Order Quantities (%)	Setup Time Reduction Factor (%)	Average Job Throughput Time (Hrs)	Pieces Completed
Initial Run	0	0	17.77	2910
Run #1	75	0	12.73	1427
Run #2	75	50 on all workstations	8.35	2281
Run #3	75	+ 75 on workstation # 4	6.96	2788
Run #4	75	+ 75 on workstation # 3	6.34	3170

of setup time reduction on the system performance, when the lot sizes are reduced by 75 percent.

By using STARCELL® in this particular case, it was determined that the design objective could be achieved by reducing job lot sizes by 75 percent if the setup times could be reduced by at least 50 percent on the machines in workstations No. 1, No. 2, and No. 5 and by at least 75 percent on the machines in workstations No. 3 and No. 4. These setup-reduction levels then became the minimal target levels for the setup team.

MANUPLAN II is a modeling package developed for designing and evaluating manufacturing systems. MANUPLAN II applies RMT based on queuing network theory. Typical MANUPLAN II outputs, from the design of a manufacturing workcell to fabricate different types of disks at a large aeronautic constructor, are illustrated in figures 7.9 and 7.10. In this case, the objective was to reduce the average flow time from more than three weeks to less than one day.

The graph in Figure 7.9 shows both the WIP waiting in the queues before each workstation (Waiting) and the WIP on the machine at each workstation (In-Pro-

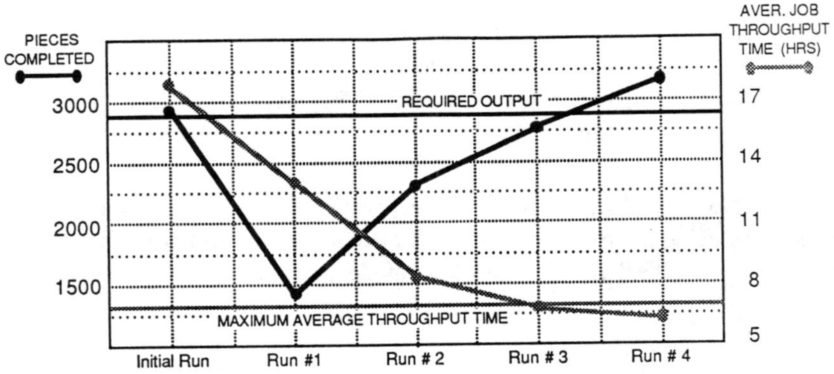

Figure 7.8. Cell Performance Under Diverse Strategies.

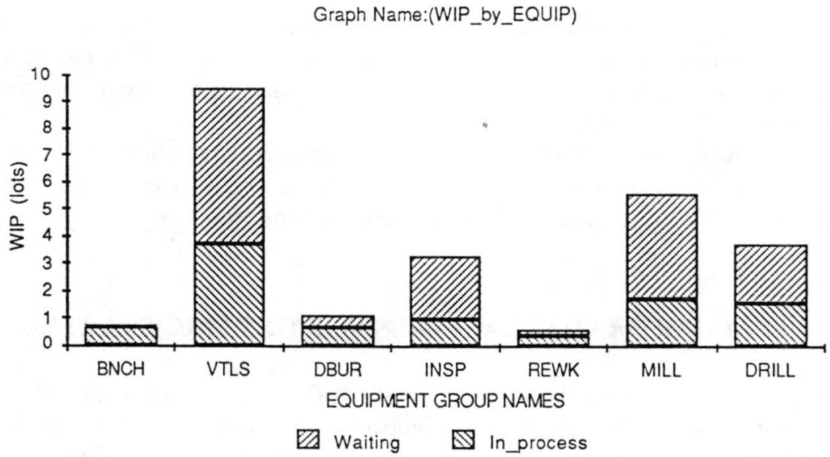

Figure 7.9. Equipment Utilization Before Setup-Time Reduction.

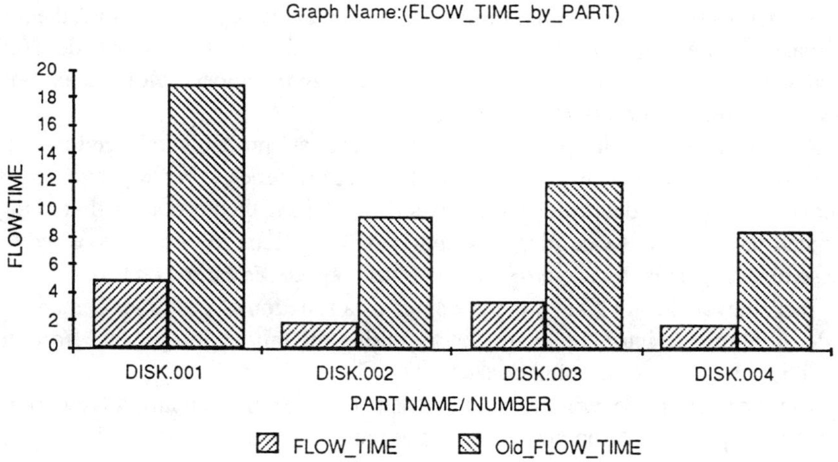

Figure 7.10. Part Flow Times Before and After a 75 percent Setup-Time Reduction on the Vertical Lathes.

cess) after reducing the lot sizes from weekly demand to daily demand, but before reducing the setup time. The graph in Figure 7.9 highlights the fact that the vertical lathes (VTLS) are setup-related bottlenecks. The graph in Figure 7.10 shows how the part flow times (or throughput times) decrease after having modeled a setup-time reduction of 75 percent on the VTLS. For example, the flow time for Disk 001 is reduced from nineteen days (Old Flow Time) to five days (Flow Time).

In this project, MANUPLAN II® enabled the quick evaluation of various alternatives; target values for lot-size and setup-time reduction could be identified in a matter of hours, thereby permitting quick feedback on the planned system behavior.

The STARCELL® and MANUPLAN II® examples show how sophisticated rapid modeling tools—simple to understand and easy to use—can be applied to quickly identify target values for lot-size and setup-time reduction.

7.3 SUMMARY AND STRATEGIES FOR GETTING STARTED

The goal for reducing inventories has traditionally been the reduction of the immobilized investment. World-class companies have understood that reducing inventories results in improving production lead times, production quality levels, customer service levels, and increasing response speed to the variation in customer demand. Such companies have demonstrated that these benefits far exceed the benefits that result from merely reducing holding costs.

Numerous indicators identify the need to reduce inventory, for example, the loss of market shares because of a lack of flexibility in adapting to customer demand, large percentages of scrapped or obsolete parts, high inventory-holding costs, and requests to expand or modernize storage facilities.

A major reason for large inventories is to fabricate products in large lot sizes; therefore, ways to reduce lot sizes must be found in order to reduce inventories. The initial step is to determine target levels for lot sizes, then obstacles that hinder lot-size reduction must be identified, and finally, actions to eliminate or reduce these obstacles must be taken so that the initial objectives can be met.

Target levels for lot sizes are determined as a function of customer demand, in order to eliminate intermediate inventory during the fabrication process from the receiving dock to the shipping dock.

The increase in receiving frequency (that results from reducing the receiving quantity of purchased components and materials) and the increase in the number of setups and material moves between machines (because of reducing setup quantities—also called batch sizes, or simply lot sizes) are two of the numerous obstacles to overcome when reducing lot sizes.

Practical methods for dealing with these obstacles are addressed in different chapters of this book. Chapter 5 shows how to rearrange machine layout and simplify part flows. Chapter 6 explains how to reduce individual setup times so that setups can be multiplied, and chapter 10 explains how to attain an improved relationship with vendors that will facilitate the increase of receiving frequencies. In addition, the cost of individual purchase orders can be further reduced by reorganizing purchasing, ordering, and receiving activities.

Simple arithmetic relations can estimate the minimum levels of setup-time reduction that will meet target levels of lot-size reduction. However, by capturing the effects of the dynamic interactions that exist in manufacturing operations, more sophisticated—yet easy-to-use—PC-based, modeling tools can more realistically estimate these minimum levels of setup-time reduction. In addition, these tools (such as STARCELL® and MANUPLAN II)® help to predict the effect of lot-size reduction on system performance, regarding such factors as throughput rate, WIP levels, and part flow times.

Lot-size reduction is one more step on the way to becoming a mean and lean world-class competitor ready for the nineties. The step that naturally follows is to adapt production planning and control systems to the new manufacturing conditions.

8

Production Planning and Control

Companies can dramatically modify their manufacturing environment by successfully adopting total quality and total maintenance, standardizing and simplifying the design of their products, implementing cellular manufacturing, developing partner-like relationships with preferred vendors, reducing setup times and lot sizes, and educating their work force to share common objectives. The functions linked to production planning and control are deeply affected (as are others) by the reorganization of the production on the shop floor, and the information systems that had been developed to support these functions now demand to be updated to reflect the organizational changes.

Since the early seventies, many industrial companies have implemented new production planning and control systems primarily based on material requirement planning (MRP). For most of these companies this implementation has required a large investment of capital and human resources. However, for the past few years, publications and articles by manufacturing authorities have provoked the question of whether these systems represent a competitive advantage or are actually a handicap for the companies engaged in worldwide competition, along with suggesting that the example of successful Japanese companies in replacing "push systems" (for example, MRP) by "pull systems" (for example, Kanban) be followed.

Do all the efforts invested and experience gained in the past twenty years then become completely worthless? Should companies redesign their information systems from scratch?

The nature of both push and pull systems is clearly explained in this chapter, and their positive as well as negative features are detailed. Often the best system for planning and controlling a company's production operations is not a pure pull system or a pure push system, but a hybrid system that combines the best of both pull and push approaches. Numerous examples in this chapter show how to modify existing MRP systems to reflect the implementation of a cellular organization, how to implement local pull systems to plan and control repetitive operations within cells, how the tasks of the traditional production planning and control system is transferred from the workstation level to the cell level, and how these different tasks can be combined in a single system. The emphasis is on the simplification of the planning and control procedures, made possible by the application of concepts developed for pull production systems.

This simplification is made possible by the streamlining of production operations, the reduction in random events (such as rejected parts or machine breakdowns), the reduction in lot sizes, and an increased reliance on vendors. Therefore, although the review of a company's production planning and control system is a major event that must be envisioned at the strategic planning stage, its full implementation is the natural capstone of the successful implementation of the other steps toward streamlined operations and WCM.

8.1 PUSH SYSTEMS VERSUS PULL SYSTEMS

The industrial companies that are competing or are considering competing in global markets usually plan and control their manufacturing operations with information systems based on MRP (materials requirement planning) or manufacturing resources planning(MRP II). These systems are called "push systems" referring to the common image of in-process inventory being pushed from one work center to the next after completion of a work order. The development of push systems has been facilitated by the continuous decrease in the cost of information processing. In the past decade, some companies—mostly in Japan, but also in the United States and Europe—have implemented "pull systems" where materials are viewed as being pulled from preceding work centers only when needed by the subsequent ones. Material flows are regulated in pull systems by cards called "kanbans," and these systems appear to be less dependent on computer technology than are the pushed systems. JIT is commonly associated with pull systems and kanbans. (Note: when written with a lower case "k" the word "kanban" refers to the original Japanese meaning: label. When written with an upper case "K", it takes the larger meaning of the production system, as in "Kanban system").

Although the respective names of these two families of systems might imply that they are based on radically opposed concepts, they do have many points in common. For example, they both attempt to schedule production so that individual parts and subassemblies are produced only when needed by the next step in the process—not too far in advance and not too late. In fact, as recently as eight to ten years ago, MRP was promoted as the system that would provide the right materials "just-in-time." Furthermore most of these systems' higher level functionalities are similar, such as long-term and mid-term planning, master-production scheduling, and bill-of-materials management.

Although it is clear to many that production control requirements can be significantly reduced with pull systems, there are a lot of misunderstandings and misconceptions about the differences between push and pull systems because they are often compared under the limited perspective of shop-floor operations, leaving the entire production planning function out of the scope of comparison. This section is aimed at clarifying the actual differences between the two families of systems; the respective planning and scheduling functions will be described and compared, examples of Kanban scheduling systems will be detailed, and the application of push and pull systems to cellular flow control will be assessed.

8.1.1 Illustrative Example of each Type of System

The general framework of a production management system usually consists of: (1) Long- and mid-term planning modules (yearly and monthly horizons), (2) short-term scheduling and shop-floor control modules (weekly or daily horizon, depending on the products), and (3) performance reporting and measuring modules. The major differences between push and pull production management systems lie in the short-term scheduling and production control modules.

When considered from the perspective of shop-floor operations, scheduling and control, push systems operate as depicted in Figure 8.1. After completion of the operations at a work center, the parts (or lots) are "pushed" to the next work center

Figure 8.1. Parts Being "Pushed" to the Next Work Center. Source: Everardo M. Hernandez P., "Western Style Just-in-Time," May 1986, p. 19, Technical Report, Manufacturing Systems Engineering, University of Wisconsin-Madison, 1513 University Avenue, Madison WI 53706

where they wait in queue. Under this system, as long as the queue in front of a work center is not empty, operator and machine are busy, thus their utilization is maximized. However, by the same token, building up inventory is encouraged and it is then common for companies to experience uncontrollable levels of inventory.

Figure 8.2 illustrates how pull-system operations are perceived from the same perspective as the push system in Figure 8.1. It is only when a work center needs a part that this part (or a lot of parts) is "pulled" from the preceding work center, which in turn, produces parts only if the number of parts waiting between the two work centers is below a predetermined limit—thereby preventing inventory build up.

From the illustrations in Figures 8.1 and 8.2 one might attempt to conclude that push systems are best suited to companies with low inventory-holding costs and high labor and equipment costs (since utilization of these assets is maximized), and that pull systems are best suited to companies with high inventory-holding costs (since inventory is limited) and low labor and equipment costs. However, to determine whether a pull or a push system (or even a combination of the two) is best suited to specific production conditions, the perspective adopted in the preceding illustrations, which is solely focused on shop-floor operation scheduling, must be enlarged. A broader picture of the functionalities of the two types of systems needs to be taken.

Push Systems

The first so-called push system, namely MRP, appeared in the United States in the late sixties. The "founding fathers" of MRP are Joseph A. Orlicky, George W.

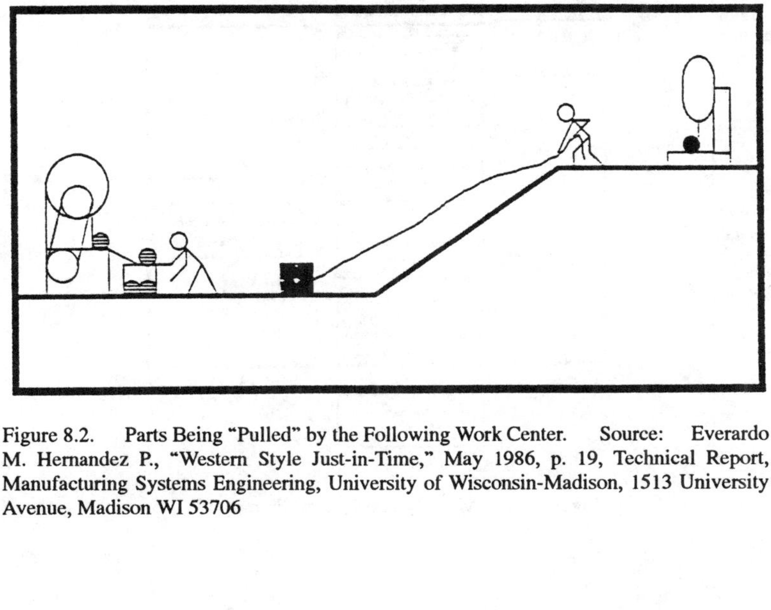

Figure 8.2. Parts Being "Pulled" by the Following Work Center. Source: Everardo M. Hernandez P., "Western Style Just-in-Time," May 1986, p. 19, Technical Report, Manufacturing Systems Engineering, University of Wisconsin-Madison, 1513 University Avenue, Madison WI 53706

Plossl, and Oliver W. Wight. The rapidly increasing popularity of MRP was due mainly to a strong support by IBM and the American Production and Inventory Control Society (APICS). Material requirement planning was not initially aimed at managing production operations, but rather at planning material requirements (MRP was originally standing for material requirement planning) as a replacement for the traditional reorder-point method, which had the drawback of independently managing the inventory levels of every single part. Quickly, new functionalities were added to MRP, which then became a true production management system. Since then the acronym MRP (or MRP II) has stood for manufacturing resources planning.

The backbone of the system is formed by the bills of materials, which regroup individual parts into subassemblies and final products. Bills of materials and their associated lead times enable one to determine the production requirements for individual parts and subassemblies from the commercial forecast of the final products.

The general framework of an MRP system is shown in Figure 8.3. The main inputs to the system are the commercial forecast input to the master production scheduling module, the process—or work centers—capacities input to the various

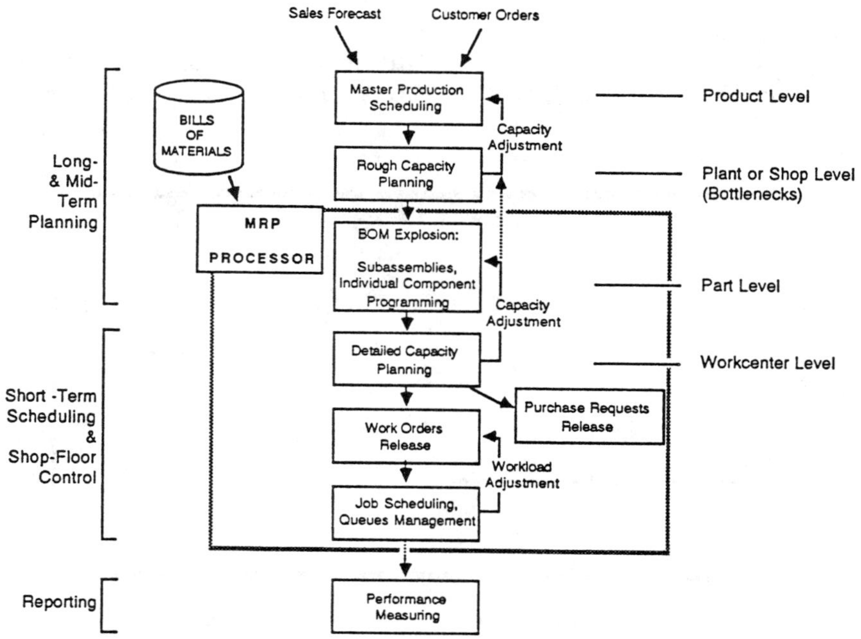

Figure 8.3. MRP System General Framework.

capacity planning modules, the bills of materials, and the various production and inventory statuses. The main outputs are the production plans that include different horizons and levels of detail, the production work orders, and the purchasing requests.

The objective of the MRP system is to release work orders and purchasing requests for all product components in a timely manner so that final products can be shipped on due dates. Theoretically, MRP is the ideal system to ensure JIT production with zero inventory. This would be the case if the system theory could be rigorously applied and if the lead times known by the system exactly reflected the actual lead times.

However, practical constraints and traditional ways of thinking have contributed to building systems that are better suited to inventory management than to ensuring smooth production patterns with minimum levels of inventory. A major constraint to the system is that the amount of information to process is enormous in most companies. Thus, for simplicity of computation, the production horizon is decomposed in time intervals called "buckets," and lead times are expressed in number of buckets. For example, if a bucket corresponds to a working week, then the lead time between two operations cannot be less than a week, even if the two operations could be completed in two days. In addition, long setup times, traditionally regarded as given constraints, have resulted in further increasing actual lead times because the parts are produced in lots larger than the quantities actually needed and the surplus is stored in stockrooms for undefined periods. These examples contribute to the explanation of why MRP systems are sometimes viewed as systems essentially aimed at "producing inventory."

Another major problem is that, since the lead times are predetermined and assumed fixed, the MRP system does not provide any incentive for improving operations on the shop floor and, thus, for reducing the manufacturing lead times. For example, improvements in assembly operations and assembly-sequence modification could reduce assembly time by 50 percent; if the information concerning the actual lead time reduction is not transmitted to the production planning department, then the only result of the improvements would be an increase in the inventory of finished products waiting to be shipped. In addition, the remoteness of production planners from manufacturing operations commonly leads to large discrepancies between the production parameters entered in the system and the actual production parameters.

On the other hand, MRP systems do present several unique advantages; for example, their ability to plan nonrepetitive as well as repetitive production and their flexibility in adapting to customer-demand variation. Customer-demand variation, as long as it can be forecasted within a time range larger than the production lead time, can easily be input to the master production scheduling module. The repetitiveness of the production undoubtedly has little affect on the performance of MRP systems; scheduling the production of home air-conditioning units (a typical seasonal product), for example, presents no more difficulty to an MRP system than scheduling the production of washing machines (which are produced all year long). This is not the case with pull systems.

Pull Systems

While almost all large companies were embracing push systems modeled on MRP, Toyota Motor Corporation was developing its own production management system—a pull system. The origin of the so-called Toyota production system can be traced to the early years following World War II at the Honsha plant, where the late Mr. Taiichi Ohno, then machining department manager, created and developed this system. He eventually spread his ideas throughout the corporation as he moved up the hierarchy to become vice president. The Toyota production system was revealed to Japanese companies when most of them went in the red after the first oil shock in 1973; while at the same time, Toyota was showing large profits. Western companies started to adopt pull systems modeled on the Toyota production system in the early eighties when they realized that they could not compete anymore with their Japanese counterparts unless they dramatically altered their manufacturing organizations. A striking example in this country is the JIT rescue of the Harley-Davidson Company, which was only a few weeks away from having to close its doors when it started to entirely—and successfully—review the way its motorcycles were manufactured. A major improvement was the switch to a pull production system.

The objective of a pull system is to attain smooth production with the lowest possible level of intermediate inventory, from material delivery to product shipment. As shown in Figure 8.4, the framework of a typical pull system is based on (1) a central system linked to two subsystems: Cell Planning (2) and Detailed Production Control (3).

Central production management system. By comparing Figure 8.4 (pull system) to Figure 8.3 (push system), it can be noted that the top portion of the framework of a typical pull system does not look much different than that of a push system. The main inputs to the system are the long-term commercial plan input to the annual production and planning module, the monthly orders input to the monthly production and planning module, the daily orders input to the detailed production scheduling module, the divisions and/or plants capacities input the rough capacity planning module, and the cells capacities input to the monthly capacity planning. Bills of materials and process lead times play a similar role in pull systems as they do in push systems. The main outputs are an annual rough-cut production schedule issued at the division and/or plant level, a monthly production schedule per cell sent to cell supervisors, and a daily detailed assembly schedule sent to the assembly lines.

At this point, however, two main differences from traditional MRP systems can already be noted for pull systems: (1) There is no capacity planning at the work

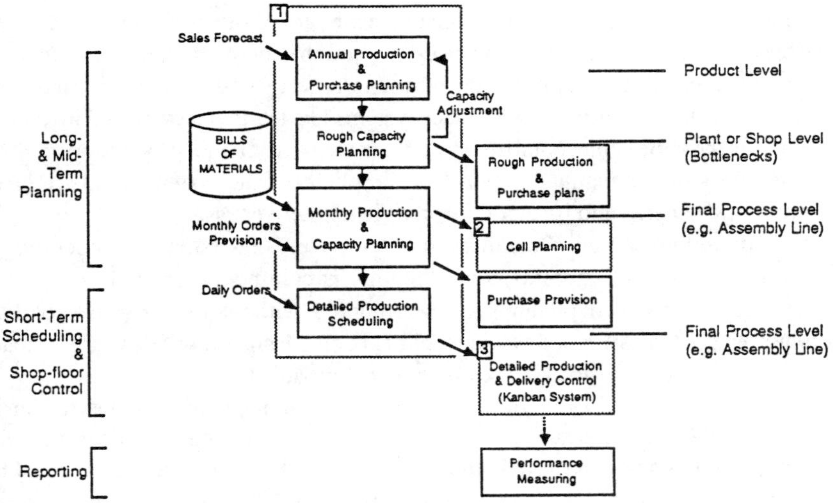

Figure 8.4. General Framework of a Typical Pull System.

center (machine) level, and (2) there is no detailed scheduling at any of the intermediate manufacturing stages; only the assembly line receives a detailed production schedule, which corresponds exactly to the actual customer demand. (The assembly schedule is established only after taking the orders.)

Another observation is that this system could not operate as described without "some" level of safety stock—unless the monthly forecast was exactly equal to actual demand. The more unforecasted variation there is in customer demand the more a safety inventory is required in order to satisfy that demand.

Note, in Figure 8.4, that the central system does not plan the capacity at the machine level or schedule the work at the cell or machine levels. Since these functions are still necessary to achieve production, two additional subsystems are required. The first one, cell planning, allows cell supervisors to estimate the number of workers needed for each cell and the number of kanbans circulating in the cell for the next time period (for example, for the next month). The second subsystem, detailed production control, synchronizes the entire production process to the assembly-line schedule.

Detailed production control. In a pure pull system, cell- or machine-detailed production is controlled by a Kanban system. The Kanban system physically links upstream operations (for example, machining) to downstream operations (for example, assembly). The kanbans (which may actually be cards, containers, or even limited buffer areas) coordinate the cells together so that the required number of individual parts or subassemblies arrive "just in time" to the final assembly stations. In addition, the kanbans regulate the amount of inventory inside each cell and between successive cells. Inventory limitation, one of the results of using a Kanban system, is the main reason why this type of system is so attractive to companies whose experience with push systems has led to uncontrolled inventory levels. Furthermore, the Kanban system as seen from the production control perspective is truly a manual system. The simplicity of the system greatly reduces control requirements and further reinforces its attractiveness.

Cell planning. While Kanban systems are well understood, another major building block of pull systems, cell planning, is commonly ignored. A possible explanation is that cell planning is meaningless without the existence of two conditions: (1) Cross-trained workers and (2) cells designed with variable capacity—both of which are not satisfied by many companies.

Cell planning consists of first determining the number of kanbans to put in circulation, thereby determining the level of inventory within and between cells. Second, it consists of determining the number of workers required for each cell by considering the desired throughput rates—assuming that cells have been designed so that their throughput rate is variable as a function of the number of workers assigned to them, and that the workers are trained to operate several machines. Usually the computations are performed monthly by cell supervisors based on the monthly production plans.

Cell planning allows the beforehand adjustment of upstream process production levels to the variations in final assembly planning, that is, to variations in customer demand. These monthly adjustments are required in dynamic market environments because the variations that can be transmitted upstream by the sole Kanban system are commonly considered as being limited to ± 10 percent. It is nevertheless important to understand that even with adequate cell planning, conditions for the smooth operation (that is, without excess inventory) of a pure pull system are a stable customer demand, a repetitive production (but not necessarily high volume), and accurate monthly forecasts.

Through this comparison of push and pull systems, it appears that long-term and mid-term planning functionalities are similar for both of them and that the main differences lie in short-term scheduling and production control. A major difference is that, contrary to triggering production by the release of work orders to individual work centers like in push systems, in pull systems part and subassembly production is regulated by flows of circulating kanban cards. The Kanban system allows a tight control of the levels of intermediate inventories. Thus, Kanban systems are often viewed as the cure to large inventories.

8.1.2 Kanban Systems

As noted in the previous section, a Kanban system is a subsystem of a larger PP&C system. The Kanban system's aim is to synchronize the production of upstream processes to the assembly-line schedule. In addition to the communication of this schedule, the main information flow from the PP&C system to the Kanban system is a periodic update of average production rates and of the recommended numbers of cards to put in circulation. Actual cell lead times are also regularly fed back to the PP&C system.

There is no obstacle to the coexistence of pull Kanban systems with current push PP&C systems. The major modifications to existing systems consist of adding a module that is able to convert the master production schedule into daily or weekly production rates and to compute the number of cards to put in circulation at the different stages of the production process. There are actually several vendors that already offer PP&C systems that are able to plan push-and-pull operations.

In general, the functioning of Kanban systems is well understood. The two-card system developed by Toyota has often served as an example to describe how Kanban systems work. However, other types of kanbans, such as kanbans for lot production or constant cycle-time kanbans, are usually not as well understood. In addition, how to compute the number of kanbans—a key to Kanban system implementation—is still unclear to many manufacturing professionals. This section will focus on these less-understood aspects of Kanban systems; the "golden rules" associated to Kanban will conclude the section.

Production Kanban with Constant Reorder Quantity

The production kanban and the withdrawal kanban are the basic elements of the two-card Kanban system. The production kanban is used to originate production of a predefined quantity of parts (constant reorder quantity) and follows the parts from the initial workstation to the final buffer; withdrawal kanbans, on the other hand, follow the parts during their transfer from the final buffer to a downstream process. The flow of production kanbans is illustrated in Figure 8.5, where the "process" can depict a single machine or, most likely, a manufacturing or assembly cell.

In order to determine the number of kanbans needed for each part and for each process, it is first necessary to compute the level of inventory that must be kept in the final buffer— that is, the average part quantity consumed by subsequent processes during the interval from part-production ordering to part withdrawal from the buffer. This interval is equal to the time it takes for one kanban to complete the entire cycle (which is illustrated in Figure 8.5), and is expressed in the following equation:

$$\text{Parts consumed during one kanban cycle} = \text{Average demand} \times \text{kanban cycle time}$$

However, since the actual demand usually is not constant, a safety coefficient (α) can be introduced at this point (the value of (α) should be kept less than 10

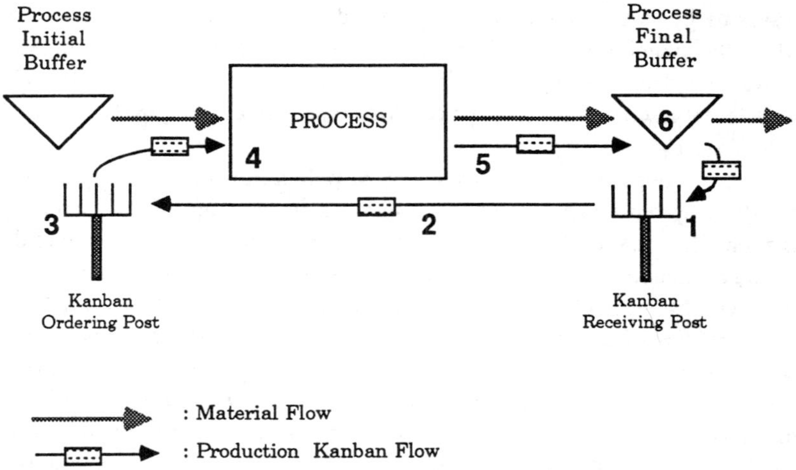

Figure 8.5. Flow of Materials and Production Kanbans.

percent and should be decreased as the demand becomes more levelled). The following equation is therefore valid:

$$\text{Parts consumed during one kanban cycle} = \text{Average demand} \times (1 + \alpha) \times \text{kanban cycle time}$$

Since there is one kanban per container of parts, the number of kanbans in circulation is then obtained by dividing the number of parts consumed during a kanban cycle time by the number of parts per container, which is expressed as follows:

$$\text{Number of kanbans} = \frac{\text{Parts consumed during one kanban cycle}}{\text{Number of parts per container}}$$

Or, more precisely:

$$\text{Number of kanbans} = \frac{\text{Average demand} \times (1 + \alpha) \times \text{kanban cycle time}}{\text{Number of parts per container}}$$

This last formula is the key for computing the recommended number of kanbans in any type of Kanban system.

The kanban cycle time is the time spent by a kanban to complete a full cycle; it is important to consider the different phases involved. (The phase numbers correspond to the numbers in Figure 8.5).

1. The production kanban is removed from the full container when the container is withdrawn from the final buffer for transfer to a downstream process; the kanban is then placed in the kanban receiving post and waits there until collection.
2. The kanban is transferred from the kanban receiving post to the kanban ordering post at the initial workstation.
3. The kanban waits at the production ordering post behind the other production kanbans (first-in—first-out queue).
4. The kanban is taken from the production ordering post and attached to an empty container, the process machines are set up, the quantity of parts to process (equal to the container size) that was indicated on the kanban are withdrawn from the initial buffer; the parts are then processed and placed in the container upon completion.
5. The full container (with its kanban) is transferred to the final buffer.
6. The container waits until withdrawn by downstream processes (the cycle is then complete).

The following formula thus sums up the kanban cycle time:

kanban cycle time =
   kanban waiting time in receiving post
+ kanban transfer time to ordering post
+ kanban waiting time in ordering post
+ lot processing cycle time
   (internal setup time + run time + in-process waiting time)
+ container transfer time to final buffer
+ container waiting time in final buffer.

There is a direct relation between the process's performance and the number of kanbans in circulation—that is, the maximum part inventory level. Some of the waiting times, such as the kanban waiting time in the ordering post, cannot be directly computed and are usually determined by experience. Therefore, the number of cards is initially estimated by applying the aforementioned formulas and then adjusted according to the process's behavior. If there is too much inventory waiting in the final buffer, one or several cards can be removed; if there is a risk of stock-out, the number of cards can be increased or the kanban cycle time decreased (through process improvement).

Because all containers should contain the same quantity of parts of the same part type, the container size significantly affects the system's behavior. Ideally, a size of one part would ensure the smoothest flow of parts throughout the entire process and the most regular production; however, this ideal situation could only be attained with zero setup times. A good rule of thumb is to set the container size to 10 percent of the daily demand requirement, thereby limiting the number of setups to ten setups per day and per part; this implies, however, that machine setup times have been reduced to single digit numbers (less than ten minutes).

The following equations illustrate how the kanban-cycle-time formula and the number-of-kanbans formula are used. First the following values are inserted:

average demand = 100 parts/hour,
container size = 100 parts,
kanban waiting time in receiving post = 5 min.,
kanban transfer time to ordering post = 1 min.,
kanban waiting time in ordering post = 30 min.,
maximum internal setup time = 5 min.,
maximum processing time = 0.25 min./part,
in-process waiting time = 10 min.,
container transfer time to final buffer = 5 min.,
container waiting time in final buffer = 10 min.,
initial safety coefficient: $\alpha = 0\%$.

Thus,

$$\text{Kanban cycle time} = 5 + 1 + 30 + (5 + 0.25 \times 100 + 10) + 5 + 10 = 91 \text{ min.},$$

and,

$$\text{Number of kanbans} = \frac{\dfrac{100 \text{ parts}}{60 \text{ min}} \times 91 \text{ min.}}{100 \text{ parts}} = 1.52.$$

The actual number of kanbans can only be an integer; thus, the initial number of kanban cards is 2, and the actual kanban cycle time is:

$$\text{Kanban cycle time} = \frac{91 \text{ min.} \times 2 \text{ kanbans}}{1.52} = 120 \text{ min.}$$

The difference between 91 and 120 minutes of 29 minutes is the additional average total waiting time for the parts in the final buffer. Because of this average waiting time equal to about 25 percent of the kanban cycle time, there is no need to consider an additional safety coefficient. The initial choice of (α) being equal to 0 percent is then justified. To maintain smooth production, the sequence of the different cards should be set so that the production of lots of the same part occur at regular intervals, every hour in the case of the part considered in this example. The smaller the setup times are, the smaller the containers can be, and the easier it is to obtain a regular production pattern. Accordingly, card sequence should be adjusted when cards are removed or added.

Withdrawal Kanban with Constant Reorder Quantity

The withdrawal kanban is the second basic element of the two-card Kanban system. The withdrawal kanban is used to withdraw a predefined quantity of parts (constant reorder quantity) from an upstream final buffer, and then follows the parts to a downstream initial buffer. Part containers cannot leave the upstream buffer without a withdrawal kanban. The withdrawal kanban is removed from the container only when the parts are needed for production at the downstream process. The flow of withdrawal kanbans is illustrated in Figure 8.6.

Using withdrawal kanbans is recommended when the distance between two processes (and the resulting lead times) requires that two successive inventory buffers be kept between the two processes. If the processes are close enough, withdrawal kanbans are not required. In the case of two adjacent manufacturing cells, for example, a common buffer might be sufficient, and therefore when upstream production kanbans are removed from the container they can be replaced by downstream production kanbans. A further simplification would be to balance

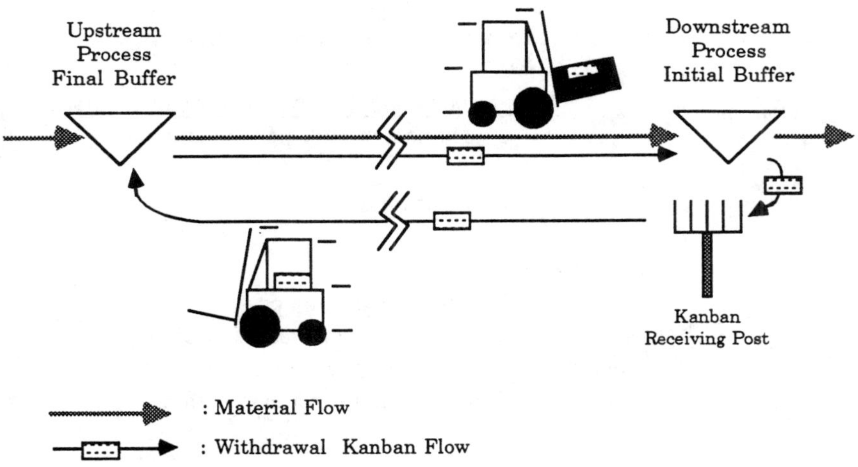

Figure 8.6. Flow of Materials and Withdrawal Kanbans.

the operations of the two processes so that the intermediate buffer would not be required at all, in which case production kanbans would follow the parts through the two successive processes—just as parts move from machine to machine in flow-line cells. However, if the two processes are not located fairly close together (for example, if they are in different buildings), two separate buffers are necessary. In addition, variations in the transportation process usually require buffers to be kept at the output of the upstream process as well as at the input of the downstream process. The flow of materials between the two processes is then regulated by withdrawal kanbans.

The number of withdrawal kanbans between two processes can be estimated in a similar fashion to the calculation of the number of production kanbans. (Note that there can be several downstream processes fed by a single upstream process, or several upstream processes feeding a single downstream process; in any case, a distinct number of kanbans has to be determined for each pair of upstream-downstream processes.) It is first necessary to compute the level of inventory that must be kept in the downstream initial buffer; that is, the average part quantity consumed by the downstream process during the time interval from container-withdrawal ordering to the actual consumption of the first part of the ordered container. This interval is equal to the time it takes for one withdrawal kanban to complete the entire cycle illustrated in Figure 8.6. As for production kanbans, the following equation is applied:

$$\text{Parts consumed during one kanban cycle} = \text{Average demand} \times (1 + \alpha) \times \text{kanban cycle time}$$

The number of kanbans is similarly calculated by the formula:

$$\text{Number of kanbans} = \frac{\text{Average demand} \times (1 + \alpha) \times \text{kanban cycle time}}{\text{Number of parts per container}}$$

The Kanban cycle time is calculated as follows:

Kanban cycle time =
kanban waiting time in receiving post
\+ kanban conveyance time to upstream buffer
\+ container conveyance time to downstream buffer
\+ container waiting time in downstream buffer.

As for production kanbans, there is a direct relation between the process's performance and the number of withdrawal kanbans in circulation (and the inventory level). In the present case, the process consists of conveying the withdrawal kanbans and the part containers.

The following equations and values illustrate how to compute the number of withdrawal kanbans:

average demand = 100 parts/hour,
container size = 100 parts,
kanban waiting time in receiving post = 5 min.,
kanban conveyance time to upstream buffer = 10 min.,
container conveyance time to downstream buffer = 15 min.,
container waiting time in final buffer = 10 min.,
initial safety coefficient: $\alpha = 0$.

Then,

$$\text{Kanban cycle time} = 5 + 10 + 15 + 10 = 40 \text{ min.}$$

and,

$$\text{Number of kanbans} = \frac{\dfrac{100 \text{ parts}}{60 \text{ min}} \times 40 \text{ min.}}{100 \text{ parts}} = 0.67.$$

The initial number of kanban cards is then set equal to one, and the actual kanban cycle time is:

$$\text{Kanban cycle time} = \frac{40 \text{ min.} \times 1 \text{ kanban}}{0.67} = 60 \text{ min.}$$

The difference between sixty and forty minutes signifies an additional average waiting time of twenty minutes for the parts in the downstream buffer. Again, one can note the impact of the container size and setup time on inventory levels. With the number of kanbans equal to one, the only way to reduce the inventory level is to diminish the container size. However, because the container size determines the lot size and the number of setups at the upstream process, diminishing the inventory between the process can only be accomplished if the upstream-process setups are short enough so that process capacity is not exceeded.

Supplier Kanbans with Constant Order Cycle

The supplier kanbans with constant order cycle serve the same function as traditional withdrawal kanbans. They permit the withdrawal of material from an initial buffer upstream (the supplier's finished-product inventory) and follow the products until they are processed at a downstream process (the company's operations). There are, however, two fundamental differences between the withdrawal kanbans with constant quantity and the supplier kanbans with constant order cycle: With the constant order cycle, product ordering and withdrawal occur at regular and fixed times, and the ordered quantity is withdrawn from the supplier inventory after a fixed delay. The withdrawal times and the frequency of deliveries and delays are negotiated between the supplier and the company.

Because product withdrawals occur at regular times and the demand is usually not perfectly constant, the quantity of kanbans can vary between withdrawals; although in the long run the average quantity of withdrawn parts is equal to the average demand. A notable joint consequence of the quantity variation and the contractual delay is that, on a given day, the number of full containers and supplier's cards conveyed from the supplier back to the company can be different from the number of supplier kanbans previously conveyed to the supplier for the company. The flow of supplier kanbans is illustrated in Figure 8.7.

Because the reorder cycle is constant, the kanban cycle time is also a constant quantity. How to determine the kanban cycle time is illustrated in Figure 8.8 and explained in the following example.

Negotiated number of deliveries = 1 daily
negoiated delay = 24 hours (from day one 1 P.M. to day two 1 P.M.)
kanban conveyance time to supplier = 5 hours (from 8 A.M. to 1 P.M.)
truck waiting time at supplier plant = 1 hour (from 1 P.M. to 2 P.M.)
material conveyance time from supplier to company = 5 hours (from 2 A.M. to 7 P.M.)

As a result,

$$\text{Kanban minimum cycle time} = 24 + 5 + 1 + 5 = 35 \text{ hours}$$

Because the cycle time must be expressed in days (daily withdrawal), the kanban cycle time in this case is equal to two days or forty-eight hours. The additional thirteen hours represent the time that the cards spend at the company plant, either attached to full containers, or in the kanban receiving post before collection just before truck departure from the plant (daily at 8 A.M.).

Therefore, the kanban cycle time is determined as the first feasible value superior to the minimal cycle time. The following formula is used to determine the minimal kanban cycle time:

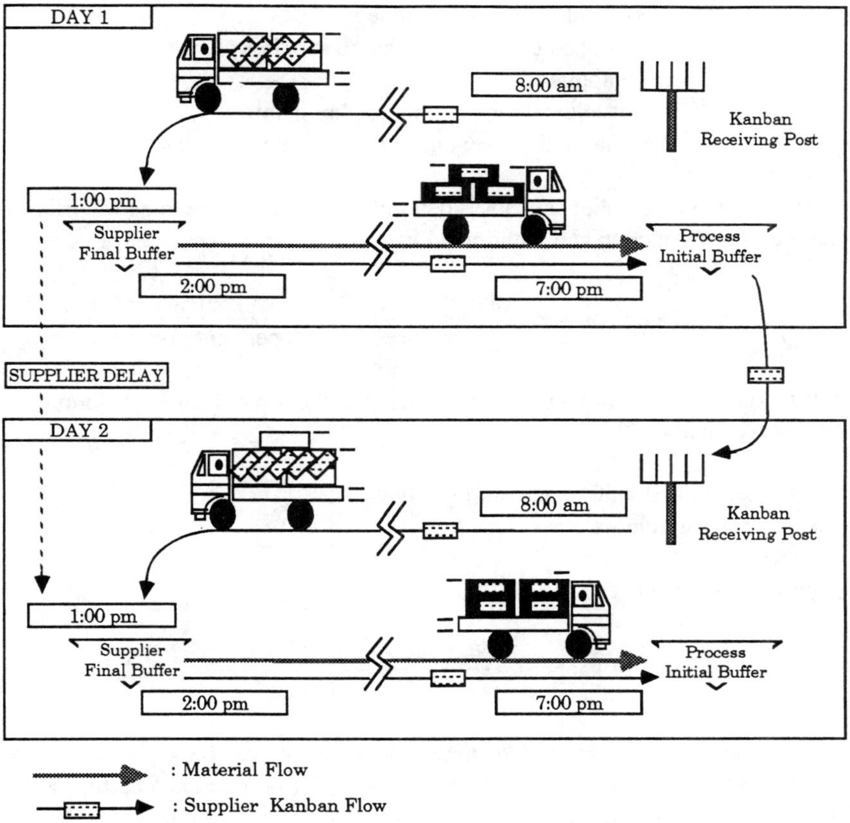

Figure 8.7. Flow of materials and Supplier Kanbans.

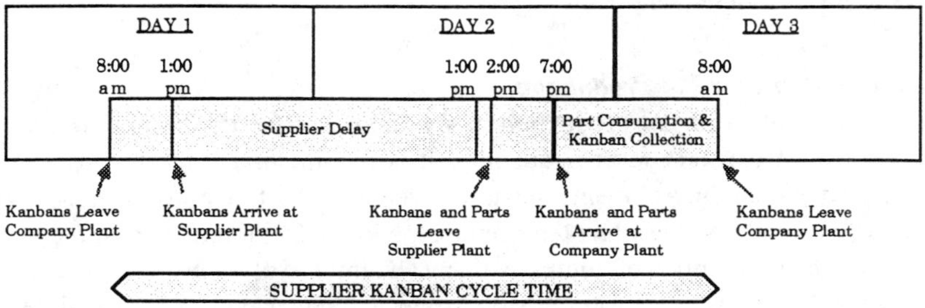

Figure 8.8. Supplier Kanban Cycle Time Breakdown.

Minimal kanban cycle time =
kanban conveyance time to supplier plant
+ supplier delay
+ truck waiting time at supplier plant
+ material conveyance time to company plant.

The number of supplier kanbans is determined by using the same formula as the one used for production and withdrawal kanbans:

$$\text{Number of kanbans} = \frac{\text{Average demand} \times (1 + \alpha) \times \text{kanban cycle time}}{\text{Number of parts per container}}$$

If, for example, the following values are used in the same formula the number of kanbans is equal to 22:

average demand = 50 parts/hour or 800/day
kanban cycle time = 2 days
container size = 80 parts
initial safety coefficient: α = 10%.

$$\text{Number of kanbans} = \frac{800/\text{day} \times (1 + 10\%) \times 2 \text{ days}}{80} = 22$$

From the supplier's perspective, the supplier kanbans can be seen as production orders, and with this perspective, maximum supplier production lead times seem to be extremely short (a single day in the previous example). Actually, the supplier kanbans are merely a confirmation that the products must be withdrawn. On the average, the withdrawn quantity is equal to the expected demand. Thus, this system works effectively for both the supplier and the company if the supplier knows the expected demand (which can be communicated monthly by the company's main production-planning system) and can adjust, in the negotiated delay, its production level to short-term demand variation.

Signal Kanbans for Lot Production

The signal kanban for lot production is a variation of the traditional production kanban with a constant reorder quantity. It is recommended for processes that produce large quantities of identical parts, have internal setup times expressed with single digit numbers (less than ten minutes), and have external setup time spans that restrain the number of setups per day. Examples of such processes in the metal-working industry are forging, die casting, and punch pressing. The minimum lot size is equal to the number of parts that can be produced during the external setup time; with the traditional production kanban, however, the lot size (order quantity) is equal to the number of parts per container. In the case of lot production, the lot size is expressed by the number of containers or pallets. Since the lot size does not change between lots of identical parts, it is not required to place a production kanban on each of the containers; only one kanban, the signal kanban, is used to order the entire lot production. In addition to the production-ordering signal kanban, a material-ordering signal kanban can be used to order raw materials from storage. The flow of signal kanbans is illustrated in Figure 8.9.

Lot size and the position of the two signal kanbans have to be determined to enable regular replenishment of the finished part inventory. The lot size is computed as a function of the average number of setups per day. For example, if on the average the total setup time (internal + external) takes two hours, then the maximum number of setups per day is on the average equal to eight, assuming a two-shift operation. The minimum lot size per part can then be computed as follows:

$$\text{Minimal lot size} = \frac{1}{\text{Average number of setups for the part}} \times \text{Demand}$$

Since the actual demand is usually not perfectly constant, a safety coefficient α can be introduced in the computation as follows:

$$\text{Minimal lot size} = \frac{1}{\text{Average number of setups for the part}} \times \text{Average demand} \times (1 + \alpha)$$

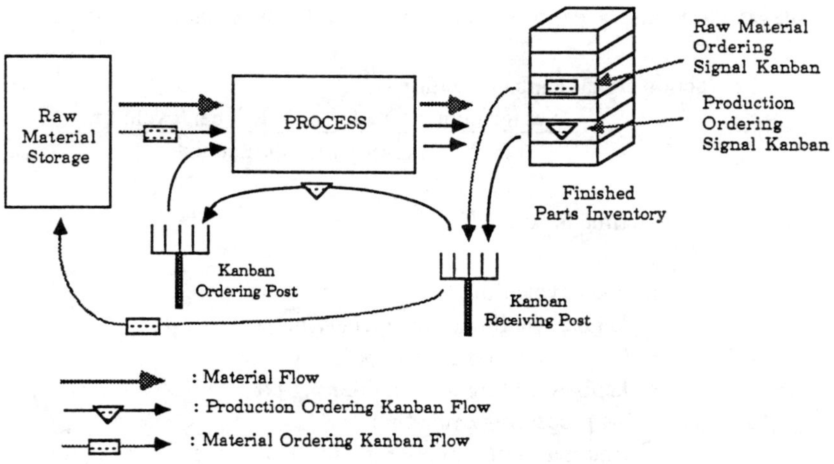

Figure 8.9. Flow of Materials and Signal Kanbans.

Using the above equations, if a machine that produces five different parts has on the average a maximum of 8/5 or 1.6 possible setups per part, and if the demand for that part is 1,000 per day, then the minimal lot size for that part, when α is equal to 10 percent is calculated as follows:

$$\text{Minimal lot size} = \frac{1}{1.6} \times 1000 \times 110\% = 688.$$

The minimal lot size is the most economic lot size (the lot size that produces the lowest level of inventory); however, since every container must hold the same part quantity the lot size must be rounded off to the nearest multiple of the container quantity. In this example a lot size of 700 parts, or seven containers holding 100 parts each, should be used. The rounded-off lot size is then the quantity indicated on the production-ordering signal kanban.

The next point of interest is the position of the production-ordering signal kanban in the lot of processed parts. This position is determined by the desired level of safety stock at the time a new lot is reordered. The safety stock is equal to the average number of parts that are consumed during the time interval from when the lot is ordered to the delivery of the final product to its stock location. The following formula uses a similar reasoning as the one used for the traditional production kanban:

$$\text{Production signal kanban position} = \text{Average demand} \times (1 + \alpha) \times \text{Kanban cycle time.}$$

This position must be expressed in integer number of containers, thus:

$$\text{Production signal kanban position} = \frac{\text{Average demand} \times (1 + \alpha) \times \text{Kanban cycle time.}}{\text{Number of parts per container}}$$

The kanban cycle time is defined as:

Kanban cycle time =
  kanban waiting time in receiving post
\+ kanban transfer time to ordering post
\+ kanban waiting time in ordering post
\+ lot processing cycle time
  (internal setup time + run time + in-process waiting time)
\+ container transfer time to final buffer.

To demonstrate how to calculate the position of the production signal kanban, the following quantities will be used:

average demand = 1,000 parts per day or 1.042 parts per min.
lot size = 700 parts
kanban waiting time in receiving post = 5 min.
kanban transfer time to ordering post = 1 min.
kanban waiting time in ordering post = 30 min.
maximum internal setup time = 10 min.
maximum processing time = 12 sec./part
in-process waiting time = 10 min.
lot transfer time to final buffer = 5 min.
coefficient: α = 10%,

when these quantities are inserted in the cycle time formula the following results:

$$\text{Kanban cycle time} = 5 + 1 + 30 + (10 + (12/60) \times 700 + 10) + 5 = 201 \text{ min.}$$

From here the number of containers can be calculated as follows:

$$\text{Production signal kanban position} = \frac{1.042 \times (1 + 10\%) \times 201}{100} = 2.30 \text{ containers}$$

The actual position of the production-ordering signal kanban will then be rounded off to the next highest whole number, in this case to three containers, as shown in Figure 8.10. In other words, as soon as the inventory level for that part becomes equal to or less than three containers, the production of a new lot is ordered. In several ways, the production signal-kanban system is similar to the traditional reorder-point method.

The position of the material-ordering signal kanban is similarly determined by examining the number of finished parts that are withdrawn from inventory during the time delay from raw-material ordering to new lot-production ordering. The value of this time delay must be set so that the raw materials are available at the machine when the internal setup is to begin. The delay is then equal to the material-ordering lead time minus the production-kanban lead time from ordering to machine setup. The material-ordering lead time is determined by the following relation:

Material ordering lead time =
material-kanban waiting time in receiving post
\+ material-kanban transfer time to raw material storage

+ material-kanban waiting time at raw material storage
+ time to withdraw material from storage
+ time to convey raw material to process.

The production-kanban lead time from ordering to machine setup is determined by the following relation:

Production-kanban lead time =
production-kanban waiting time in receiving post
+ kanban transfer time to ordering post
+ kanban waiting time in ordering post

Thus, the time delay between material ordering and production ordering is calculated as follows:

Time delay =
material-kanban waiting time in receiving post
+ material-kanban transfer time to raw material storage
+ material-kanban waiting time at raw material storage
+ time to withdraw material from storage
+ time to convey raw material to process
- (production-kanban waiting time in receiving post
 + production-kanban transfer time to ordering post
 + production-kanban waiting time in ordering post)

Note that the value of the delay can be equal to zero or even be negative. If this is the case, raw materials should be ordered at the same time as lot production and the two cards should have the same position. When this is not the case, the difference in positions of the two signal kanbans is given by:

$$\text{Kanban position difference} = \frac{\text{Average demand} \times (1 + \alpha) \times \text{Time delay}}{\text{Number of parts per container}}$$

To determine how to calculate the difference in positions of the two signal kanbans, the following quantities will be used:

material-kanban waiting time in receiving post = 5 min.
material-kanban transfer time to raw material storage = 10 min.
material-kanban waiting time at raw material storage = 15 min.
time to withdraw material from storage = 10 min.
time to convey material to machine = 10 min.

production-kanban waiting time in receiving post = 5 min.
production-kanban transfer time to ordering post = 1 min.
production-kanban waiting time in ordering post = 30 min.
average demand = 1000 parts/day or 1.042 part per min.
container size = 100 parts
coefficient: α = 10%

When these quantities are inserted in the time-delay formula, the following results:

$$\text{Time delay} = [5 + 10 + 15 + 10 + 10] - [5 + 1 + 30] = 14 \text{ min.}$$

This quantity (14 min.) can then be used in the kanban position difference formula as follows:

$$\text{Kanban position difference} = \frac{1.042 \times (1 + 10\%) \times 14}{100} = 0.16 \text{ Container.}$$

Since the difference can only be expressed by an integer number of containers, the material-ordering signal kanban is placed one container above the production-ordering signal kanban as shown in Figure 8.10.

Variations
There is practically no limit to the variations of physical implementation of Kanban systems. Cards can be replaced by containers or carts with limited capacity, or squares painted on the floor (Kanban squares) can be used to limit the amount of inventory between two processes, or limited space on a conveyor or in a gravity chute can constrain the inventory level between two machines. As long as an upper limit to inventory is clearly defined and lot sizes are kept small enough to allow

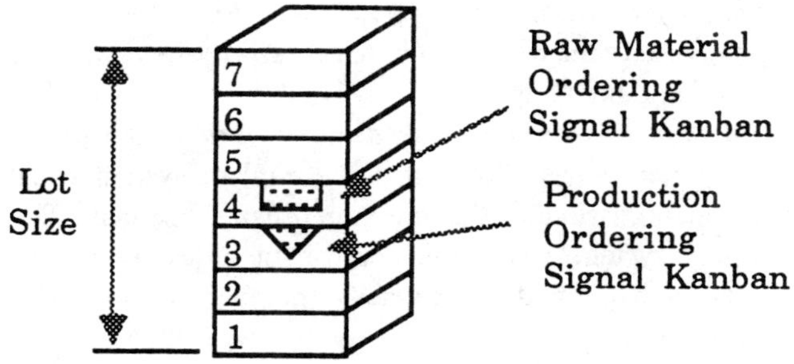

Figure 8.10. Production Lot Size and Position of Signal Kanbans.

smooth production, the best system is the simplest one. The system should be simple enough so that everyone using it understands how it works.

The upper inventory limit is computed by using the various formulas given in this section. It should be remembered, however, that these formulas only indicate a recommended number of kanbans and the number that best suits production conditions should be adjusted as these conditions change. Typically, the number of kanbans is set higher than the recommended number and gradually decreased as problems emerge and are solved. In this regard, the Kanban system is not only a production scheduling system, but also a means to improve production conditions.

Kanban Golden Rules

The following is a set of rules that must be followed to ensure that a Kanban system can be started and can function at all:

1. *Do not move nonconforming parts to a downstream process.* No bad part should exit a process toward another process. The worst alternative is to detect the bad parts before they are moved to the final buffer. The best alternative is to improve the process so that *no bad part is produced.*
2. *Ensure that downstream processes withdraw parts from upstream processes in the correct quantity and at the right time.* Parts are "pulled" from upstream to downstream only as the need arises, contributing to the smooth operation of the whole system.
3. *Do not let upstream processes produce more than the quantity of parts withdrawn by downstream processes.* This is the basis for inventory limitation.
4. *Ensure that production is leveled.* The last process, which pulls production through the entire manufacturing process, must have a leveled schedule to ensure smooth production patterns throughout the plant. Setup times must be reduced to permit small lot sizes and to synchronize production to customer demand. An underlying rule to ensure leveled production is that the *kanbans should always be processed on a first-come-first-serve basis.* Altering the card order will result in unnecessarily speeding up some parts and in delaying other parts.
5. *Do not attempt to transmit large demand variations with the Kanban system.* Experience has shown that the kanban system usually cannot transmit demand variations that are larger than 10 percent. The mid-term cell planning system serves the purpose of adjusting inventory levels (number of kanbans) and cell capacity (number of workers) to demand levels variations.
6. *Balance cycle times for smooth production, and constantly improve cells and workstations.* To ensure smooth production, workstations and cells

should have similar cycle times. In addition, workstations and cells must be constantly improved to reduce cycles times, setup times, and breakdown times, thereby also increasing capacity and flexibility.

The constraints imposed by these six rules indicate that a Kanban system is only a tool to be used in combination with other techniques, such as continuous quality checks, leveled assembly scheduling, cell planning, and continuous equipment improvement, if the company wants to achieve continuous-flow manufacturing with reduced inventories and progress toward WCM.

When this global approach is used, it is important to understand the differences between the various Kanban systems. Although the idea of a Kanban system and its circulating cards looks rather simple at first, the many possible variations tend to confuse the understanding of how Kanban systems work and, more importantly, of how to implement such a system. The next section will address these concerns and describe how companies can get started in the application of a Kanban system to control cellular flow.

8.1.3 Controlling Cellular Flow

All companies that move from traditional manufacturing to cellular manufacturing (CM) sooner or later are faced with the need to assess their control strategies. At that time, some important questions should be asked: Can the traditional shop-floor control system be used for cellular flow control? Should a Kanban or a kanban-based system be used? Should within-cell flow be controlled in a different manner than is between-cell flow? How is material flow to be controlled on workstations that are not integrated into cells? And, how is the transition to be made from the current system to a different system? This section will answer these questions by depicting the typical case of a company that has been using an MRP system to control its job-shop type production operations and yet wishes to adapt its control system to the newly designed manufacturing and assembly cells.

The aforementioned cellular flow takes place within and between manufacturing workcells; yet, what is a manufacturing workcell? As defined in chapter 4, a manufacturing workcell is "*a group of closely linked, dissimilar workstations (automatic or manual) that are dedicated to performing a sequence of operations on families of similar parts or products.*" As seen from the perspective of part flow and as shown in Figure 8.11, a cell usually consists of an input buffer where parts wait for processing, a series of workstations between which parts are transferred in small batches, intermediate buffers with allowances for a limited amount of WIP, and an output buffer where processed parts wait for transfer to the next cell.

The control of flow within the cell has to be organized first. Then, the cells are integrated in a larger flow-control scheme.

The control of flow of parts within cells

Machines or manual stations grouped in a manufacturing workcell, process sequentially the same lots of parts. To control the flow of parts, the first change to implement is to reduce the number of work orders released by the production system by simply releasing one work order for each cell and for each lot instead of one for each machine. When designing policies for the control of part flows within cells, a particularly critical parameter is the size of the transfer batches; that is, the number of identical parts that must be processed by one workstation before they can be transferred to the next (refer to Figure 8.11). The transfer batch size is equal to a fraction of the lot size; thus, the largest possible size is the lot size itself, and the smallest possible size is a single part. Large transfer batches result in long waiting times between workstations and thus long flow times; small transfer batches result in a large number of material handling operations Furthermore, there is no simple formula to apply to derive the transfer batch size. The transfer batches should be transferred by dedicated material-handling means (not by means common to an entire shop such as bridge cranes and forklifts) and possibly even by hand. For example, in a cell that processes shafts for lawn mower engines, the transfer batch size could be between half a dozen and a dozen shafts, which could be manually handled on dedicated trays or carts. To maintain balanced cell

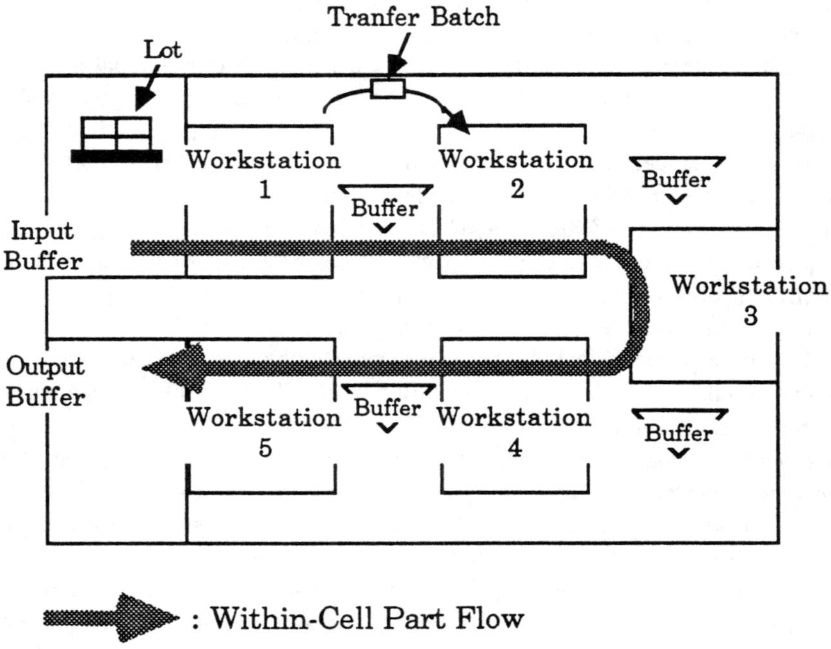

Figure 8.11. Manufacturing Workcell Diagram.

operations, the transfer batch size should be consistent for a given lot of parts across the cell. Using dedicated containers helps in counting the parts (possibly at a glance) and assists in keeping the size of the transfer batches accurate. (The effects of transfer batch size on cell performance is addressed in chapter 5.)

There is no need for using kanban cards between workstations within a cell. In general, parts and products are "pushed" within a cell. Requirements for production arrive at the first workstation in the form of work orders or production kanbans. When a number of parts equal to the transfer batch size has been processed by the first workstation, the transfer batch is moved to the second workstation, where it waits until it can be processed; from there it moves on to the third workstation, and so on until the entire lot of parts has been processed by the cell. In a cell with balanced operations, the flow of transfer batches throughout the cell is systematic without significant accumulation of WIP between workstations. When the operations in a cell are not balanced, for instance, if the third workstation cycle time is twice as long as the preceding workstation's, then WIP will accumulate in front of the slowest workstation until the upstream workstations slow down. To prevent this accumulation, the processing rate of every workstation should not exceed that of the bottleneck workstation. This can be achieved by the dynamic assignment of operators within a cell so that the bottleneck machine always runs while the other machines are operated only when needed. For example, one operator could work on workstation 3, while another operator works on both workstations 1 and 2. Limiting the size of the buffer before a workstation can serve as a signal to indicate when upstream machines should stop sending parts to a downstream machine that does not have the capacity to absorb them. For example, if the transfer batch is a container of five parts, it can be decided that no more than four containers (or twenty parts) should be present at any time in front of any workstation; thus, WIP accumulation over this limit would show production problems that need to be solved before normal production can resume. Either the upstream machines are producing at a rate that is too high, and this rate must be reduced, or the downstream machines are producing at a rate that is too slow (because of repeated jams or quality problems), and this rate must be increased (by finding and eliminating the reasons for the jams or the quality problems).

The sizes of intermediate buffers can be determined by using a trial-and-error process on the actual system or, more efficiently, by experimenting with a simulation model of the cell as discussed in chapter 5. There are various ways of limiting the WIP between workstations that are quite simple. Some of them are shown in Figure 8.12, such as kanban squares painted on the floor between the workstations (one container or pallet is allowed for each square), a limited number of carts, short conveyors that hold only a small number of parts, and gravity chutes with limit switches that signal when a predetermined numbers of parts is present in the chute.

The control of within-cell part flow, as it is described here, does not depend on

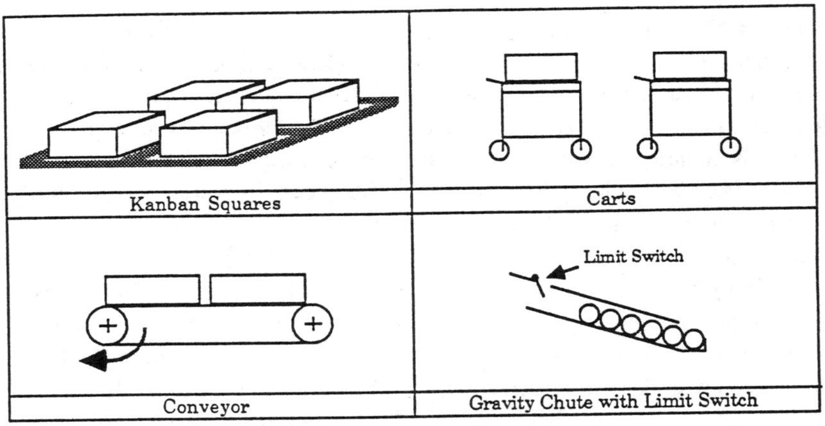

Figure 8.12. Diverse Ways to Limit WIP.

which type of PP&C system (push, pull, or hybrid) is used. In the case of a push system, cell production is triggered by work orders that are released to the cell, and in the case of a pull system, by production cards that are sent back to the cell's ordering post. Therefore, within-cell part flow control is a step that can be implemented as soon as the new cells are ready to enter into production. The next step is to adjust the control of the flow of parts between cells in an integrated system.

Cell integration and the Control of the Flow of Parts Between Cells

Cell integration involves shifting from a push system to a hybrid push-pull system or to a pure pull system for controlling the flow of materials between several cells. Cell production is triggered by individual work orders sent to each of the cells or by the need to replenish downstream buffers. Cell integration links together, as shown in Figure 8.13, manufacturing and assembly cells as well as individual workstations that have not been integrated into the cells because (1) they remain in a "machine-shop" area, or (2) they are shared by several cells. Expensive multipurpose CNC machines and heat-treating equipment are two examples of such individual workstations.

The key condition permitting part control in pull modes is *repetitiveness*—which does not necessarily mean high volumes. A typical example is the company that every few months produces products that will be sold over several months to a year. By reducing these monthly batches to weekly batches for every product, the same production pattern can be repeated every week with smaller production quantities. These repetitive production patterns can be easily regulated by Kanban systems.

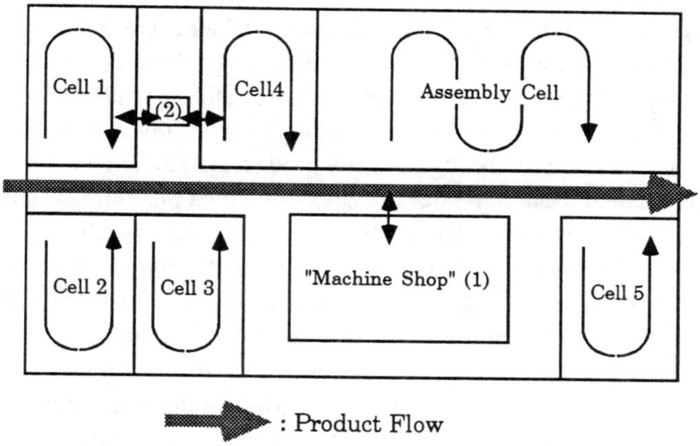

Figure 8.13. The Scope of Cell Integration.

It is clear that pull production control systems are not applicable to all production environments and that in some cases manufacturing and purchasing operations are best controlled with a push system with accurately monitored lead times. Some companies whose products are highly custom oriented or which sell one-of-a-kind products will have difficulty in identifying repetitive patterns in their production operations. However, most manufacturing companies offer a mix of generic and custom products and therefore experience some level of repetitiveness at various production stages. Furthermore, more and more products are designed to be modular, so that even though the final products are different, each is a result of a specific combination of modular elements, which when viewed separately can be repetitive. Some of the company attributes that indicate either a high or low likelihood of success for the identification of repetitive operations—and therefore for success in scheduling in the pull mode, are listed in Table 8.1.

TABLE 8.1. Company Attributes Affecting the Success of Pull Production.

High Probability of Success	Low Probability of Success
- Stable customer demand - Standardized design - Modular design - Frequent shipments are required by customers (several per month)	- Highly customized products (one-of-a-kind) - Very low volumes (less than ten of same type per year)

Once repetitive production operations have been identified, a pull system can be implemented to regulate the flow of materials through these operations. The cells (or isolated workstations that are not grouped into cells) to which these production operations are assigned can be linked together by withdrawal-kanban loops. The production in every cell can be regulated by production-kanban loops. The number of kanbans and the production rates are calculated by a planning system based on customer demand and production lead times, as explained in section 8.1.2.

Therefore, in the modified production system, cells and isolated workstations receive both production kanban cards (for pulled jobs) and work orders (for pushed jobs). The challenge for company supervisors and cell managers is to schedule these various jobs so that all due dates are met. One of the rules used could be to produce on a first-come-first-serve basis (FCFS), or another rule could be devised that would permit some degree of job combination to minimize the number of setups. (Job scheduling is further detailed in section 8.3.2 of this chapter.)

At this stage of the company's evolution, various applications of the Kanban system described in section 8.1.2 can be applied, such as signal kanbans for lot production, production kanbans with constant reorder quantity for individual cell production, and withdrawal kanbans with constant reorder quantity for regulating the flows of parts between distant cells. The next stage is to start integrating vendors in either the hybrid or pull system by controlling repetitive deliveries of purchased parts using supplier kanbans. The new challenges that are created by this integration are discussed in chapter 10.

Control procedures are the final link of the PP&C system. How well these procedures work is a result of how well the production is planned: The number of released work orders and the number of circulating kanban cards are direct consequences of the production planning. In turn, production planning follows master production scheduling (MPS) and the definition of the manufactured products described in the bills of materials that are associated with the production operations are also described in part routings. The following section will deal with the modification of these functions (MPS policy, bills of materials, and part routings) in order to adapt traditional MRP systems to world-class operating strategies.

8.2 MODIFYING MANUFACTURING RESOURCES PLANNING (MRP II) SYSTEMS

The modification of an MRP system that is imposed—or allowed—by organizational changes (such as cellular manufacturing, teamwork, and decentralization of support departments) must start with the adaptation of the major system inputs to the new operating conditions; namely, the adaptation of the MPS and the manufactured products' technical data, which are recorded in the bills of materials and the part routings. This section will describe how each of these inputs is affected by the organizational changes and how they should be modified in order to implement a PP&C system that is best able to take advantage of the new organization.

8.2.1 Adjusting the MPS Policy

The MPS is a formal production plan, established jointly by the marketing, manufacturing, and purchasing departments and approved by management. It is a periodic statement that indicates when and in what quantity final products will roll off the final stations. Consequently, the MPS dictates when and in what quantity assembled parts will be manufactured and raw materials or nonmanufactured components will be purchased. Closer and better relationships between departments—a characteristic of world-class companies—reinforce the view that the MPS is an agreement between marketing and manufacturing. In this agreement both the customers' needs and the production capacity requirements are considered.

Different time horizons cover the different functions of the MPS, as is shown in Figure 8.14. On a long-term basis a flexible schedule describes the planned production for which no action, such as material purchasing or special tooling design, has yet been taken. This long-term schedule is based on forecasted sales, and its main function is to enable the planning of future capacity. Updating this schedule has no effect on the current manufacturing or purchasing operations because it covers future production for which no purchasing or manufacturing action has been initiated. This long-term schedule is composed of periods that range from months to years, depending on the nature of the business, and on the proximity of those periods, (for example, months within a one-year horizon, and years beyond).

On a short-term basis a firm production schedule describes the production of items for which some action has already been initiated. It is based on both customer orders and forecasted sales. Updates to this firm production schedule have a direct effect on current operations. A "frozen schedule" is usually added to the firm

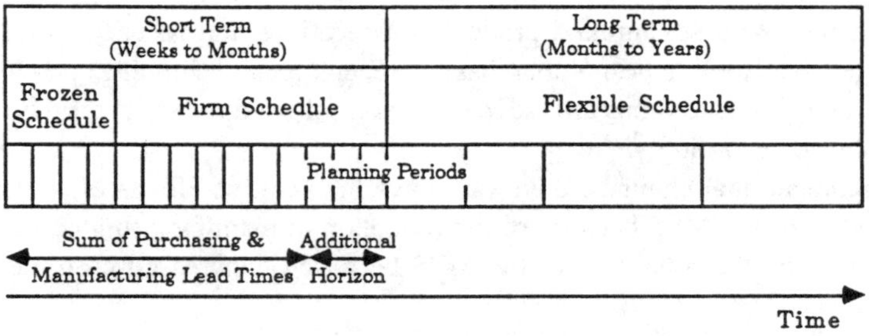

Figure 8.14. An MPS's Time Horizons.

schedule in order to limit disturbances to ongoing operations. Production or purchasing events planned in the frozen schedule cannot be modified. The length of the short-term schedule (frozen plus firm schedules) is traditionally the sum of the purchasing and manufacturing lead times with an additional horizon that allows for part grouping (such as "economic lot sizing"). The lengths of the periods that compose the short-term schedule are mainly a function of the nature of the business; commonly each period is equal to a week.

The MPS is established by first considering bottleneck resources for the planned production. Examples of bottleneck resources are a critical CNC machine, the capacity of the heat-treating facility, or the rate of delivery of a crucial purchased material.

The organizational changes (such as shortened lead times), which result from applying techniques described in the other chapters of this book, directly affect the inputs to the MPS. These changes also affect how the output of the MPS can be used to establish production plans. Although the principles of the MPS are not altered by the transition from traditional to WCM techniques, the MPS policy has to be adjusted to reflect and exploit these organizational changes.

How organizational changes affect the inputs to the MPS

Improved relationships with vendors result in a smoother flow of purchased products because the quantities received are smaller and they are received more frequently. Similarly, the improvement of relationships with customers results in a smoother demand: Shipped quantities are smaller, shipping frequency is higher, and demand is more predictable. Therefore, a more stable demand enables the extension of the horizon of the short-term schedule (including both the frozen and the firm schedules).

On the other hand, changes such as setup-time reduction, cellular-manufacturing implementation, and decentralized support departments enable the horizon of the short-term schedule to be reduced. With short setup times, parts needed across several planning periods do not need to be grouped in a single period to reach the "economic lot size"; as a result, the additional horizon for part grouping can be eliminated. With streamlined production operations and focused support operations (such as tooling design, purchasing, receiving, and shipping) purchasing and manufacturing lead times are reduced several fold and the short-term horizon can be reduced accordingly.

Organizational changes, therefore, have the positive effects of increasing the flexibility of the MPS because of the reduction in manufacturing constraints and of increasing the reliability of the MPS because of increased customer-demand stability.

In addition, preparing the MPS becomes easier because bottleneck resources are more easily identified in a cellular environment than in a traditional manufacturing environment. Work loads can be determined for each cell's bottleneck workstation, and capacity can then be adjusted accordingly for the planned productions.

How organizational changes affect the utilization of the MPS outputs and the scheduling horizons

In a traditional push system, the MPS outputs are transferred to the MRP processor, which then computes for each period the manufacturing and purchasing requirements for every level in the bills of materials. Having reorganized their manufacturing facilities and stabilized the customer demand, companies with push systems that implement pull systems need to use the MPS outputs differently. As explained in the first section of this chapter, and as shown in Figure 8.15, the outputs of the production plan are used (1) to establish a final assembly schedule (FAS), and (2) to compute the number of kanban cards required to pull the products through the different production stages (cell planning). The FAS is based on actual customer orders and is established with a very short horizon (in the Toyota example given in Section 8.1.1 the horizon was days). Cell planning is usually based on sales forecasts and occurs several planning periods before the actual customer need is known. (Again, in the Toyota example, cell planning was computed monthly, two weeks before the beginning of the month.)

Consequently, the short-term schedule is primarily affected by the organizational changes. Improvements in customer-demand prediction and the reduction of lead times allow the short-term schedule horizon to be set so that production requirements over the short-term schedule range can be converted into production rates that cover every planning period of the short-term schedule. The maximum range of the short-term schedule is limited by the time horizon for which precise customer-demand predictions can be obtained. The minimum range of the short-term schedule is limited by the manufacturing and purchasing lead times. Inside the short-term schedule, the frozen-schedule horizon is reduced to match the FAS

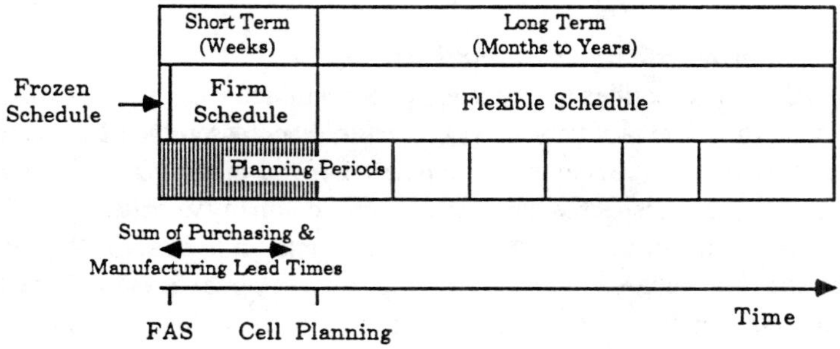

Figure 8.15. Time Horizons for Planning Pull-Production Mode.

range, both being set equal to the assembly lead time. Limited changes in the assembly of the planned end product are permitted only until the finalization of the FAS.

In addition, the range of the planning periods of the short-term schedule must be decreased as much as possible in order to ensure smooth production. For example, if the schedule covers a four-week range, planning periods could be set equal to a day. The demand for every product during the four-week range is converted to daily production rates so that the same production sequence is repeated every day. The number of kanban cards can then be set according to the production rates.

How to adjust the MPS when implementing WCM techniques

Companies that implement WCM techniques need to adjust their MPS policy. Primarily, the various schedule time horizons must be updatable as their reduction is made possible by the reduction in lead times. In addition, companies that desire to evolve from an MRP system to a pure pull system or to a hybrid pull-push system need to modify the MPS in order to (1) implement specific horizons for each product, (2) compute production rates of the products that are scheduled in the pull mode, and (3) frequently edit FASs.

8.2.2 Modifying Bills of Materials

The MPS represents planned production in terms of end items. Bills of materials describe the end item's structure in terms of subassemblies and individual components. The organizational changes described in other chapters also affect the nature of the information contained in the bills of materials, and actually enable the simplification of the bills of materials' structure. After a brief description of bills of materials and their purpose—including the utilization of modular bills—this section will explain how organizational changes affect the bills of materials structure and how the bills of materials should be accordingly modified in push, pull, or hybrid production environments.

What bills of materials are and what their purpose

A bill of materials describes the structure of a manufactured product by listing the sequence in which the products are manufactured and assembled from raw materials and individual components to end items. Bills of materials are often considered from different perspectives by the different company departments. The marketing department views the billing of materials in terms of the product's characteristics and subassemblies, which represent optional or variable features in the commercial catalog. For example, the marketing department of an office-supply manufacturer will consider the desk bills of materials in terms of table size, type of wood, color, number of drawer sets, et cetera. For the design department, the bills of materials are the lists of individual product components; a desk bill of materials would then list all the wooden parts, fasteners, coating products, and other individual components that compose a complete desk. As seen by the manufacturing department, the bills of material are even more detailed: In addition to giving the complete end item's structure, they provide information pertaining to the condition of the components and subassemblies at different stages of the manufacturing process. For example, a desk bill of materials could differentiate between a raw drawer board and the same board being machined and ready to be assembled.

From the manufacturing perspective, the bills of materials serve two purposes: (1) To give the complete end-item composition so that raw materials and individual components can be purchased and their inventory controlled, and (2) to provide an intermediate stage description for control purposes. As shown in Figure 8.16, for example, a drawer-subassembly bill of materials could be comprised of the following levels:

- Purchased-components level (the lowest level)
- Machined stage before assembly (wooden parts only)
- Intermediate assembly stage (wooden parts only)
- Painted or coated drawer
- Final assembly stage (complete drawer with wooden and metallic or plastic parts).

Each level shown in Figure 8.16 corresponds to one or several records in the data base of the MRP system. Each record serves an inventory-control purpose, and therefore the bill of materials includes only the levels at which some control must be exerted, such as where inventory is stored or where a counting point in

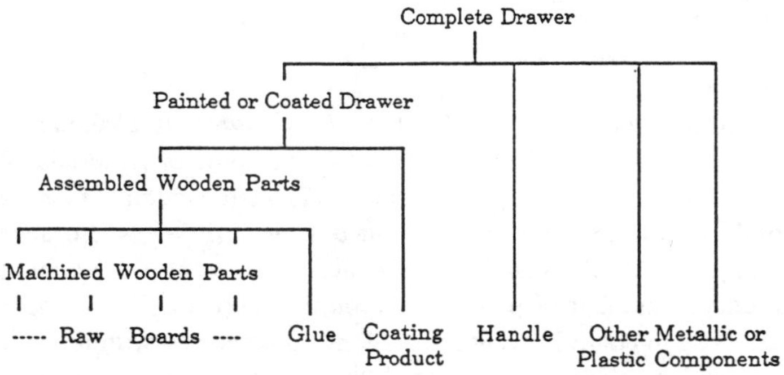

Figure 8.16. Drawer-Subassembly Bill of Materials.

the production process is required. Because the cost of implementing, maintaining, and operating the MRP system increases with the number of records, and because information accuracy is difficult to maintain when extensive data entry is required, a trade-off between the degree of control and cost of control must be established. The number of levels in a product's bill of materials is the consequence of management decisions on the degree of detail necessary in tracking inventory and WIP. The partial bill of material shown in Figure 8.16 allows very detailed control on the intermediate inventory stages, but all intermediate levels could be eliminated if no control was required on the intermediate inventory, only the lowest level (individual components) and highest level (complete drawer) could be kept.

The highest level of description in the bill of materials corresponds to the end product that appears in the MPS. However, companies that sell products with many optional or variable features do not plan their production in terms of final products, but in terms of major subassemblies, because the number of possible combinations of end items usually exceeds the total number of products that will be sold during the product's life. For example the office supply company could offer desks with a choice of many different colors, table sizes, wood quality, number of drawer sets, et cetera; however, it would not be efficient, or even possible, to maintain a bill of materials for every possible configuration. In addition, any new optional feature (for example, a PC-monitor stand) would double the number of bills. A different approach is to use modular bills for the main subassemblies (table, drawer sets, table support, and optional features). The number of modular bills is independent of the possible combinations, and the addition of new features results in minor updates of the bill of materials' data base. In addition, the sales forecast of a limited number of subassemblies is more easily established and the resulting MPS is more accurate when expressed in a limited number of subassemblies. When this approach is used, the different subassemblies are temporarily stored and the final products are assembled when the FAS is edited on receipt of firm customer orders.

How the organizational changes affect the bills of materials' structure

The main changes that will have an impact on the bills of materials' structure are (1) the cellular manufacturing organization, (2) the reduction of manufacturing and assembly lead times, and (3) the reduction of WIP levels. Because of the simplified part flows and because of the reduced time spent by parts on the shop floor and reduced number of parts present on the shop floor at any given time, there is less need to control inventory. As long as parts stay within a cell, visual control can replace the control exerted through the MRP system: The cell operators or cell manager can continuously monitor the completion stage of parts inside the cell, keeping track of the flow of parts entering and leaving the cell. In addition, since the transfer batch size inside a cell is often equal to one, it is not feasible and, furthermore, not necessary to update the MRP data base each time a part completes a production stage. Using the drawer example described in the preceding section, if all operations are now performed sequentially in the same cell that contains a saw, a planer, a sander, one or two assembly benches, and a coating station with a dryer, then the lead time to produce a drawer is equal to minutes or tens of minutes depending on the coating drying time, and different types of drawers are processed in the cell in small lots as needed by the next workstations. The only inventory information of interest is, therefore, the number of initial components available for drawer production and the number of complete drawers available for the next production stage. The bill of materials does not need to list more levels than these two stages. A primary effect of the organizational changes is a reduction in the number of levels in the bill of materials.

The reduction of final-assembly lead time combined with design standardization also affects the definition and number of modular bills. Modular bills describe modular subassemblies and optional features that are produced for stock and later assembled when firm customer orders are received. The choice of the modular bill's highest level (end item) depends on the time it takes to assemble all of the modular end items and optional features into the final product to be shipped to a customer; however, this assembly time must not exceed the delay from order taking to product delivery that was negotiated with the customer. When assembly times are reduced it is possible either to reduce the delays negotiated with the customers or to modify the end item of the modular bills in order to decrease the number of references kept in temporary storage as well as decrease the number of modular bills to be maintained. For example, consider a manufacturer of medical radiographic equipment who offers the equipment on the commercial catalog in modules (such as, patient table, X-ray source, electronic analysis module, et cetera). This manufacturer produces the modules for stock, assembles the optional features to the modules when customer orders are received, and then ships the ordered configurations. The modules thus represent the highest level of the modular bills (end items). However, if our manufacturer has been able in the past few years to highly improve the degree of design standardization, then as a result, many subassemblies internal to the modules have become common across the modules and even across product lines; for example, electronic circuit boards, electric drives, mechanical components. Today, the number of different internal subassemblies is less than the number of modules, and the final assembly time is also less than the delay accepted by the customers. These internal subassemblies can thus replace the modules as the highest level of the modular bills, and, as a result, the number of references stored in inventory is reduced along with the number of modular bills to be maintained. Thus, a second effect of the organizational changes is a reduction in the number of modular bills of materials.

How to modify the bills of materials

To reduce the number of levels in the bills of materials, careful attention must be paid to how much control is actually necessary. As a rule of thumb, as long as a flow of parts is continuous (that is, parts do not accumulate in buffer areas for long periods of time) then no control is required. In the example of the drawer subassembly if the flow of drawers can be considered continuous from the initial gathering of components to the completion of a drawer, no inventory control is required between these stages and the bill of materials does not need to include more than two levels, which is shown in Figure 8.17.

The number of levels could be further reduced if, for instance, the "drawer cell" was part of a larger "desk cell" where only the number of drawers required for a specific desk would be produced just-in-time for final desk assembly. In this case, the desk bill of materials should not include the complete drawer level since no control is required on the drawer inventory. On the other hand, if the drawers are kept uncoated (because of the numerous types and color of coating products) until the time when firm customer orders are received, then a temporary inventory of uncoated drawers (assembled wooden parts) has to be kept and inventory transactions have to be entered in the MRP system when drawers enter or leave the inventory, and then the drawer bill of materials would include an intermediate inventory stage as shown in Figure 8.18.

A way of modifying the bill of materials less radically than simply deleting records is to transform intermediate records into "phantom records." A phantom record is useful when the production process does not justify maintening a specific record in the bill of materials (such as when the part flow is continuous), yet some inventory must be maintained (such as in the case of service parts or of customer returns). For example, an electronic manufacturer might keep track of the circuit-boards inventory through phantom records if no buffer inventory is required for production yet some inventory must be kept for service purposes. The main differences between a phantom record and a regular record is that the system's treatment of the phantom record is simpler than that of a regular record (no

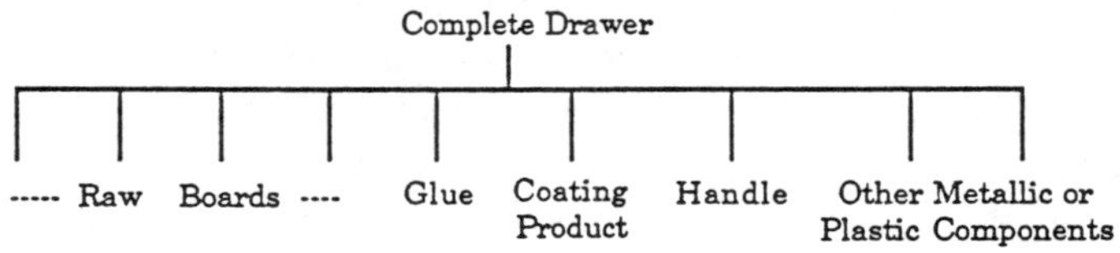

Figure 8.17. A Simplified Bill of Materials of a Drawer Subassembly.

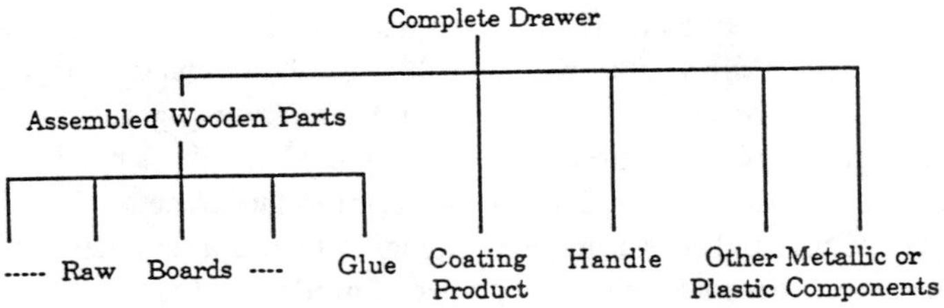

Figure 8.18. A Bill of Materials of a Drawer Subassembly with an Intermediate Inventory.

individual requirement is edited for item production) and that the lead time associated with the phantom record is set equal to zero (no delay between part completion and availability for the next production stage).

Another simplification of the bills of materials structure is possible when the supplier base has been reduced and when several parts within a product are purchased from the same supplier. One master record can be assigned to this group of parts, and individual components can be maintained in the supplier's bill of materials rather than in the company's.

These principles for determining the number of levels of the bills of materials in push scheduling systems (such as MRP) can also be applied to determining the number of levels of the bills of materials in pull scheduling systems (such as the Toyota production system). In the context of a pure pull production system, or in that of a hybrid push-pull production system, the bills of materials, along with sales forecasts, are used to plan the demand rate of individual components and subassemblies for products that are scheduled in the pull mode. The bills of materials are the basis for computing the component demand rate at every stage of the production process. They are therefore one of the inputs required to compute the number of kanban cards needed at each production stage (production kanbans), between the production stages (withdrawal kanbans), and for controlling the flows of purchased items (supplier kanbans). The number of records in the bills of materials is dictated by the same principles for both pull production systems and push production systems: A record for each interruption in the flow of parts. In both types of systems these interruptions are represented by temporary inventories: In push systems, items coming in and leaving these inventories are tallied and transactions are posted in the MRP system; in pull systems, these buffer inventories are controlled by the number of circulating kanban cards. In practice, the bill of materials includes a record for each individual part or subassembly, from individual components to end items, that enters or leaves a manufacturing or assembly cell.

The number and structure of modular bills should be determined by carefully examining existing bills of materials, which detail the products' composition, and by identifying the appropriate level of part and subassembly groupings that results in the smallest number of groupings, and, as a result, the smallest number of modular bills. The diagram in Figure 8.19 illustrates this precept. The curve shows the number of part and subassembly groupings, which is a function of the number of individual components per grouping (size of grouping). On the left-hand side of the figure, each grouping is composed of only one individual component, and the total number of groupings is equal to the total number of individual components; on the right-hand side of the figure, each grouping represents an assembled product ready to be shipped; and accordingly, the total number of groupings is equal to the total number of possible end products. The objective is to find the optimal intermediate level of grouping so that the number of groupings is the smallest, as illustrated in the figure. In the example of the medical radiography equipment manufacturer, the optimal level could be catalog modules (patient tables, for example), or internal subassemblies (electronic circuit boards, for example), depending on their design modularity and assembly times.

When modifying the structure of the bills of materials, it should be kept in mind that the groupings defined by the modular bills will be temporarily stored and later assembled as customer orders are received; therefore, the level of detail should be such that the assembly time does not exceed the delay negotiated with customers. If the delay is exceeded, a trade-off level might have to be chosen as shown on Figure 8.19.

Modifying the bills of materials primarily involves reducing the number of bill

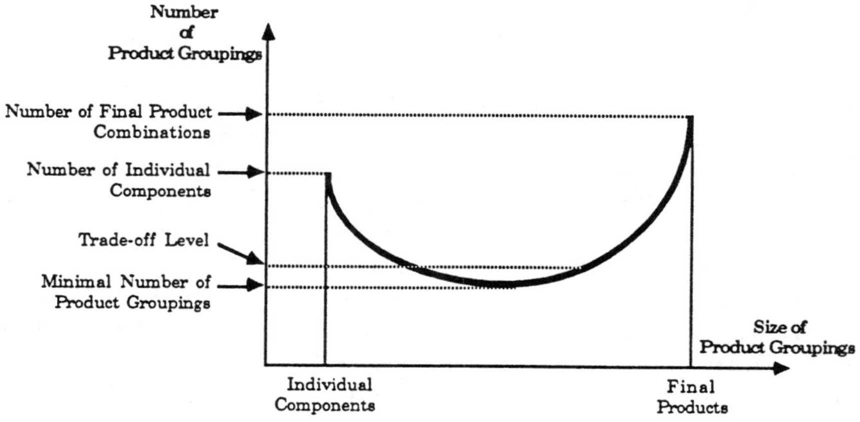

Figure 8.19. Number of Part Groupings as a Function of the Grouping Size.

levels and the number of modular bills that cellular reorganization and the resultant trimmed production lead times permit.

8.2.3 Simplifying Part Routings

Although the information contained in the bills of materials completely describes product structures, it does not describe how the components and subassemblies are progressively modified to become the final products. For example, bills of materials provide no indication of the processing time that specific manufacturing operations require, or what equipment is to be used to execute those operations. This information is given by the part routings. The part routings describe how individual components and subassemblies can evolve from one child level in the bill of materials to the next parent level. In particular, part routings include information such as which machines to utilize to transform or assemble the product components, the procedures to be used, and the time these procedures take. Together, bills of materials and part routings constitute the complete information package a company needs to define how to manufacture the products offered to its customers.

Parts routings traditionally are inputs to many different functions: To production planning, execution and control, to product costing, and also to evaluate individual worker's performance. Companies striving to achieve world-class status have modified and improved each of these functions and, hence, have reviewed and modified the structure and content of the part routings to allow these modifications. For example, when teamwork training commences, following the implementation of workcells, workers on the same team are taught to set up and run a variety of machines inside their workcell; thus, the workers can no longer be evaluated on the basis of machine-dedicated standard times that are traditionally specified on part routings. Therefore, the existing standard times have to be applied differently or new standards need to be defined. The remainder of this section will describe the effect of organizational changes on the part routings' structure and content, and will show how the routings can be simplified to conform to the company's new organization and new objectives.

How the organizational changes affect part routings

The organizational changes that have the most impact on a part routing's structure and content are the cellular manufacturing organization and the manufacturing organization's focus on products. With a traditional job-shop or functional layout, the part routings define the flow of parts through the plant by indicating for each part or subassembly the succession of operations from machine to machine until their completion or assembly with other parts or subassemblies. With a cellular organization, the flows of parts are drastically simplified by easily distinguishing two types of flows: (1) The flow of parts from cell to cell, and (2) the flow of the parts inside the cells. The flow of parts between cells needs to be planned and controlled centrally. The central production planning system must know production rates and lead times for each cell and for each part or subassembly in order to be able to prepare realistic production plans. However, each cell can be perceived as a black box by the central system, and within-cell planning and control can be performed locally, within the limits of the production conditions known by the central system. The routing structure should be modified to adhere to these different requirements.

In addition, cellular manufacturing organization casts a new light on the time standards that accompany the description of operations in the routings. Primarily, time standards have a planning purpose: They help to estimate production rates and machine capacities and, as a result, plan production output. For planning to be accurate, it is important to verify that the actual times comply to the standards; consequently, elaborate systems to verify production rates or production times have been implemented. These systems have often deviated from their initial planning objective and also have been used for control purposes, mainly to control individual workers on individual machines. This deviation has several negative consequences:

(1) Workers who are requested to follow standard times and can easily do so are not willing to exceed them because they know that eventually this would result in raising these standards.
(2) Because the standards must be "realistic" they are often based on the slowest worker's performance, which results in underutilizing the capabilities of the rest of the work force; therefore, using the standard times for control purposes deters improvements of production processes and encourages poor worker performance.
(3) When an incentive system is linked to the time-control system, workers are rewarded for producing as many parts as they can regardless of the number of parts actually needed, resulting in a double waste for the company: The waste of parts produced when not needed and the waste of the bonuses given to the employees for unproductive operations.
(4) Traditional time standards can reward only the isolated performance of individual workers as opposed to the collective performance of teams of workers.

World-class companies understand that individual piece-rate incentive systems cannot enhance their global performance and have devised different ways of achieving this purpose. Time standards should return to their original purpose—production planning. Yet, traditional, machine-dedicated time standards cannot be used in their current form to plan the duration of linked operations in a manufacturing cell. Some of the new questions that emerge in the cellular manufacturing environment are: What is the production rate of a cell when the production rates of the machines in the cell are different? How is cell setup time defined and measured? How is the variation of cycle time and setup time with the variation in the number of cell operators to be accounted for?

The following example shows how the cell setup time for a new job can fluctuate depending on the number of operators assigned to the cell, the cell's setup policy, and the processing times of the initial and new jobs. Consider a simple cell dedicated to the discrete metal cutting of small lots of cylindrical parts that is made up of three automatic machines: a lathe, a milling machine, and a drill. All parts first go to the lathe, then to the milling machine, and finally to the drill. For each of the three machines, Table 8.2 shows: The processing times per part for Job 1 and Job 2, the part unloading and loading times, and the machine setup times from the first to the second job.

The machines are automatic; therefore, manual operations involve only unloading and loading parts and setting up the machines. The transfer batch size is one, and the unloading/loading time includes the transfer of each part to the next machine. The setup time indicated in the table is the internal setup time; the external setup is performed while the machines are processing the last parts of the first job. The definition of an isolated machine's setup time, as given in chapter 6, is "the time elapsed from the production of the last part of one lot to the production of the first good part of the next lot." The same definition can be applied to manufacturing cells; therefore, the cell setup time is measured by the time elapsed from the production of the last part of job 1 by the last machine of the cell (the drill) to the production of the first good part of job 2 by the same machine. Consider, now, how this time fluctuates depending on the conditions listed in Table 8.3.

The two possible setup policies are (1) the three machines are set up when the entire cell is empty of parts, that is, after the last part of job 1 is unloaded from the last machine (drill) and before the first part of job 2 is loaded on the first machine (lathe); and (2) the three machines are set up sequentially, that is, as soon as the last part of job 1 leaves a machine, that machine is then available for setup, and as soon as a machine is set up, it is available for production. When two workers are assigned to the cell, one of them is assigned to the lathe and the other one to both the milling machine and the drill.

TABLE 8.2. Time Data.

Machine	Job 1 Processing	Job 2 Processing	Unloading/Loading	Setup
Lathe	3 min.	4 min.	10 sec.	10 min.
Milling Machine	2 min.	3 min.	10 sec.	8 min.
Drill	4 min.	6 min.	10 sec.	4 min.

TABLE 8.3. Cell Setup Conditions.

Setup Policy:	Setup Empty Cell	Sequential Setup of Machines in Cell
Number of Cell Operators:		
1	Case 1	Case 2
2	Case 3	Case 4

The cell setup time in each of the four cases is indicated on the diagrams in Figure 8.20.

Note that in Figure 8.20, the cell waiting times are a consequence of the interaction between individual machine setup times and job processing times. The cell setup times are summarized in Table 8.4. Depending on the setup policy and on the number of cell operators, the cell setup time can vary by as much as 100 percent.

This example clearly shows that although isolated machine standard times must still be mentioned on the within-cell routings for local cell planning, they cannot be used directly to determine lead times on between-cell routings. Individual machine standards cannot be used directly to estimate these lead times.

There are other effects of the cellular manufacturing organization on part routings and on how part routings are established. First, the routings become simpler because operations that were previously performed in several departments now take place in the same cell. Second, the routings are easier to establish because of the classification of parts into part families and the assignment of part families to specific cells. (Once a new part is classified in a part family, its assignment to cells and machines is straightforward, and routings for a new part can be established by modifying existing routings of a similar part from the same family.) Third, the time spent establishing the within-cell routings is reduced because the same manufacturing engineer or process planner produces routings that cover all the machines in a cell even though the machines are functionally different (as opposed to traditional routings prepared sequentially by several different people).

When the manufacturing organization is focused on products and when the focused manufacturing departments are physically moved onto the shop floor next to the cells, these effects become even more apparent. In addition, faster identification and rectification of routing errors occurs when the routings producers are moved to a location next to the users of the routings, and, as a result, the quality of the part routings improves.

How to simplify parts routings

Because within-cell routings and between-cell routings each hold different information and serve different purposes, distinguishing between the two routings simplifies each regarding its information and purposes.

The information contained in within-cell routings is similar to the information

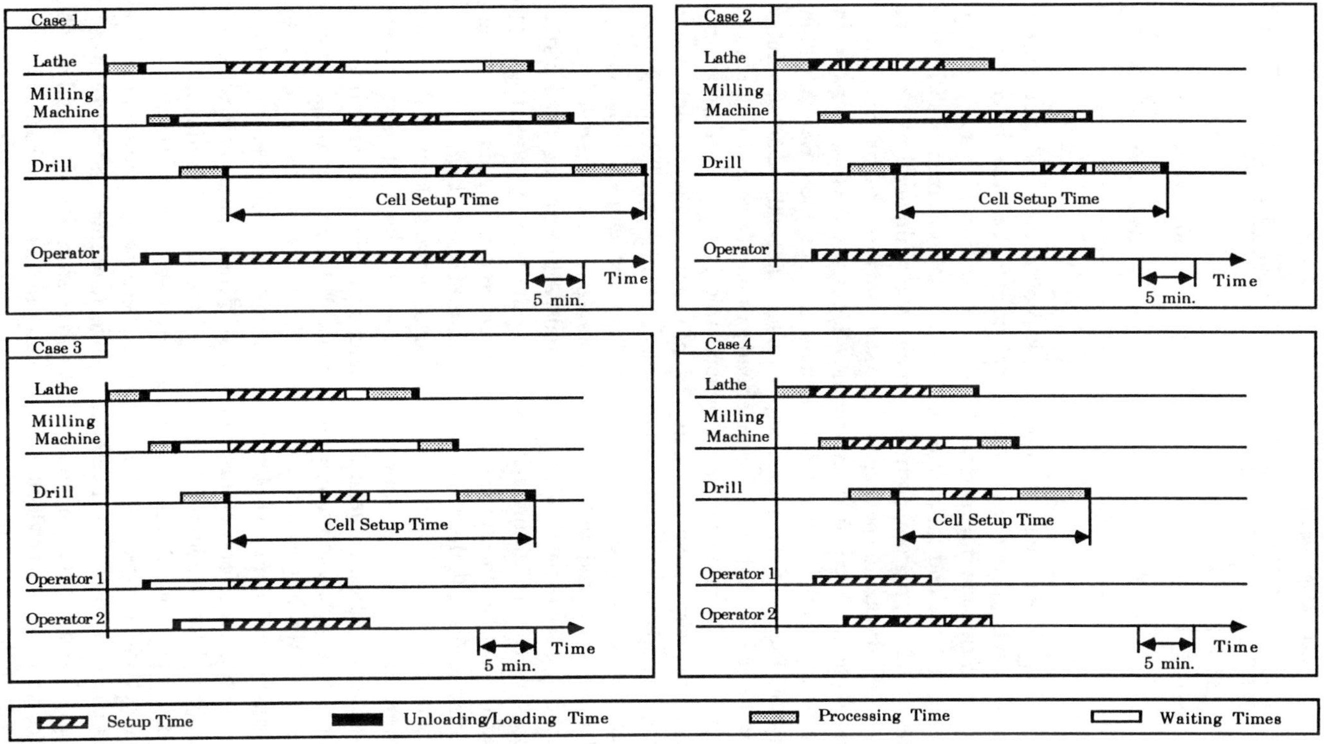

Figure 8.20. Cell Setup Time Under Four Different Setup Conditions.

TABLE 8.4. Cell Setup Times.

Setup Policy:	Setup Empty Cell	Sequential Setup of Machines in Cell
Number of Cell Operators:		
1	Case 1: 35 min. 30 sec.	Case 3: 22 min. 30 sec.
2	Case 2: 25 min. 30 sec.	Case 4: 17 min. 10 sec.

contained in traditional routings; however, it is structured and used differently. This information is created, used, and maintained at the cell level by cell managers and focused manufacturing personnel. The description of manufacturing and assembly operations and the reference to additional documents (specifications, setup sheets, CNC program) form the core of the within-cell routings, and are targeted toward production personnel. Time standards are used by the cell manager to plan the within-cell operations. In within-cell routings, the first operation should be the cell-entry operation and the last one the exit operation. Hence, within-cell routings should start at the point where parts enter a cell and end at the point where parts exit the cell. Recall that these enter and exit points were also used to define records in the bills of materials. A general rule is that a within-cell routing corresponds to every parent record in a bill of materials. For example, the three-level bill of materials for a wooden drawer, depicted in Figure 8.18 is composed of two parent records: The complete drawer and the wooden-parts subassembly, two levels that are associated with temporary storages. Two within-cell routings are consequently associated with this bill of materials: A first routing that describes the operations involved in cutting and assembling the wooden parts, and a second routing that describes the coating operation and assembling the remaining purchased components with the coated wooden-parts subassembly.

Between-cell routings, which link within-cell routings, describe the groups of macro-operations that are performed across several cells. Each macro-operation covers an uninterrupted sequence of operations performed at an individual cell thereby creating a one-to-one correspondence between macro-operations and within-cell routings. Between-cell routings are centrally used to plan cell capacity and to estimate end-item completion time; therefore, they need to include the following information for each macro-operation: The required cell capacity (machine hours and labor hours) as a function of the desired production rate, the estimated cell lead time (setup and processing times), and the estimated waiting time in the cell input storage location. However, because this type of information is a result of such variables as the interaction between machine and worker availability, the mix of products being manufactured, and lot sizes, it is also highly dynamic. Therefore, how are cell production rate, setup time, and processing time determined? Although a cell's production rate can sometimes be estimated by the production rate of the cell's bottleneck (if the bottleneck is fixed), there is no clear-cut answer to any of these questions. One approach to realistic production planning is to identify the lowest and highest bounds for these variables as a function of factors such as product mix, lot size, and worker assignment. By using simulation, for example, and by recording the resultant information in the between-cell routings, actual production rates and lead times can be estimated at the time of production as a function of product mix, lot sizes, and operator assignment for each cell. Furthermore, this information could be used by the cell production planner to assign more or fewer operators to cells relating to the final output. In any case, this variability shows an imperative need for an information system that is flexible in the way the information can be used and also in the way the information can be updated.

The structure of a between-cell routing is similar to the structure of a bill of materials because the between-cell routings link together within-cell routings, and there exists one within-cell routing for each parent record in the bills of materials. Initial levels of a between-cell routing are the lowest parent levels of the bills of materials, which then correspond to the first macro-operations performed on purchased materials. Figure 8.21 is a graphical representation of the structure of the between-cell routing of the drawer subassembly previously considered.

In between-cell routings, lead times are associated with the links (waiting times) and with the macro-operations (setup and processing times). The total lead time of an end item is represented by the longest lead time of any of the paths from one end of the routing to the other.

The modification of the part-routing structure mirrors the physical reorganization of the activities on the shop floor. Differentiating between within-cell and between-cell routings enables each to include only the information pertinent to local cell planning or to central production planning respectively and thus results in routing simplification. The onset of teamwork leads to the obsolescence of individual standards being used for worker evaluation, and standards specified on part

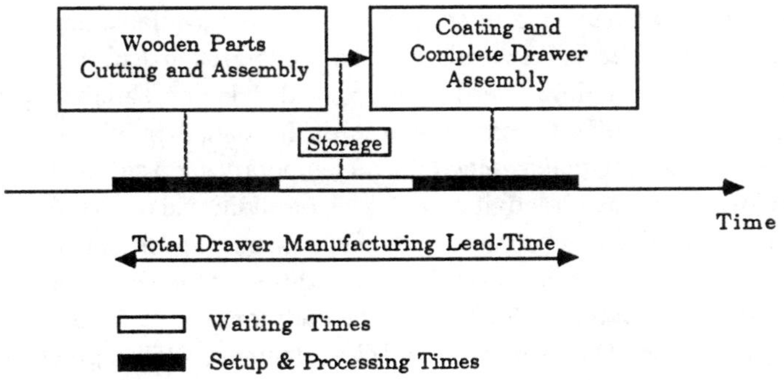

Figure 8.21. A Between-Cell Routing Structure of a Drawer Subassembly.

routings return to their primary planning function. The shop-floor activity reorganization also enables a simplification of the structure of the bill of materials: The number of records are reduced and the number of modular bills to maintain is also reduced. Bills of materials and between-cell routings have corresponding structures. The MPS policy needs to be adapted to the improved production environment, time horizons should be easily updated as lead times are continuously reduced, production rates should be often computed for each product and part scheduled in the pull mode and the FAS should be edited frequently. The modifications of part routings, bills of materials, and the MPS policy are prerequisites to the implementation of simplified production-planning and scheduling systems.

8.3 SIMPLIFIED PLANNING AND SCHEDULING SYSTEMS

Having adapted the major inputs of the PP&C system to the new manufacturing organization, the next phase consists of adapting the way the system actually plans the production (namely, the way the system adjusts work load and production capacity and the way individual part production and assembly is scheduled) so that the final products' assembly dates conform to the master production schedule.

8.3.1 Planning Capacity

Capacity planning follows the preparation (or update) of the MPS. In this section, the role of capacity planning will be discussed, along with its major inputs and outputs and then how capacity planning can be modified to best take advantage of the organizational changes will be described.

The Role of Capacity Planning, its Inputs and Outputs

Planning capacity is an important management function of any production planning system. Its purpose is to ensure that the load resulting from the customer orders and the sales forecast is in accordance with the available capacity of the production system. If the load exceeds the capacity, then such decisions as building a new plant, subcontracting, overtime work, and delaying shipments, must be made. On the other hand, if the capacity exceeds the load, then other choices must be made, such as closing plants, scheduling a temporary shop shutdown, reducing the number of subcontracted activities, or taking on additional orders. As suggested by the numerous and diverse choices that can be made, capacity planning is done with different levels of precision associated with different time horizons depending on which path is taken. Figure 8.22 shows the traditional three levels of precision in capacity planning: (1) long-term capacity planning, (2) rough-cut capacity planning, and (3) detailed capacity planning.

Long-term capacity planning is performed on the basis of the sales forecast with

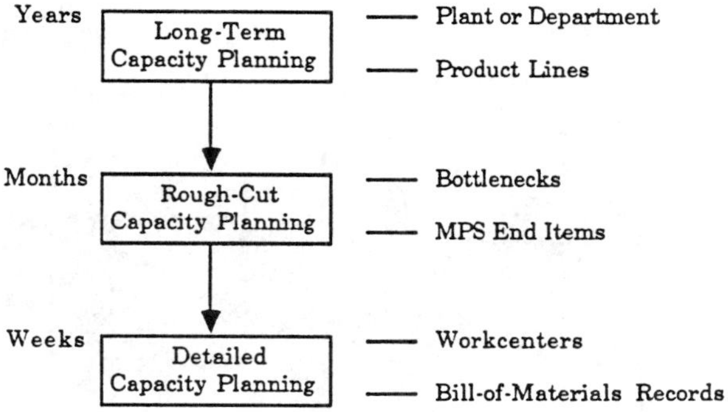

Figure 8.22. Traditional Levels of Capacity Planning.

a time horizon of one to three years at the production-unit or large-department level for each major product line. Its output can determine, for example, whether a plant should be remodeled to increase or reduce the number of departments or whether a major activity should be subcontracted. Rough-cut capacity planning is based on both customer orders and the sales forecast. Having a time horizon of several months to a year, it is performed at the department level for each end item on the MPS. In rough-cut capacity planning, the load resulting from the prevision in end-item production is compared to the available capacity of bottleneck equipment. This comparison results in the need to make decisions regarding the delays that can be negotiated with customers, the need for subcontracting or for obtaining more orders, and the size of the work force that will be required. Detailed capacity planning scrutinizes the load for each record in the bill of materials for every work center and typically has a time horizon of weeks. Its output shows the resulting load for each work center. For example, the diagram on Figure 8.23 shows the load resulting from the production of four different items at a work center. Again, decisions need to be made when work centers appear to be overloaded or not loaded enough. Rigorous capacity planning ensures that due dates can be met for all products upon the arrival of work orders on the shop floor.

Long-term capacity planning is not affected by the organizational changes. Although the composition of departments might evolve when manufacturing activities are focused on product lines, the process of planning long-term capacity and the precision in that planning are unaffected by the reorganization. Therefore, the remainder of this section will focus on rough-cut and detailed capacity planning functions.

For rough-cut capacity planning, the information traditionally maintained is a list of bottleneck processes or machines, and for every item on the MPS, a time-phased load profile on these different bottlenecks (see Figure 8.24). The information normally maintained for detailed capacity planning is a list of the

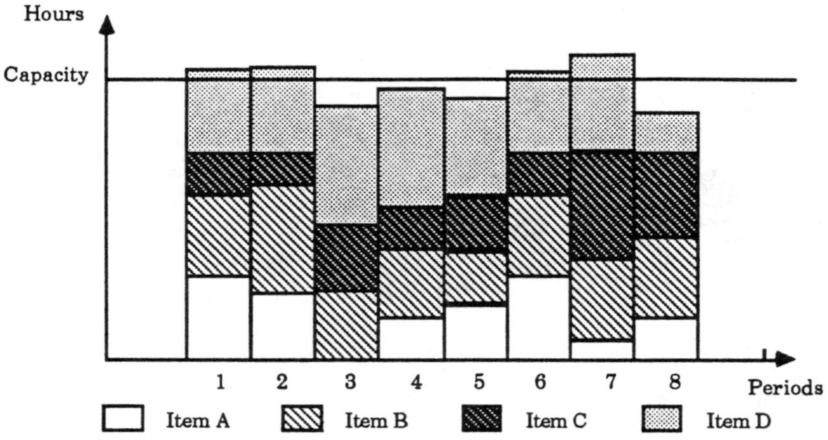

Figure 8.23. Comparison of a Bottleneck's Load and Capacity.

specific work centers used in each production process, detailing the equipment and the manual stations involved in each production process and a time-phased load profile on these work centers for every record in the bills of materials.

These load profiles are then superimposed in order to easily compare the load capacity of the work centers and bottlenecks, as shown in Figure 8.23. The challenge for production planners is to correctly identify bottlenecks and define work centers so that the partial information provided by the synthetic analysis of bottlenecks and work centers is representative of the entire production system. The major problem is that in a traditionally functional or job-shop organization, bottlenecks shift depending on the current product mix in the shop. An additional difficulty is that if the definition of the work centers—the grouping of several machines into single work centers—is too broad, it can mask possible bottlenecks, and if the definition is too restrictive it can create artificial bottlenecks. Incorrect

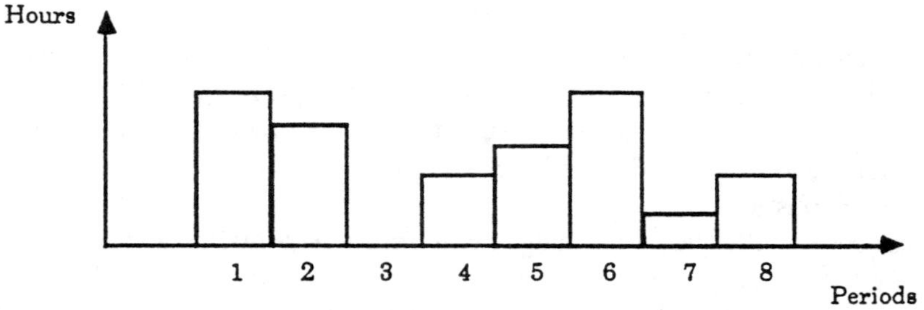

Figure 8.24. A Time-Phased Load Profile of an MPS Item.

bottleneck identification or inappropriate work center definition lead to the over- or underestimation of the shop's capacity.

The move to a cellular manufacturing organization and the redefinition of the bills of materials and part-routing structure demand a revision of the input information to the capacity-planning function eventually leading to an overall simplification of this function, along with increased precision in planning.

How the Organizational Changes Affect Capacity Planning

The first effect of cell implementation is that rough-cut capacity planning is simplified by being performed at the cell level. The time-phased load profiles for MPS items that were established for each bottleneck are now established for each cell on the basis of the information contained in the between-cell routings, as illustrated in Figure 8.25. Remember that the between-cell routings describe for every MPS item its time-phased progression from cell to cell (or to a work station not included within the cells), with indication of the load on the cell. Identifying bottlenecks on which the load is measured is a much simpler task for a cell with a limited number of machines than for a machine shop because cell bottlenecks are generally identified when the cells are designed. A second effect of cellular organization, which is attributed to the reduction of lead times, is the increased accuracy of the load profiles, which in turn, results in improved capacity planning.

Another effect of cellular manufacturing and of the reduction in lead times is

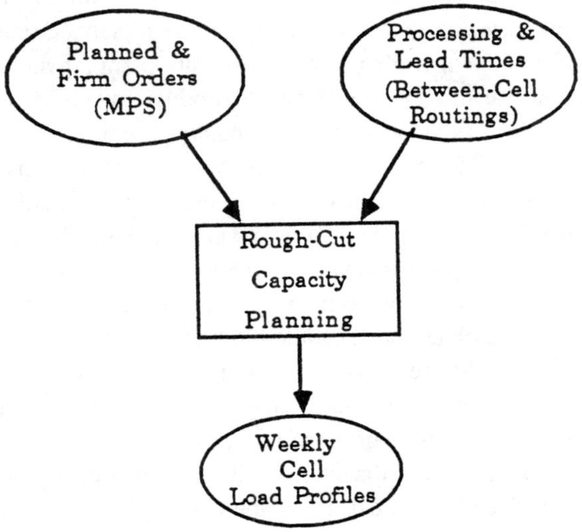

Figure 8.25. Load Profile Preparation.

the obsolescence of detailed capacity planning at the work center level. Several reasons are in favor of suppressing detailed capacity planning. The first reason is that cell capacity is determined by the capacity of the cell's bottleneck; the processing rates of the other machines in the cell are adjusted according to the bottleneck's processing rate. Therefore, as long as the capacity is sufficient on the bottleneck it is also sufficient on the other stations in the cell. A second reason is that, because of the very short within-cell lead times (hours to days), the amount of information that would be entered to the production planning system regarding the orders currently in production (open orders) would be enormous and not worthwhile for the little additional insight provided on how to plan current production. Furthermore, if the planning period is kept equal to a week, as in almost all PP&C systems, then the information provided by the system would be of little use for planning daily or hourly operations. Another difference between traditional and cellular manufacturing is that in cellular manufacturing batches of parts are simultaneously processed by several machines (it is not necessary to wait for the entire batch to be processed by one machine before transferring the first processed parts to the following machines), thus the load profiles overlap. Taking into account this overlap would further complicate detailed capacity planning, again without providing any additional value to planning cell operations. There is no need, therefore, for detailed capacity planning in cellular manufacturing.

Does this mean that only the rough-cut capacity planning performed several months before production and partly based on the sales forecast is sufficient to ensure that due dates on firms orders can be met? Most likely not. The rough-cut capacity planning must be updated as planned orders become firm orders, as new firm orders are taken, and as customers cancel or report planned orders. Depending on the nature of the business, this updated capacity planning should be prepared sometime between one and four weeks before production and be based as much as possible on firm orders. Like the rough-cut capacity planning, the updated capacity planning is performed at the cell level on cells' bottlenecks or on individual stations that are not integrated into cells. If necessary, or determined valuable, an additional update of the capacity plan can be performed when all planned orders have become firm orders. In hybrid systems, capacity plans integrate items that are pulled through the plant and are scheduled in daily or weekly rates with those items that are pushed through the plant and scheduled in lots.

Another significant difference between traditional and cellular organization is that the cell cycle times can be adjusted to the desired production output by changing the assignment of operators. For example, a cell composed of five machines could produce 150 shafts each day if manned with two operators, or 200 shafts could be produced if the cell is manned with three operators. Thus a new view of capacity planning is made possible by cellular manufacturing: Not only can the load be adjusted to the capacity, but the capacity can be adjusted to the load.

In practice, being able to adjust capacity to the load means that a range of processing rates, which is dependent on the number of cell operators, is associated with each cell in the between-cell routings and that this information can be dynamically used by the capacity planning system to ensure that due dates are met—the only constraint being the amount of available work force. Production supervisors and cell managers use this information to assign the adequate number of operators to each cell, to plan trainings, and so forth. Cellular manufacturing and cross-training operators (within cells and across cells) increase the flexibility of capacity adjustment to the required output from that of traditional manufacturing. Figure 8.26 summarizes the main evolutions in capacity planning.

Improvements in manufacturing operations and their effects on lead times must be fed back regularly to the production planning system so that these improvements can result in a reduced overall lead time for future production and not in an increased part or product waiting time between operations or between cells. However, production plans must be realistic; using unrealistic, unmatchable due dates as a means to place pressure on the shop-floor workers, does not motivate them to improve the production process—it only achieves to degrade their working conditions. Sound capacity planning paves the way to simplified cell scheduling.

8.3.2 HOW TO SIMPLIFY SCHEDULING

Scheduling entails solving the following problem: Given a set of jobs and a group of machines, determine job to machine assignments that enable the job due dates to be met. A subproblem of job/machine scheduling is job sequencing: Given a queue of jobs waiting in front of a machine or workstation, determine a sequence that enables the job due dates to be met. Scheduling and sequencing are often undifferentiated; yet, although they aim at the same objective of meeting job due dates, their variables are different, such as the job and machine assignments in the scheduling problem, and the job sequences on predetermined machines in the sequencing problem.

Sophisticated mathematical techniques exist to solve these problems, such as network analysis, linear programming, et cetera. These techniques have been applied with success to entirely automated production systems such as flexible manufacturing systems (FMS). However, in almost all companies, the people who schedule and sequence jobs are schedulers, production supervisors, foremen, and machine operators: People who tend to rely on their experience of previous conditions in the shop. For example, a group of similar jobs is sequenced on a machine because this sequence reduces the setup time, or a job is assigned to machine A rather than to machine B because production problems were encountered the last time it was assigned to machine B or because the current operator on machine A is the one who worked on the last lot of similar parts. In some companies, scheduling and sequencing algorithms are embedded in dedicated

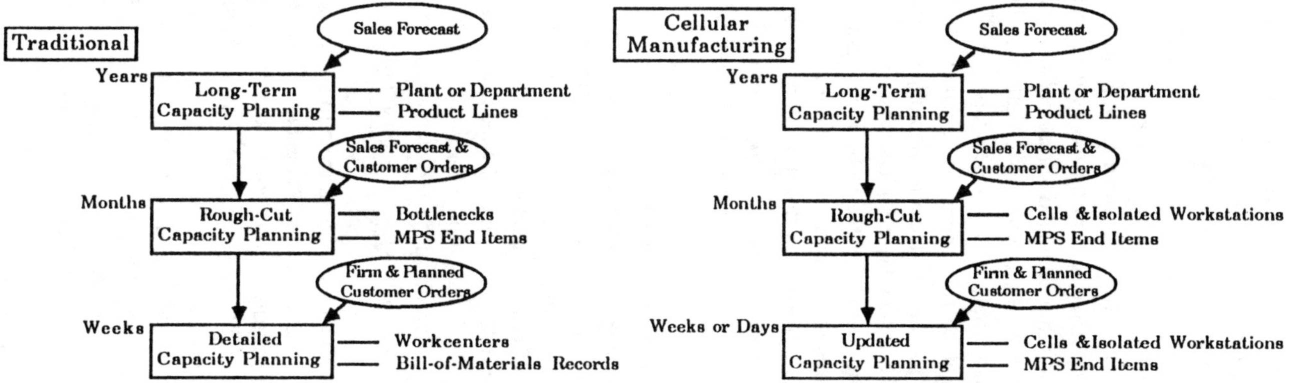

Figure 8.26. Evolution in Capacity Planning.

modules of the central MRP system. The system's outputs then consist of sequences of jobs for each machine. However, the observation of actual situations shows that these sequences are rarely followed on the shop floor as jobs are often resequenced, even manually reassigned to different machines. There are two main reasons for this inconsistency between the central information system and the actual operations, both of human origin. First, when a computer gives solutions that do not lend any flexibility to the users (shop-floor personnel), and the users do not understand how these solutions are obtained and do not find them satisfactory, then the workers might modify the solutions to correct what they consider unsatisfying. (For example, the job sequence A-B-A given by the computer would require the machine to be setup twice for job A, thus the sequence A-A-B is chosen by the machine operator, which actually results in missing job B's due date.) Second, and this is often a more severe reason, the inconsistency in output results from the inconsistency in input. When the management of input data (setup and processing times, lead times, lot sizes, machine capacities) is left to remote support departments, it cannot be kept up to date directly by the users (shop-floor personnel). Progressively these users lose confidence in the information system, thus leading the flow of information from the shop floor to data processing to become thinner and thinner, and the gap between the information in the system and the real conditions on the shop floor to grow larger and larger, until finally the system is simply not used on the shop floor.

The organizational changes that most affect scheduling and sequencing are the implementation of cells with streamlined part flows and increased visibility, the minimization of setup times, and the decentralization of support operations.

Implementing cells almost completely eliminates the scheduling problem. The job/machine assignment problem is solved for the most part when cells are designed, which consists of assigning groups of parts to groups of machines (the cells). The assignment of new jobs to cells is straightforward because it is based on part similarities, such as when cells are designed to process GT families of parts or when cells are designed to process specific products. The assignment of jobs to machines that are inside cells simply follows the within-cell routings, and the machines are set up as they become ready to process the transfer batches waiting in front of them. Similarly, the scheduling problems for assigning jobs to machines that are not grouped into cells is drastically reduced and in most cases completely eliminated. A typical reason for which a machine could not be integrated into a cell is that it is a single multipurpose machine on which must be processed parts that are assigned to different cells. Because this type of machine is usually unique, it does not cause a scheduling problem. In addition, even if similar isolated machines originally had similar processing capabilities, it is common, in order to reduce setup times, to constrain their capabilities to specific types of parts. (For instance, each of five punch presses could be assigned specific dies depending on the die height; thus parts would automatically be assigned to one of the presses depending on the height of their die.) When some scheduling problems remain (such as several cells or isolated machines processing a given job), the reduction in the number of possible options drastically simplifies the problem. Because scheduling cannot compensate for missing capacity, the most beneficial time to solve the job/machine assignment problem is not after the jobs have stagnated on the shop floor, but at the capacity-planning stage. Furthermore, part routings must indicate the specific cells and isolated machines to which to send the jobs. Thus, the scheduling problem is not only simplified, but also eliminated at the execution level. This simplification is illustrated in Figure 8.27.

Sequencing is also simplified because instead of defining appropriate job sequences at each machine, sequencing is executed only at a cell's first workstation and at isolated machines. Remember that lots wait only in front of cells or isolated workstations, and within cells the parts are moved in small transfer batches. An example of a sequencing problem is shown in Figure 8.28. Arriving consecutively in front of a cell, jobs A1, B1, A3, and A2 wait to be processed by the cell. A1, A2, and A3 belong to part family A, and B1 to part family B. The cell setup time is

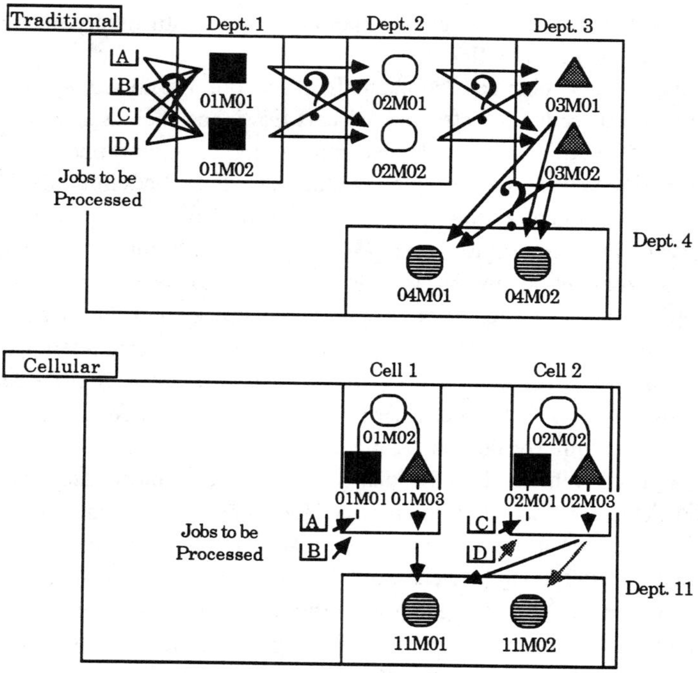

Figure 8.27. Job/Machine Assignment in Traditional and Cellular Environments.

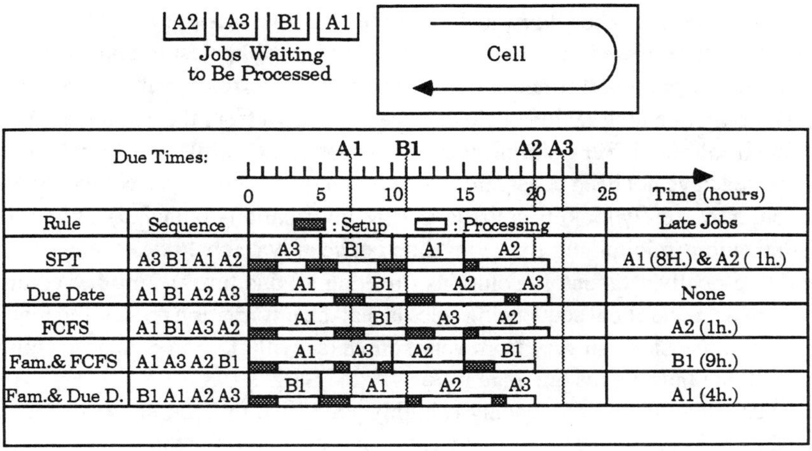

Figure 8.28. Example of a Sequencing Problem.

equal to two hours when changing jobs from family A to B (or B to A) and is equal to one hour when changing jobs within families. Job processing times by the cell and due times are given in Table 8.5. Consider that the job being finished by the cell belongs to family C and that, as a result, an initial setup time equal to two hours is required, regardless of which sequence is chosen.

Various sequencing rules are shown on the diagram in Figure 8.28. The sequencing rules dictate the choice of the next job to process in the cell according to the following criteria: shortest processing time (SPT), earliest due time (Due Date), FCFS, grouping all jobs from the same family (Fam.).

The various job completion times, that depend on the chosen sequencing rules, are also illustrated in Figure 8.28. In addition, the lateness of the various jobs, if applicable, is shown.

In this example, only one of the rules enables all due times to be met: the

TABLE 8.5. Time Data per Job.

Job	Processing Time	Due Time
A1	4 hours	In 7 hours
A2	5 hours	In 20 hours
A3	2 hours	In 22 hours
B1	3 hours	In 11 hours

due-date rule. All other rules, including the ones that attempt to minimize setup times, result in missed due times. The rule that is the simplest to apply—FCFS—shows rather good performance compared to the other rules; only one job is late, and by only one hour. Two conclusions can be drawn from this example: (1) This problem, although very simple (new jobs that arrive while the initial jobs are processed have not been considered), does not have an immediate solution that can be found by merely looking at the jobs' time information (Table 8.5), and (2) local optimization (minimizing the setup time between two jobs) does not necessarily lead to globally satisfactory solutions (meeting all due times). Another comment that can be made about sequencing rules is that the information processing required to make a decision can vary from very simple (as with the FCFS rule) to complex (as with the family plus due-date rule).

Thus, although the sequencing is highly simplified by grouping machines into cells, it still must be accomplished at the cell level and for those isolated workstations that are not integrated into cells. The most advisable rule is still FCFS. This rule ensures the smoothest part flows across the entire production process because the part flows are not disturbed by reassigning job priorities. It is obvious that this rule can only be applied effectively if there is enough capacity to allow the setups to be performed; that is, if sufficient time for setups has been reserved at the capacity-planning stage. The FCFS rule is the primary rule for pure pull systems where part flows are strictly regulated by kanban cards. Because the cards are taken from the ordering post for production in the same order as they were placed on the post, following this rule guarantees that downstream inventories are regularly replenished.

The most favorable conditions for applying the FCFS rule are a stable production plan; small lot sizes; a reliable production process, both in terms of part quality and machine reliability; and short setup times. If this last condition is not met, some degree of family grouping can be performed to avoid exceeding the machines' capacities. However, it is more efficient in the long run to attack and resolve the problem of long setup times rather than to struggle with implementing sequencing rules.

When the production plan is neither repetitive nor stable, creating the need for sequencing to be determined dynamically, there exists an interesting alternative to relying on a central scheduling system. The alternative is to provide production supervisors or cell managers with local computer-based tools to enhance their decision-making process. An example of computer-based tools are cell simulators that can be fed with the daily job schedule issued from the central system (MRP, for example) and allow the various job sequences or job sequencing rules to be tested with the current daily conditions (operator assignment, machine conditions), and the most feasible sequences to be identified. In the case of an unfeasible production schedule, the information should be "fed back" to the central system for an update of the estimated completion dates. Using these tools presents several advantages. First, they apply state-of-the-art techniques to deal with the remaining complexity of the sequencing problem. Second, this type of tool shows whether sequences chosen by the user are feasible, instead of merely providing "the solution" to the sequencing problem. The user can thus select the most satisfying sequence or, if no sequence appears to be feasible, modify the production conditions (operator assignment, for example) to find a feasible sequence. Third, the control of the production schedule and operating conditions is left to local production personnel, who are the most aware of current production conditions. Note that using these simulation tools at the shop-floor level is realistic only if the problem to be solved is simple enough to keep simulation times low,thereby allowing fast decision making. It is, therefore, only after grouping machines into cells and having simplified any resulting sequencing problems, that these local tools can be used for real-time decision making (in minutes). The simulation of an entire machine shop is too complex to be performed on tools simple enough to be used by production personnel on the shop floor.

Scheduling and sequencing have the common objective of meeting due dates assigned to jobs. Job due dates are based on available capacity and on the final item's due date. The overall performance of the production system depends not only on the ability of isolated cells and production departments to meet assigned due dates, but also on the ability of the production-planning system to assign intermediate due dates that will result in the least amount of intermediate inventory and will enable the final items' due dates to be met. This is the problem of coordinating final assembly to the upstream cell's operations.

8.3.3 How to Coordinate Final Assembly to Upstream Operations

Company A is an important manufacturer of large industrial electric motors; for example, motors used to continuously rotate ovens in cement making plants. Company B is a supplier of kitchen appliances; its products can be found in almost every household. In company A, products are made to order and every design is different in order to fulfill the specific customer needs. The motors come in many different sizes and specifications; the major subassemblies are not modular; only some purchased parts, such as electrical connections, are found in more than one design. In company B, the products are shipped to department stores and small retail stores across the country. Customer demand is predictable, and, although the products are offered in several different models, a high level of component modularity allows for a stable production process. These two examples show extremes in repetitiveness of production operations: From no repetitiveness to a high level of repetitiveness. It is clear that very different systems are used to plan the final assemblies in these two companies. These examples will be used to illustrate two approaches for matching intermediate production operations to final assembly. One approach is called backscheduling. The second is the mixed-model assembly. A third possibility is to combine the two.

Backscheduling

Backscheduling is the traditional approach used in the MRP system. A due date is assigned to every record in the bill of materials by subtracting the estimated assembly or production time of the next parent record from the parent due date. For example, consider Figure 8.29, which illustrates the assembly of the rotor and stator of an electric motor. Because the assembly time is three days and the final due date is day 20, then each of the subassemblies and components are due on day 17.

Depending on the degree of the product's customization, the backscheduling scheme can include tests and controls, assembly, production, transfers, subcontracting, purchasing, tooling design and manufacturing, process planning, and product design. The accuracy of the backschedule and, eventually, customer satisfaction, depends only on the correct estimations of the various lead times. Underestimated lead times result in subassembly and final-assembly delays. Overestimated lead times result in wasteful inventories or poor productivity, both of which result in higher costs than necessary. To keep abreast of its world-class competitors, company A—the industrial motor company—needs to tightly control all lead times, in particular the production-related lead times that are included in

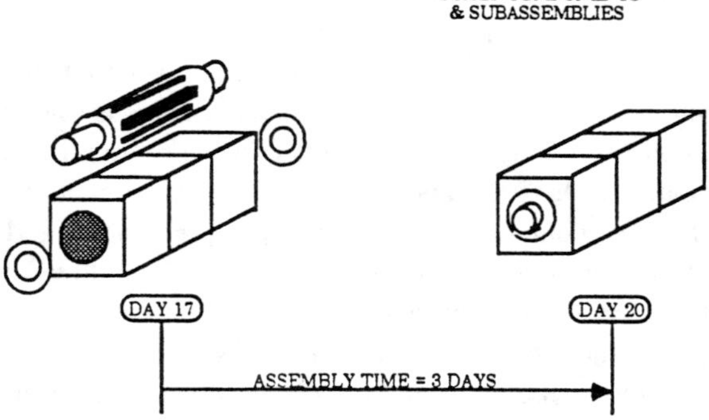

Figure 8.29. Backscheduling a Subassembly Using Due Dates.

between-cell routings. To enable detailed planning and control of production operations, a daily time bucket is the minimum range for most of the businesses. Some companies that offer nonrepetitive, customized products that have very short production lead times need to increase their planning precision by using smaller time buckets, such as one shift or an hour. Note that backscheduling does not necessarily generate large intermediate inventories and long waiting times. If correctly planned, the different production operations should be chained without provision for waiting times between operations. This absence of intermediate inventory is facilitated by the very small lot sizes resulting from a shipping quantity of a single product.

Company B can also plan its production of kitchen appliances using backscheduling. However, because tens of products roll off the assembly lines everyday and because every single one of these manufactured or purchased parts would need a different reference number, it would neither be useful nor realistic to backschedule the production operations for every product. Furthermore, the timing of every single production would have to be tightly followed; for example, body number 6789 is due on day 1 at hour 7 to be assembled with other components to form the final product number 16789 which, in turn, is due on day 1 at hour 7.5; similarly, body number 6790 is due on day 1 at hour 7.1 to be assembled with other components to form the final product number 16790 , which in turn, is due on day 1 at hour 7.6. Not only is it unrealistic to try to backschedule such a quantity of products, it is not even useful. Because the operations are repetitive, it is not important to know which specific body is used in final assembly of product number 16790 or to know when that body was made, as long as a body is available when product number 16790 has to be assembled.

For this reason, the traditional approach has been to assemble and produce in lots and to backschedule the production of lots. Consider the food-processor line at the kitchen appliance company. There are eight models of food processors, labeled A through H, and a lot of each model is assembled every month. Because model A's monthly production is 1,000 units, production and purchasing of its components and subassemblies could easily be backscheduled every month for a batch size of 1,000 and thus have 1,000 motors, bodies, switches, electric connections, and so forth, ready on D Day for assembly. Clearly, this approach is against all precepts for WCM: Lot sizes would be large, inventories high, lead times long, et cetera. An evident solution is to reduce the final-assembly batch size, shifting from monthly lots to weekly lots. Model A could then be assembled in lots of about 250. A better solution is to produce every product continuously and thereby replace the backscheduling system by a replenish system, which is controlled by the rate of consumption at the assembly line. When sales volumes do not justify dedicated assembly lines for each model, the mixed-model assembly line is an option.

Mixed-Model Assembly.

The basic idea that supports mixed-model assembly is that even if customer orders are erratic and unpredictable, the assembly sequence can still match the customer orders. To ensure the smoothest flow of components through the plant, the assembly sequence can be adjusted so that the rates of consumption of individual parts are as constant as possible (for example, one model A can be assembled every twenty minutes, one B every twelve minutes, one C every thirty minutes, one D every three minutes, and so forth). Thus, the assembly sequence would be: DBDADBDC...and not AAABBBBBCCCDDDD....

Contrary to the backscheduling approach, which enables the scheduling of each upstream operation such that the components arrive just in time to the next downstream production process, the mixed-model assembly approach creates an assembly pattern that automatically adjusts the rate of repetitive production operations to the rate of product consumption by the customers. Note that the mixed-model assembly requires some level of intermediate inventory. Figure 8.30 illustrates, for example, that parts for each of the eight models must be continuously available at the assembly line. Knowing, checking, and reducing the lead times is crucial when applying the mixed-model assembly pattern because the shorter the lead times, the smaller the intermediate inventories. Note also that the mixed-model assembly is more advantageous than small-batch assembly for keeping smooth production flows through the plant, even if finished products are shipped in batches (for example, even in the case of a truck leaving the plant every four hours with twelve model A's, twenty B's, eight C's, and sixty D's). Companies

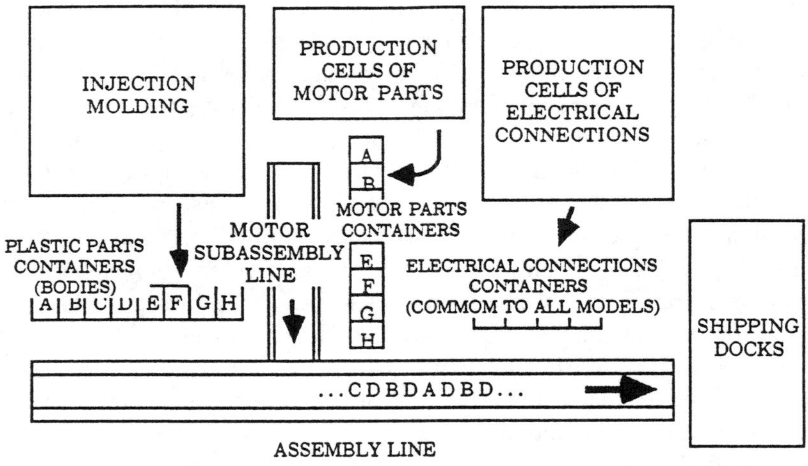

Figure 8.30. Layout for Mixed-Model Assembly.

should strive to extend the mixed-model assembly to subassemblies, and even further to manufacturing operations when repeated efforts in setup-time reduction eventually lead to negligible setup times.

Mixed-model assembly is typical of a pull production system (for example, the Toyota production system); backscheduling is typical of a push production system (for example MRP). When various levels of repetitiveness coexist for similar products, forcing the production to be planned and controlled by a hybrid push-pull system (a system where kanban cards coexist with work orders), production operations are matched to assembly operations through the combined use of mixed-model assembly and backscheduling.

Mixed-Model Assembly Combined with Backscheduling

In a hybrid system that combines mixed-model assembly with backscheduling, the repetitive components and subassemblies are assembled following the mixed-model assembly and the production of optional and specific features are back-scheduled. Consider, for example, the assembly of radiographic equipment for hospitals and clinics. These products are highly modular, and each subassembly comes with many optional features that are too costly to be temporarily stored before orders are received. However, because every model has a large number of standard components (mechanical components, cabinet parts, etcetera), the major subassemblies can be assembled on mixed-model assembly lines. The specific options are then backscheduled to be made available at the time of assembly, as shown in Figure 8.31: Upon reception of customer orders (1) and determination of the planned assembly date, the lead times for producing the options (2) are used to determine the time when the option production (or purchasing) must start (3).

The time between order taking and assembly is then mainly a function of the optional features' lead times. Figure 8.32 summarizes the various ways to coordinate final assembly to production operations depending on product characteristics.

Note in Figure 8.32, that an intermediate possibility between pure mixed-model

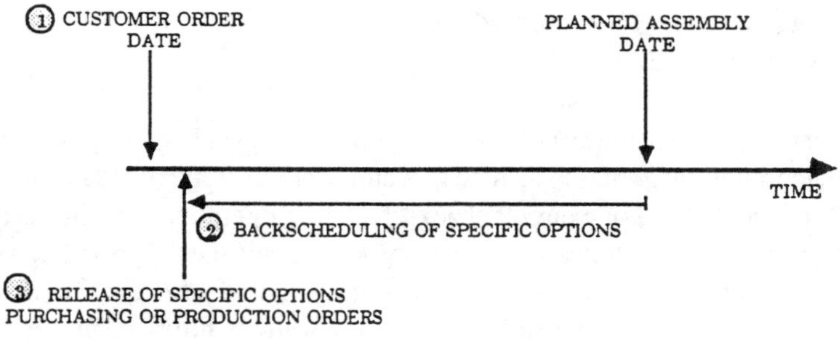

Figure 8.31. Backscheduling Specific Options.

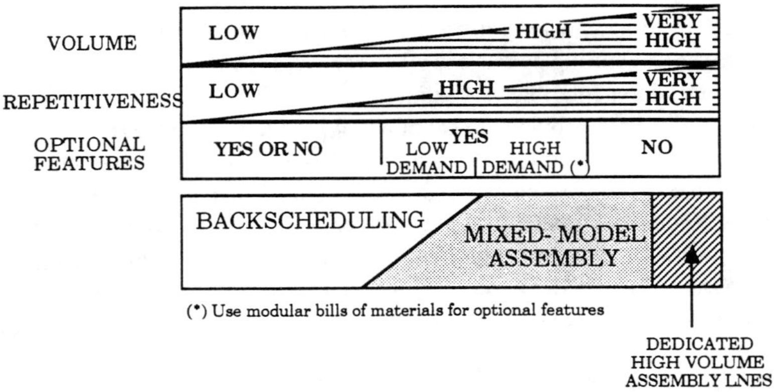

Figure 8.32. Coordinating Assembly to Production as a Function of Product Characteristics.

assembly and hybrid mixed-model assembly/backscheduling is a pure mixed-model assembly that uses modular bills of materials when the variable and optional-features demand is high and can therefore be predicted with enough precision to allow for short, temporary storage before assembly. In addition, note that if production volumes are large enough, the mixed-model assembly can evolve toward dedicated assembly lines where each model is continuously assembled.

The organizational changes that pave the way to world-class manufacturing enable companies to simplify their planning and scheduling systems, spanning manufacturing and assembly operations. A cellular organization and the delegation of the central scheduling responsibility to production personnel are the two major changes that allow this simplification. The same changes also have a significant effect on the control of operations and on the management of information on the shop-floor.

8.4 SIMPLIFIED INFORMATION AND SHOP-FLOOR CONTROL

The precision (and feasibility) of a production plan depends greatly on the accuracy of such input information as actual customer orders, available capacity, and production lead times. For example, backscheduling operations can be accurately performed only if lead times are precisely known by the information system. Controlling production operations by recording manufacturing and assembly times, within-cell lead times, and between-cell waiting times gives production managers and production planners the information that enables them to build precise production plans. In order to record useful information the purpose for recording that information must first be discerned.

In world-class companies, most machine operators and cell team leaders would state that the purpose for recording information is to make sure that due dates and production rates are met, to follow improvements in scrap reduction and product quality, and to know the capabilities of team members. Production supervisors and managers, on the other hand, would reason that the purpose for recording information is to improve customers' satisfaction regarding shipment dates and product quality, to improve plant or shop profitability, and to enhance the capabilities, potential, and safety of personnel.

Nonetheless, in companies that want to reach world-class status, control must be reoriented toward decision making for action; that is, the physical performance measures that are key factors in improving productivity must be closely followed by the plant managers as well as by the cell operators: "Eyeball" control systems must be introduced that allow cell team members and production managers to take immediate action, and the complexity of existing computerized systems must be dismantled and adapted so that they provide the information needed to manage production operations.

8.4.1 Physical Performance Measures

World-class companies are differentiated from traditional companies, not only by their organizational structure, but also by their culture. In a world-class organization, more autonomy and responsibility are given to the production personnel, thereby lessening control that needs to be exerted by the central production planning department on ongoing operations. Figure 8.33 shows the plan-execute-control cycle as it still is practiced in many companies and as it should be practiced.

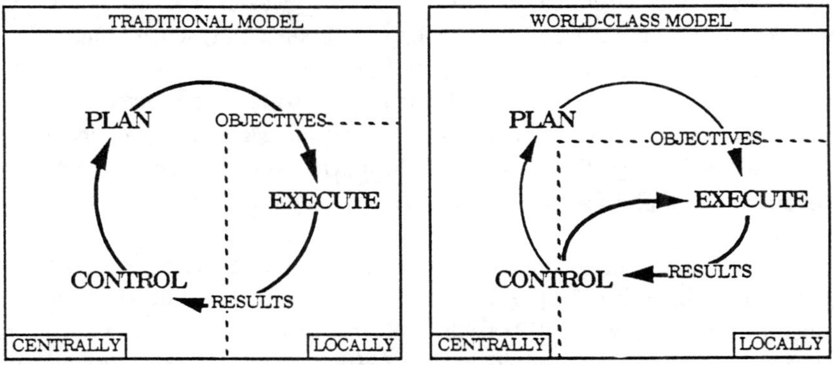

Figure 8.33. Traditional and World-Class Models for the Plan-Execute-Control Cycles.

Traditionally, production personnel have been in charge of only the actual execution of processes, while the central production department planned the execution, controlled how the execution conformed to the plans, and modified subsequent plans accordingly. In the world-class model, the central production planning department still plans and controls, but the type of control it exerts is different because the production personnel not only execute the process, but also control the execution, and even adjust the execution according to current conditions and given production objectives.

Operators need to know how their team is doing and what changes need to be made so that their objectives can be met, while what the central planning department needs to know is what production units or cells can manufacture the ordered products and how long production will take. The traditional indicators followed by the production control systems are (1) the state of completion of current jobs, (2) the levels of WIP and of inventory, and (3) the setup and processing operation times. However, these traditional measures must be reconsidered from the new perspective of a world-class organization and new measures must also be added.

These three traditional measures will take on a different meaning under the new manufacturing organization. In cellular manufacturing, a job's state of completion is clearly to be measured and interpreted differently than it is in a functional organization. Within cells, the job's state of completion can be followed visually and does not have to be reported. Job tracking is therefore reduced to measurement points at the cell level—when jobs enter and leave the cells—thereby enabling the within-cell job lead times and between-cell waiting times to be followed. Reporting this information to the central system allows the lead times indicated on between-cell routings to be verified and updated. This information feedback is capital for the correct planning of backscheduled parts (pushed production). For parts that are not backscheduled and are therefore regularly produced for intermediate inventory replenishment (pulled production), job tracking is replaced by counting parts to measure production rates and by measuring kanban-card cycle times. These cycle times are also fed back to the production system to improve the quality of the cell planning module outputs. In addition, job tracking measurements should be communicated to the production teams so they can assess their own performance. Monitoring production rates and reducing the within-cell lead times are among the first objectives given to production teams in world-class companies. They are the first physical indicators to display on the cell billboard (see Figure 8.34).

Cellular manufacturing implementation and lot-size reduction can lead to as much as a 90 percent reduction in inventory and WIP. When these amounts of reduction can be achieved, WIP tracking becomes simple and almost unnecessary. For example, when the transfer lot sizes between machines are equal to one part, there is clearly no need for within-cell WIP tracking. Inventory transactions are only necessary when parts must be stored for a period that can be considered significant when compared to the total manufacturing lead time. A product (say a machine tool), for example, has a total manufacturing lead time of fifteen weeks, if modular components are stored on the average for two weeks before final assembly, then the modular components inventory should be recorded; however, there is no need to record the inventory of gears stored between machining and heat-treating operations if the gears wait there for only a few hours. The inventory stored in intermediate buffers between cells can easily be monitored visually. Inventory levels are important performance measures for production managers, although they should not be used as physical performance indicators at the cell level since cell operators cannot directly reduce inventory.

As discussed in section 8.2.3, setup and processing times must be considered differently for individual machines than for cells. Knowing individual machine setup and processing times enables cell manager to assign operators to machines accordingly to achieve balanced cell operations. At the cell level, the sum of a lot setup time and processing time is equal to the within-cell lead time, which depends on the lot size and on the number of operators in the cell. The central production planning system needs to be informed of the within-cell lead times, but detailed machine setup and processing times are of little use for central planning. It is, however, important to keep measuring individual machine setup and processing times for local cell planning and for the improvement of cell operators. Both measures can be consolidated at the cell level (summed over all the cell's machines). For example, displaying the evolution of the cell setup times and processing times enables the effects of a setup-reduction campaign to be followed and also motivates equipment improvements for reducing processing times.

Cell lead times, production rates, and setup and processing times are some of the key performance indicators that should be followed to assess the productivity of cell teams. A variety of solutions exist for data entry, processing, and output edition. These solutions range from entirely manual to completely integrated into the production control system. Since the main objective of this control is to provide feedback information to cell operators, it is crucial that the way the performance indicators are produced is clearly understood by all production personnel. The performance indicators should be charted against time as illustrated in Figure 8.34, and the charts should then be posted inside every cell and regularly updated (say weekly). The evolution of the performance indicators should be discussed during work-team meetings. This type of control, based on physical performance indicators, is oriented toward action, and the cell operators must suggest appropriate actions to correct unsatisfactory situations, such as suggesting a different way of handling materials inside a cell, a specific improvement on a machine, different scheduling rules, and so on. The performance indicators discussed so far can be compiled by adapting information provided by most of today's production control

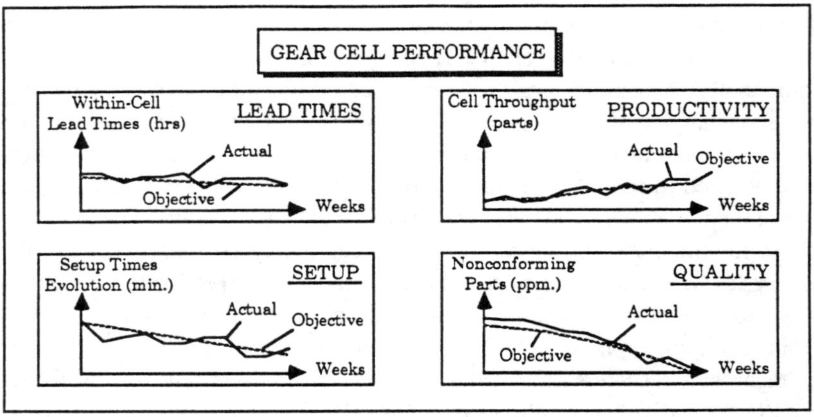

Figure 8.34. Communicating a Cell's Performance.

systems. These indicators are necessary but not sufficient to follow every aspect critical for assessing a company's competitiveness.

Additional critical aspects are: Product quality and production equipment uptime, for which the corresponding performance measures are considered in the chapters where each of these particular aspects is discussed. Examples of physical performance indicators that enable the assessment of these aspects are given in Table 8.6.

Although some of the indicators in Table 8.6 may seem distant from the actual production control, their evolution evidently reflects the evolution of production conditions on the shop floor; therefore, these indicators have their place in a production control system. Posting the indicators, which show information traditionally considered "sensitive," is part of the necessary culture shift that creates better communication between operational personnel and managers and is en-

TABLE 8.6. Examples of Additional Physical Performance Indicators.

Competitive Aspect	Possible Performance Indicators
Product Quality	Scrap Rate Rework Cost Customer Satisfaction
Equipment Uptime	Equipment Uptime Mean Time Between Failures Mean Time to Repair Cost of Curative Maintenance

hanced by the formation of production teams. Another major organizational change that follows the implementation of cells and the resulting simplification of operations is the option of replacing "management by numbers" with "management by sight."

8.4.2 Eyeball Control Systems

As in any management situation, management by sight addresses the issue of preventing problems and of solving problems as they surface. The eyeball control systems are simple physical control systems that clearly show potential problems and provoke those problems to surface.

A typical eyeball control system in the Kanban system. Although receiving a card for producing a lot of parts is not essentially a problem, not receiving cards or receiving too many cards (as indicated by an almost empty or a full Kanban ordering post), definitely shows the presence of production problems that require action from machine operators or from production managers. The advantage of eyeball control systems is that they show the problems as they occur, or even show critical situations before they actually become problems, differing in that from traditional computer-based control systems where problems are reported *a posteriori.* Observing a full storage area in front of a machine is a much more obvious and quicker way of detecting that the machine has capacity problems than by reading periodic capacity reports.

Eyeball control systems can be informal, being simply an interpretation of the current state of a system like the Kanban system, or they can be formal systems specifically designed to make problems as visible as possible. An example of a formal system is the use of warning lights on machines or manual stations. Consider a manual or automatic assembly line, each workstation being equipped with two lights: a yellow and a red light. If a workstation is stopped for any reason (jammed, next station full, et cetera), then the yellow light is switched on and the team leader or an idle worker comes and takes appropriate action. When the problem is serious and is better solved if the assembly line were stopped, then the red light is switched on, the line is stopped, and all the available help rushes to the site of the problem. Another example of a formal eyeball control system, which is even more simple than the yellow-red light system, is to place, next to the performance indicator board, another board that itemizes, in large characters, all of the current unsolved problems, as shown in Figure 8.35.

Everyone working in a cell, including support personnel such as maintenance workers, can report problems, and the reported problems stay on the board until solved. Small group activities are ideal occasions for identifying and solving these problems. Note also that displaying the performance indicator charts is an additional form of an "eyeball" control system. By simply consulting a set of charts, a

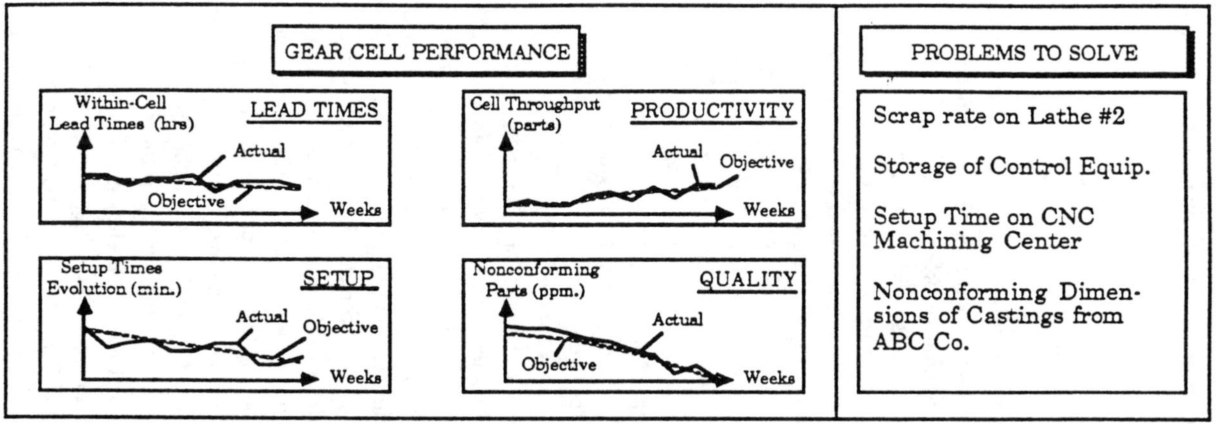

Figure 8.35. Cell Performance and Problem Posting.

worker, a production manager, or a top manager can assess the evolution of a cell or a production unit's performance,and thus originate appropriate action.

A strength of world-class companies is to have designed systems that prevent production problems from remaining hidden. There exists a human bias to remain silent on anything that goes wrong, and this tendency is even more pronounced when the person who signals a problem may be blamed for it. But hidden problems are like unhealed wounds, the longer they remain untended the worse they become. Thus it is crucial that problems get solved as soon as they occur, or better, are prevented. Few production problems remain unresolved when eyeball control systems are implemented and when workers are confident that signaling them will bring help and recognition rather than blame.

The success of eyeball control systems demonstrates that simplification of control systems should follow the simplification of the production environment. Similarly, the complexity of computer systems, which have been designed to manage complex flows of information, should be dismantled to reflect simplification on the shop floor.

8.4.3 Dismantling Computerized Complexity

Organizational changes, such as the implementation of cellular manufacturing call for a complete review of the manufacturing information systems. Figure 8.36 conceptualizes the view that the central production system has of the shop floor, before and after the organizational changes. Before the organizational changes, each machine or work center and each storage location is considered separately. In job scheduling, for example, multiple combinations of job/workstation assignments are compared in order to find a combination that enables all the jobs' due dates to be met or one that maximizes machine utilization; hence,jobs are tracked

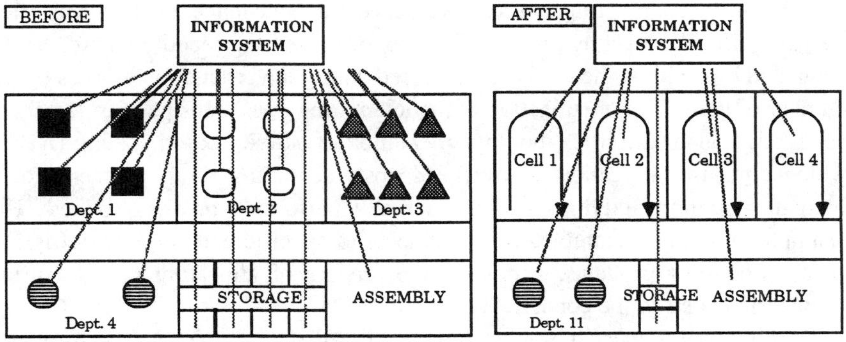

Figure 8.36. View of the Shop Floor from the Central Production System, Before and After the Organizational Changes.

at every workstation to determine the rate of completion and to check the conformance to the production schedule. After the organizational changes, workstations now grouped into cells or within the assembly line are not distinguished anymore. Instead, the cells are seen as black boxes by the system, jobs are sent to cells and are not tracked by the system within the cells, repetitive production is simply tracked by counting the number of parts entering and exiting cells. In addition, when lead times are short, counting parts can be further simplified by *backflushing* parts consumption: Subassemblies, or complete products, are counted when they exit assembly stations or assembly lines, and the number of consumed parts is determined by going through the different levels of the bills of materials.

The information needs required by the production managers (shop or plant level) for central control are thus reduced to a consolidation of the main performance indicators. These indicators should cover conformance to schedule, quality levels, equipment maintenance, safety, personnel training, and the effects of ongoing efforts such as setup-time and operating-time reductions. Furthermore, to assess the profitability of the production unit (shop or plant), production managers need to follow financial indicators. Reviewing a company's financial system when implementing WCM operations could be the object of a separate chapter, or even of a completely different book. It is beyond the scope of this discussion on PP&C systems to cover in detail the resulting review of financial systems. However a few of these consequences of replacing complex, traditional systems by more effective, less costly, and simpler ones can be addressed here. Financial indicators and other relevant information should be transmitted from the reporting module of the production information system to the financial control system, or a financial module could be added to the PP&C system. Both options are illustrated in Figure 8.37.

The financial information to report consists of the labor and material cost per part, as well as indirect costs, which can be evaluated for each cell. The indirect costs are the cell supervision costs—the costs of programming NC machines, of manufacturing engineering, maintenance, et cetera. Note that cellular organization also simplifies the costing process. The cost of processing repetitive parts can be evaluated by dividing the total costs incurred during the production process by the number of parts. The organization of manufacturing operations per part family or per product also simplifies the implementation of activity-based costing (ABC), which aims the allocation of overhead costs according to the percentage of particular support activities that are actually performed on the products. For example, assigning manufacturing engineers to specific cells and, therefore, to specific parts or products, simplifies the process of allocating manufacturing engineering costs to the concerned products.

Another consequence of implementing cells is that the employees' performance is no longer evaluated individually. In manufacturing cells, employees are part of a team, and the performance is evaluated and rewarded at the team level. This

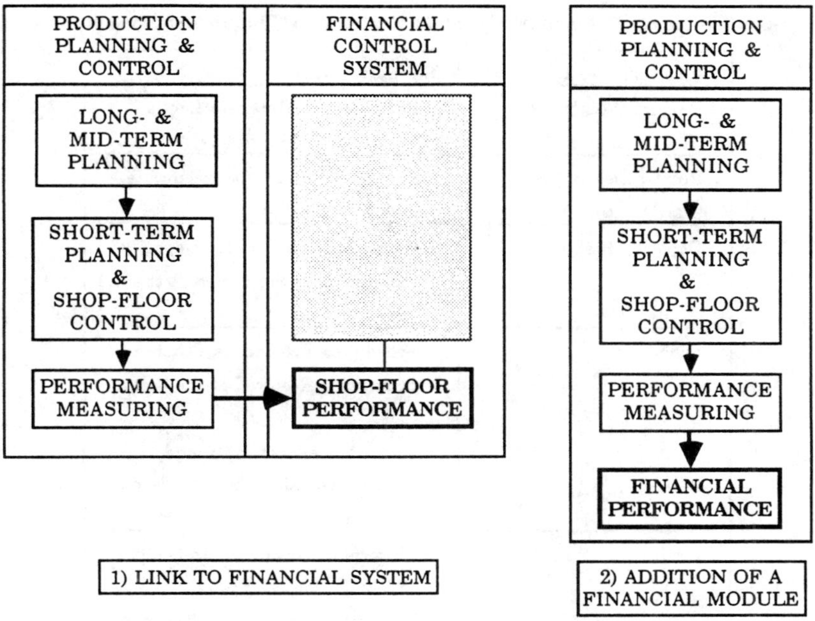

Figure 8.37. Two Options for Reporting the Shop Floor's Financial Performance.

change can dramatically simplify the performance evaluation system. In addition, WIP inventories often disappear, and their control becomes meaningless and should be eliminated leading to additional savings. On the other hand, lead times become meaningful and could be used as a cost driver in an ABC system. As a general rule, many traditional financial performance indicators are eliminated, the remaining ones simplified, and simple nonfinancial performance measures are added. The simplifications of the central information system used for PP&C are summarized in Table 8.7.

Clearly, the simplification of the central information system is made possible by delegating increased responsibilities to local structures on the shop floor and the resulting reduction in the amount of information to be centrally processed. The amount of information to be processed locally should be handled manually, but could also be supported by a local information system running on a microcomputer, for example. The applications implemented on this system could consist of a simple simulation software to model the cell behavior and resolve, if necessary, sequencing problems and an electronic spreadsheet with graphic capabilities that would generate the physical performance indicator charts. A next step in decentralizing the PP&C system is to transfer some of the planning and control functions to the local system, such as computing the number of kanban cards and determining

TABLE 8.7. Simplification of the Central PP&C Information System.

Function	Simplifications
Product Description (Bills of Materials)	- Fewer Bill of Materials Levels - Fewer Bills of Materials
Process Description (Part Routings)	- Bi-level Structure (between-cells and within-cells)
Capacity Planning	- Rough-Cut and Detailed Capacity Planning only Performed at the Cell Level
Scheduling	- Scheduling Problem Is Eliminated - Sequencing Resolved at the Cell Level - Mixed-Model Assembly and Kanban System Eliminate Need for Backscheduling Repetitive Parts
Production Control	- Physical Performance Indicators - "Eyeball" Control Systems - Reduction in Volume of Information Controlled Centrally - Backflushing - Simpler Costing Process, ABC

the number of cell operators needed to meet centrally determined cycle times as well as parts counting. The system described here leads to the implementation of a tree architecture, as shown in Figure 8.38, where local systems are in charge of cell planning and control, and the central system communicates with each local system to send production requirements and to receive information concerning local process capacities and the current status of production. The production control function is then transferred to the shop floor. Local within-cell control is placed under the responsibility of the cell leader and the cell production team. Central control should also be physically moved to the shop floor to reduce the number of production control problems and speed up the resolution of the remaining problems.

Integrating local systems and a central system into a companywide network marks a step toward CIM. Computer integrated manufacturing, however, is *not* one of the steps toward WCM. Many companies have attempted and failed to integrate exiting manufacturing information systems because they have not attempted first to review and simplify these systems. Computer integrated manufacturing is the ultimate consequence of a series of initial steps: From employee involvement to

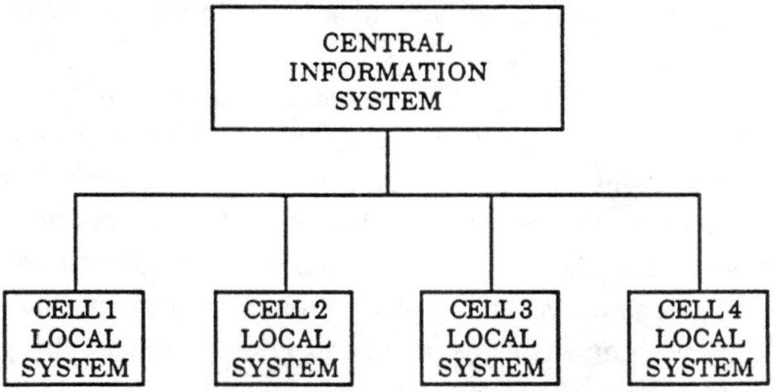

Figure 8.38. Possible Evolution of the PP&C System Architecture.

cellular manufacturing, to simplified production and planning systems; therefore, only simplified systems should be integrated. Describing in further detail this integration exceeds the scope of this book.

8.5 SUMMARY AND MANAGEMENT GUIDELINES

In the last twenty years many Western companies have implemented MRP-based PP&C systems. However, recent successes of Far Eastern competitors have been attributed, in part, to their ability to control production with JIT systems. (The Toyota production system is the typical example of a successful JIT production system.) JIT systems regulated by kanbans are commonly defined as Kanban systems or pull systems; MRP-based systems are commonly defined as push systems. The definitions of these two types of systems appear to be contradictory and the question arises of whether western companies should abandon push systems (MRP) and substitute them with pull systems (Kanban) in order to enjoy the same success as their Far Eastern counterparts.

Push systems use estimated lead times to backschedule the various production operations from final assembly to the initial processing of raw materials. Push systems are appropriate for planning the production of items with a variable demand but long planning intervals; the assumption of fixed lead times usually results in high inventory and WIP levels. Pull systems rely on the regular replenishment of intermediate inventories by upstream processes as the parts are consumed by downstream processes. Pull systems schedule only final-assembly operations, the upstream operations' rates are adjusted to the final-assembly rate by cyclic flows of kanban cards. Pull systems present the enormous advantage of fostering improvements in production operations by allowing tight monitoring of intermediate inventories. In addition, pull systems significantly reduce the need for control requirements. However, they are mostly appropriate for controlling repetitive production.

Although the two systems appear to be contradictory, their structure is essentially similar. Both rely on information contained in bills of materials and part routings, and both systems possess a strong capacity-planning function. In fact, the main difference between the two systems lies in the way production is scheduled on the shop floor. Since the production of repetitive products and parts is best regulated by pull mechanisms and the production of varying products and parts is best scheduled by push mechanisms, companies that produce both types of products should implement hybrid systems that combine the best of both approaches. There is, therefore, no reason for Western companies to abandon their current push systems to replace them with pull systems. To reduce their handicap in competing with Far Eastern companies, Western companies simply need to adapt their existing MRP PP&C system to take advantage of the organizational changes that occur on the shop floor, such as, implementing manufacturing cells and the drastic reduction in lead times.

First, the master scheduling policy is modified, as this is made possible by an increased flexibility in the firm schedule horizon: FAS's are edited with higher frequency and can rely on firm customer orders, and the MPS output provides production rates for repetitive end products. Then, the bill of materials' structure is updated to conform to the reorganization of the shop-floor activities, and accordingly, the number of levels and records is reduced so that the remaining records match the actual storage points that remain to be controlled. Records that must be kept for indicating the product structure can be modified into phantom records that do not trigger any MRP processing. In addition, the decrease in final-assembly times and the standardization of the subassemblies' internal components enable the number of modular bills to be reduced. The part-routing structure is profoundly altered by the implementing manufacturing cells: Between-cell routings, containing information used by the central planning and control system, are distinguished from within-cell routings, which contain information used locally by cell teams. In addition, world-class companies often abandon the use of standard hours, indicated in part routings, for individual worker performance evaluation.

Detailed capacity planning at the individual work-center level is replaced by an update of the rough capacity planning performed at the cell level at the time when planned customer orders are confirmed. The need for scheduling groups of parts on groups of individual machines is eliminated; parts are assigned to cells according to each cell's processing capabilities. Jobs are sequenced within the cells by the cell team, which is also responsible for meeting due dates and production rates. Assembling repetitive products is sequenced using the mixed-model assembly pattern to smooth the demand of parts produced in upstream cells. Traditional backscheduling is maintained to schedule nonrepetitive production operations. Work orders are sent to cells and to individual workstations that could not be grouped into cells.

Physical performance measures posted on shop-floor information boards make public each team's performance. Similarly, current production conditions and unsolved problems are made visible by implementing eyeball control systems. Delegating decision-making responsibilities to shop-floor personnel reduces the amount of information to be centrally processed. In addition, when focus and simplicity is achieved through decentralization and making things smaller and self-contained, support personnel such as production control people and material planners (that should be in charge of purchasing the materials consumed locally) can be housed on the production floor easily and, as a result, speed up their actions. The PP&C system is then simplified accordingly.

Reviewing and adjusting the PP&C system is certainly one of the last steps to be taken by a company on its way toward WCM. However, this step must not be neglected. Having understood the necessity for such an adjustment, the company's top management should launch the preliminary system design once the first cells have been mapped. Since the system's new design depends, for a good part, on the amount of local authority delegation, it is crucial that the extent of delegation is properly defined and understood. As much as possible, the system design should be kept flexible to enable further operating-condition modifications, such as forming new cells, increasing part and product repetitiveness, and extending the hybrid system to include external vendors.

9

Total Maintenance

World-class manufacturing companies cannot afford unexpected delays in their operations. From customer order processing to product shipment, information and product components must flow as smoothly and predictably as possible through the organization. Customers now expect on-time delivery of quality products. Not meeting those expectations might send valued customers to the competition.

A critical factor in the timely completion of products is the dependability of the production operations. Companies engaged in worldwide competition can no longer afford the luxury of including equipment breakdown delays in their production lead time; therefore, lengthy or repeated breakdowns must be prevented from occurring on the shop floor. Preventing breakdowns can only be achieved by a drastic change in the way maintenance has been traditionally considered and is still practiced in many manufacturing organizations. A whole new maintenance culture must be implemented: total maintenance.

Total maintenance (TM) regroups a set of strategies aimed toward eliminating unscheduled equipment downtime. These strategies involve personnel on a companywide basis. This chapter details the concepts involved in TM and addresses practical issues concerning the implementation of this new maintenance culture. The first section addresses some of the root causes of poor equipment reliability and maintainability, and why traditional attitudes and policies regarding maintenance are no longer acceptable. Section 2 details strategies for evolving from traditional attitudes and practices to TM. It discusses the new roles and responsibilities of both machine operators and maintenance workers, and addresses information and organizational systems necessary for eliminating equipment failure problems or making them easier to solve. The third section then presents a systematic procedure for implementing the strategies for TM detailed in section

2. The chapter concludes with a summary of the key concepts and strategies for TM, along with some guidelines for successful implementation.

9.1 TM: A NECESSITY FOR WCM COMPANIES

In conventional manufacturing environments, when a machine breaks down, high levels of in-process inventory generally allow downstream operations to run for days or even weeks at normal capacity, and the parts that are produced upstream are simply stored until the machine resumes production. Thus, a machine breakdown is considered more as a nuisance than as a problem that should be eliminated. However, implementing of WCM methods and practices results in drastic cuts in inventory levels; consequently, a single breakdown has an almost immediate and detrimental effect on the entire system's performance. Because improving maintenance is mandatory for WCM operations, maintenance policies and practices must be considered from a different point of view.

9.1.1 The Need for a Different Maintenance in WCM Operations

In any production environment, the failure of a bottleneck machine—even with comfortable levels of in-process inventory—quickly becomes a major problem for any production manager. Long breakdowns might require more costly overtime work to meet the production requirements or might even result in delayed shipments to customers. Therefore, bottleneck equipment traditionally has been the object of much attention from maintenance departments, typically under pressure from production managers to keep bottleneck equipment operating.

On the other hand, because breakdowns of nonbottleneck equipment usually are not critical to plant operations and, therefore, not a major concern of production managers, maintenance departments have a tendency to let the condition of such machines slowly deteriorate. If a nonbottleneck machine breaks down, there is no immediate need to fix the machine because production can either be transferred to a similar machine, or inventory levels are high enough to allow production on downstream machines to continue.

World-Class Manufacturing and the implementation of JIT operations have shaken-up this lax way of operating in manufacturing organizations. The implementation of manufacturing workcells in focused factories, typically linked together by pull production control systems, has slashed levels of in-process inventory and finished goods inventory. In a JIT environment, when a machine breaks down, the operation cannot simply be shifted to another machine, and there is no safety stock to keep the other machines running—in effect, *all machines become bottlenecks.*

Consider the following scenario that could happen in a typical continuous-flow manufacturing environment:

1. One of the machines in a workcell breaks down.
2. The downstream machines in the workcell stop one after another upon processing the last part that went through the stopped machine (downstream machines are starved).
3. The upstream machines in the workcell stop production because they do not receive any production requests from the stopped equipment downstream (upstream machines are blocked).
4. The entire workcell is now stopped.
5. Because the workcell is stopped, downstream workcells or assembly lines are not being supplied anymore, and after consuming (a low level) intermediate inventory, these processes will also have to stop if the first stopped workcell does not restart rapidly.
6. Because the downstream operations are stopped, they do not send production requests to the upstream workcells that are parallel to the first stopped workcell, causing these workcells to stop also.
7. If the failure of the initial machine is not fixed, the entire plant might be stopped. Customers will suffer delayed deliveries, and suppliers will not receive production requests. Thus, the effects of the breakdown of one machine can be very costly.

To avoid this extreme situation, a last-resort solution would be to "outsource" the production to either another workcell (if the capacity is available) or to outside plants. In any case, such a breakdown would be extremely costly; it would downgrade the performance of the entire production system and therefore must be prevented.

This simplistic example demonstrates why companies that have tried to implement pull production control systems (such as JIT), without understanding the range of the issues that should be addressed, have been confronted with unsolvable problems and have had to fall back to traditional manufacturing techniques. The way maintenance is considered, carried out, and organized is one of these issues.

9.1.2 The Need for Measuring Savings Generated by Improved Maintenance

By dramatically intensifying the consequences of problems that occur in production systems, WCM operations implementation has forced production personnel and managers to confront these problems and eventually eliminate them. When the necessity to eliminate problems is recognized and addressed, production costs are reduced several fold.

Traditional manufacturing can tolerate deficiencies in the production systems; however, when such deficiencies are not eliminated, the same problems, such as equipment breakdowns, occur again and again. When there is no incentive to prevent equipment failures, equipment will keep on failing. The question asked by many maintenance and production managers is *why spend money on preventive maintenance if only the extra cost can be measured and not the benefits?* When no system is available to measure the benefits that result from increased preventive and predictive maintenance—and from improved equipment maintainability—the incurred costs cannot be justified. Justification is even more difficult when the costs are incurred by a maintenance department and the benefits are realized by production or marketing departments.

Many world-class companies have developed measurement systems; other companies that aspire to WCM also need to create such measurement systems. When breakdown frequencies, repair times, cost of preventing and repairing breakdowns, and other such physical and financial indicators are heeded, companies realize that trying to save on maintenance is like trying to save on education: Short-term savings lead to long-term losses.

9.1.3 What has to Change to Evolve Towards Total Maintenance

Mean time between failure (MTBF) and mean time to repair (MTTR) are widely used indicators of performance for measuring the level and the quality of equipment maintenance. MTBF indicates the dependability of the equipment (more dependable equipment has longer times between failures). MTTR indicates the maintainability of the equipment (equipment with better maintainability has a shorter time requirement for repair). To aim toward total availability of equipment when it is needed (zero breakdown), companies must increase the quality of maintenance along two parallel paths: (1) Increase the MTBF, and (2) decrease the MTTR. That is, both the equipment dependability and maintainability must be increased. A preliminary step in reaching this objective is to identify the causes of traditional problems on which to focus the company's efforts. The remainder of this section discusses first the main causes for high frequency of breakdowns and second the major reasons for long repair times.

The main causes for the high frequency of equipment breakdowns can be grouped into four categories: (1) The way the equipment is used, (2) the level of preventive maintenance, (3) the equipment complexity, and (4) the robustness of the equipment design. Each category will be discussed in the following.

1. *The way the equipment is used.* The frequency of breakdowns can be attributed to whether a machine is often used either below or above the manufacturer's specifications. In some cases production imperatives may require a machine to be used beyond its specified characteristics. The

condition of a machine that is often used beyond the recommended limits must be monitored more closely than another machine that is used normally under the recommended conditions.

2. *The level of PM.* Some production and maintenance department managers still advocate that PM is too burdensome and expensive, and a better solution is to wait until components break down before replacing them, rather than replacing the components before they break down. However, it is unquestionable that the amount of PM has a significant impact on breakdown frequency. If a machine is thoroughly and regularly lubricated and parts (such as bearings and other components subject to wear) are changed on a timely basis, chances for machine stoppage are significantly reduced.
3. *The complexity of the equipment.* Manufacturing equipment composed of hundreds or thousands of components are more complex and more likely to break down than simple machines with fewer components. Higher breakdown rates do not result directly from the increased level of automation but rather from the higher level of complexity. For example, the rate of breakdowns of older machines has been greatly reduced when moving parts with control functions (such as cam mechanisms) have been replaced by simple electronic controllers. But some of the newer manufacturing systems, such as the FMS, which integrate CNC machines, robots, and automatic guided vehicles (AGV) are extremely complex. The integration of numerous complex components into a single system multiplies the risk of system failure.
4. *The robustness of the equipment design.* How well electronic circuits perform in a humid environment is an example of robustness. Simple as well as complex mechanisms can perform well in given temperature or humidity ranges, but are at risk for malfunction outside these ranges. It is important for the user to know the extent of equipment design robustness, and under which conditions—not necessarily thought-out by the manufacturer—the equipment might fail.

Repair times are also affected by machine design (for example the time it takes to access parts, or to disassemble and reassemble machine components). But the major reason for long repair times is poorly organized maintenance activities. The multiple causes for poor maintenance organization are summarized in the cause-and-effect diagram in Figure 9.1 and discussed in the following paragraphs.

- *Communications.* After the observation and report of a machine breakdown by a machine operator, it is not uncommon to see two or three waves of people intervene in the process until, eventually, a qualified maintenance worker is

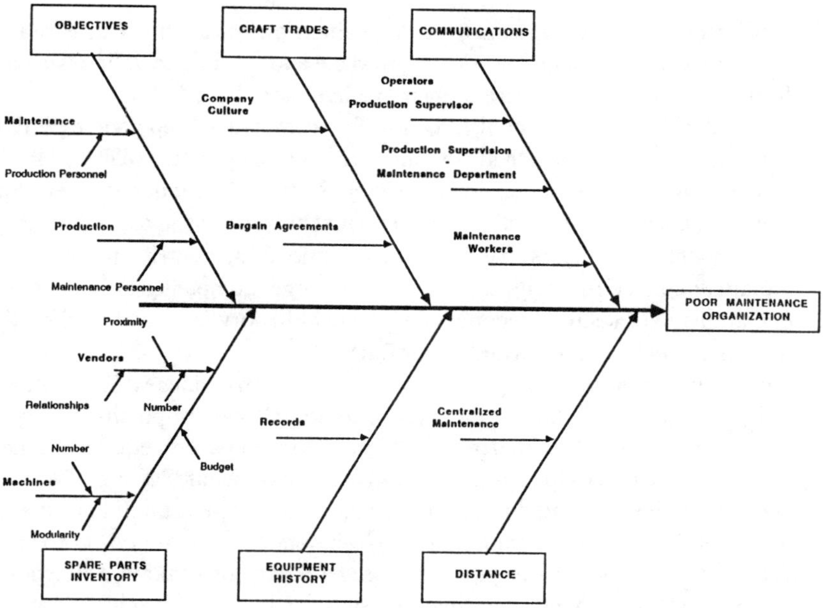

Figure 9.1. Maintenance Organization Cause-and-Effect Diagram.

notified of the problem. Similarly, in many companies operators cannot even directly call the maintenance department. These and other outdated and inflexible organization procedures prevent the fast diagnosis and resolution of problems

- *Craft trades.* Rigid craft trade classifications hinder maintenance flexibility and lengthen repair times. Specializing workers is supposed to speed up the resolution of complex problems and to foster productivity, but most of the breakdowns that occur routinely are not complex to repair; they do not require deep knowledge in any specific field but, rather, require a good knowledge base in every field. Having several specialized people work on a simple problem reduces everyone's individual interest in the work and is therefore counterproductive. Furthermore, repairs take longer.
- *Objectives.* Many companies (still) have a functional organization. With this type of organization the sense of common objective—say, meeting the customer's satisfaction—does not exist; production workers are not encouraged to prevent breakdowns, and maintenance personnel do not feel the need to consider production priorities and are not encouraged to speed up the repairing of critical equipment.
- *Distance.* The physical distance that exists between the central maintenance

department and the equipment to be maintained is often another cause for long repair times. Maintenance workers can waste a lot of time traveling back and forth between the machines and their "maintenance base."

- *Equipment history.* It is astonishing how few companies keep good records of equipment breakdowns, changed parts, last check-up dates, MTBF, MTTR, PM operations, and other equipment history events. Consequently, every time a maintenance worker is called to repair a machine, he or she has no idea what the frequency is of that problem on that machine; thus, recurring problems can remain unnoticed for months, particularly in larger companies or in multishift operations. Not keeping records of equipment history is a costly policy that increases breakdown rates and repair times.
- *Spare part inventory.* A large number of different machines with low modularity (having few or almost no parts in common) that are purchased from a wide variety of vendors hinders long-term relationships with equipment manufacturers and causes the spare parts inventory to be needlessly high. Not only does the budget allocated to the maintenance department inhibit storing the minimum level of spare parts, but also vendors can't be relied on for rush deliveries of parts. Vendor proximity or spare part distribution efficiency is seldom considered when the final decision for buying a machine is made. Equipment purchasing policies and their impact on the availability of spare parts, eventually also have an effect on repair lengths.

9.2 STRATEGIES FOR TM

The previous section has highlighted some of the root problems of poor equipment reliability and maintainability. Although solving these problems is neither straightforward nor easy, there are valuable strategies that can be applied to achieve success:

- Record information pertaining to the condition of the equipment (rate of breakdowns, length of repairs, frequency of component exchanges), and analyze the records for performance and policy evaluation.
- Clean and lubricate the machines (first-level maintenance).
- Implement effective PM programs.
- Involve the *entire* work force—up to and including top management—in resolving maintenance problems.
- Reorganize the maintenance structure and assign new responsibilities to the maintenance department.
- Implement predictive maintenance programs.
- Design (or purchase) more robust and easier-to-maintain equipment.

The rest of this section details these strategies. Some of them are illustrated with examples from the evolution toward TM at the Electromechanical and Nuclear Equipment Division of Jeumont-Schneider Industries, a French company that started implementing a WCM strategic plan in 1987. The Electromechanical and Nuclear Equipment Division manufactures high-power electrical rotating machines and nuclear power plant components. Faced with competition from other Western as well as Far Eastern companies, the division management team defined objectives of 20–30 percent improvements in productivity, quality increase, manufacturing time reduction, and cost reduction. In order to eliminate delays because of inefficiencies in the original organization, the Manufacturing Department and related support departments were broken down into five autonomous production centers. Originally centralized production support departments, including the Maintenance Department, were therefore, decentralized, and their personnel distributed into the different new production centers. The aim of this decentralization was to equip the production centers with the resources they needed to allow them to take full responsibility for organizing and controlling their own production operations. In its strategic plan, management recognized that since production equipment and machines were critical components of the future WCM operations, it was necessary to control their condition and important not to suffer their breakdowns. Hence, the choice of evolving toward TM and implementing the following strategies in order to reach this goal.

9.2.1 Information Management

Information is known to be expensive to gather and to maintain. However, free information does exist but is commonly overlooked; every process produces information (at no cost), which can be used to improve the process itself. For instance, counting the frequency of machine breakdowns helps to identify critical equipment and the need to focus on these machines for immediate improvements.

In the context of equipment maintenance, information management has three objectives: (1) Plan maintenance operations, (2) know the equipment, and (3) measure the effectiveness of the new policies.

1. *Plan maintenance operations.* A machine maintenance plan should be established for each machine. This plan is the skeleton that defines the first-level, preventive, and predictive maintenance schedules for the machine's entire working life. From this plan, a list of maintenance operations and their frequencies are identified. For new machines, the maintenance plan can be established on the basis of information from the manufacturer (technical manuals or direct input from service or design engineers). For older machines, manufacturer based information can be enriched from machine-breakdown history. (For example, the service fre-

quency should be increased on a component that is known to fail often.) Maintenance plans should be established by the maintenance workers that are assigned to a specific production area and assisted by that area's machine operators. Remember that maintenance plans are not cast in concrete and should be modified as new problems are discovered and other problems eliminated. The first-level and preventive maintenance schedules are first detailed or explained to machine operators and then posted at the workstations for easy reference.

2. *Know the equipment condition, identify problems, and take appropriate actions.* Breakdowns should be recorded by the machine operators on a machine log book kept at the workstation. (The log book is essentially a record of all the breakdowns occurring during a machine's work life.) The date, time, breakdown description, and breakdown symptoms should all be concisely recorded in the log book. These breakdown records can then be analyzed in order to evaluate any changes in the equipment's condition and the effectiveness of implementing new policies. The information reported in the log books is essential for improving the reliability of the equipment. Repairs can also be reported in the machine log books. The repair information that should be recorded by the maintenance workers consists of:
    - Brief description of the problem
    - Root cause(s) of the breakdown
    - Steps taken to solve the problem
    - Length of the repair time
    - Exchanged parts (main components)
    - Whether the repair was complete
    - Other problems noticed on the machine
    - Problems encountered during the repair

   Analyzing this information leads to improving the equipment maintainability. Even in the absence of breakdowns, machine operators can regularly record information on the equipment condition that can be used to monitor the evolution of the equipment condition.

   Figure 9.2 shows an example of a maintenance sheet that is filled out weekly by machine operators at Jeumont-Schneider. In this company, the use of a microcomputer dedicated to each decentralized maintenance group is helpful in storing and analyzing the recorded information.

   Figure 9.3 shows the printout of all maintenance operations performed on a large punching press over a six-month period. Observing a high frequency of repairs on the clutch and on the straightener, the production center maintenance group could justify to the production manager the clutch replacement and straightener modification in order to reduce the number of adjustments.

OPERATOR MAINTENANCE SHEET

Work Position N° .... Week ....

	GOOD	BAD	OBSERVATIONS
. UNACCEPTABLE CLEARANCES OR ABNORMAL NOISE	......	.....	....
. CLEANLINESS	......	.....	....
. LUBRICATION	......	.....	....
. PRESENCE OF PROTECTIONS	......	.....	....
. INSULATION Connection of electric cables	......	.....	....

COMMENTS ....

....

....

....

....

Figure 9.2. Maintenance Sheet to be Filled in by Machine Operators. (Courtesy of Jeumont-Schneider Industrie).

3. *Measure the effectiveness of the new implemented policies, and motivate (and reward) the people who are involved in equipment maintenance.* Management cannot undertake a major maintenance operations change if any improvements in operation performance cannot be measured. In many plants no one knows the current average values of MTBF and MTTR; however, when new policies are implemented, people have to be given precise objectives. Quantitative objectives, such as a monthly reduction in breakdown rates by 10 percent, can only be set if performance can be measured. The information recorded by machine operators and maintenance workers—in log books, on PM report sheets, and in inspection reports—is the source used to measure performance. Typical performance indicators are:
   - The number of breakdowns per period (week, month, year)
   - The maintenance costs (hours and parts, scheduled or unscheduled maintenance)
   - Equipment availability (the percentage of time that the equipment is available when required for production)

   Performance indicators are set by the production managers and are then explained to the workers. These indicators are computed weekly or monthly.

OPERATION SUMMARY

WORK STATION: 42PM85 3406 WEINGARTEN Press
Issue date: January 24th 90 PERIOD: 6 MONTHS

DATE	OBSERVATIONS	TIME
150689	Remove conveyor belt tables, recondition jacks ejection and table end stops	21.00
150689	Adjust unwinder straightener	01.50 ↙
170689	Overhauling clutch and discharge conveyor belt	47.50 ↙
170689	Rewire force limiter	11.00
190689	Lowering problem during slide adjustment. To be monitored	01.50
220689	Installation of a safety device for opening the press side door	02.00
220689	Connection and adjustment of the load limiter	07.50
260689	Reinstall motor on the press	02.50
300689	Remove motor, change bearings	15.00
300689	Touch commutator up	04.00
300689	Rotor balance	03.00
071289	Remove clutch for replacement of a rod	06.50 ↙
071289	Disassemble clutch	10.00 ↙
081289	Clean the sump by removing oil and plate rejects	07.50
081289	Reassemble the clutch	04.50 ↙
111289	Check the straightener motor	02.50
131289	Adjust the magnetic conveyor belt	02.00
141289	Repair the 2 tables (reattach) and adjustment following a breakage in the linkage. Welding to be programmed	05.50
141289	Check straightener motor	01.50 ↙
181289	Drill tap replacement of the rear end stop bolt	02.00
181289	Remove the bolt broken by the rear end stop	01.50
221289	Unwinder and straightener adjustment	01.00 ↙

Figure 9.3. Summary of Maintenance Operations Performed on Punching Press. (Courtesy of Jeumont-Schneider Industrie).

For example, Figure 9.4 shows the evolution of the weekly number of maintenance hours in one of the cost centers at Jeumont-Schneider. The thick curve indicates the number of hours of curative maintenance (remedial) and the bold curve the number of hours of preventive maintenance. The forty-four week period covers the implementation of TM at Jeumont-Schneider. There is clearly a decreasing trend in the number of curative maintenance hours, as well as an increasing trend in the number of preventive maintenance hours. This information is useful to managers to assess the usefulness of current actions, and should be communicated to employees.

9.2.2 First-Level Maintenance

First-level maintenance operations are an integral part of a machine's daily operations and should be routinely performed by machine operators. These operations consist mainly of cleaning the machines and workstations, lubricating the machines, and replacing machine filters. Having first-level maintenance operations performed directly by the operators serves the following purposes:

- Some imminent breakdowns can be detected before they occur (such as leakages detected during machine cleaning).
- Operators gain a better knowledge of the equipment they use.
- Lubrication is likely to be done more regularly by machine operators than by people not involved in the machine's daily operations.
- Operators gain a sense of ownership of the equipment they care of and use.
- Operators feel more responsible for their working environment.

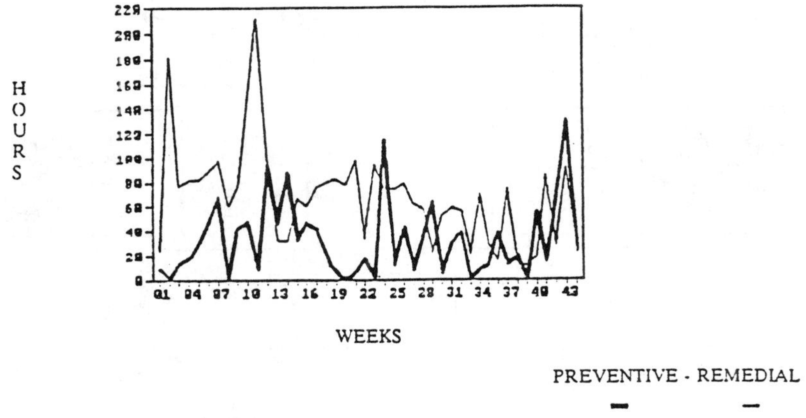

Figure 9.4. Evolution of the Number of Maintenance Hours over the Implementation of TM. (Courtesy of Jeumont-Schneider Industrie).

Educating, training, and involving machine operators in first-level maintenance activities is cost-effective since it contributes to increased plant productivity and customer service, which in turn yields higher levels of competitiveness and profitability. The cause and effect relationships and benefits associated with operator involvement in first-level maintenance is shown in Figure 9.5.

In general, the transfer of first-level maintenance operations to production workers does not encounter strong resistance if sufficient time is designated to execute these operations. Giving personnel the opportunity to work in a clean and ordered environment typically has a profound psychological effect by reviving a sense of dignity among workers. For this reason, first-level maintenance has been well accepted in many companies.

A prerequisite to successfully implementing first-level maintenance is to properly organize when and how the operations should be performed. First-level maintenance schedules should be established and posted on the workstations (next to the PM schedules). Because maintenance personnel are the most familiar with machine maintenance requirements, they should take primary responsibility for establishing the first-level maintenance schedules, on the basis of their

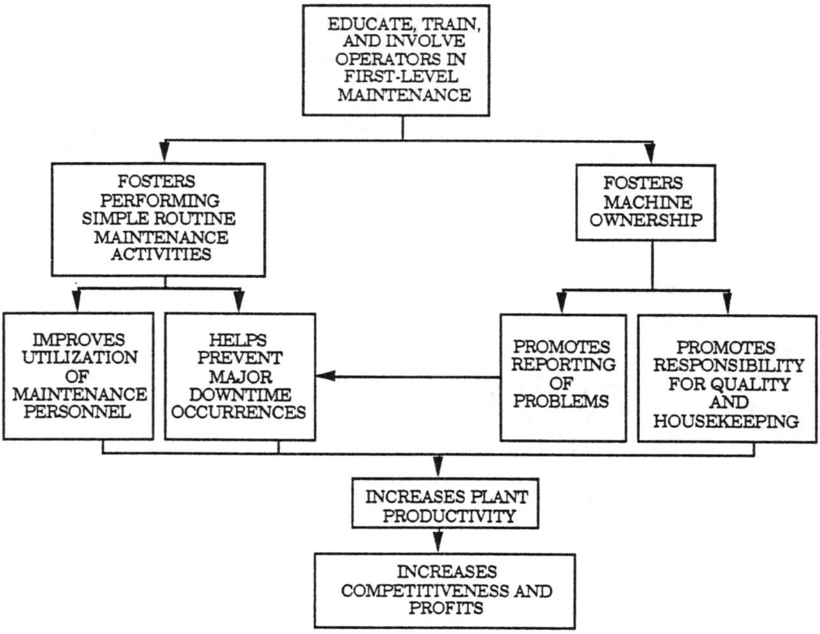

Figure 9.5. The Benefits Associated with Operator Involvement in First-Level Maintenance.

experience and also of the equipment's manufacturer recommendations. Each maintenance operation should be listed on the schedule with the following information:

- Operation description (with drawings or pictures if necessary)
- Lubricant to be used
- Reference of filters to be changed
- Estimated operation time
- Operation frequency

Machine operators should also be involved in establishing the schedules because they can provide valuable input on how the operations should be described. As a result, schedules will be more readily accepted on the shop floor—and readily followed.

Performed operations should be reported by the operators on the posted schedules. In addition, any problems that are noticed when performing the operations should be reported. Production supervisors should regularly check the records, thereby showing their commitment to implementing first-level maintenance.

All reported problems must be discussed and appropriate actions taken. Problems such as insufficient operation description, inappropriate operation frequency, and lack of time to perform operations should be addressed in small meetings where most of the problems can be solved. In addition, maintenance workers should be available to answer operator questions and help them solve startup problems.

At Jeumont-Schneider having the operators perform first-level maintenance operations is also viewed as a way to foster operator ownership of the equipment. In this regard managers observed that first-level maintenance is more difficult to implement on machines with frequent operator change because of multipurpose usage. In this case a different approach could be used, such as a team approach. For example, it is important that operators assigned as a team to a manufacturing workcell gain a sense of ownership of all the machines in the cell, so that they continuously monitor the condition of these machines.

9.2.3 Preventive Maintenance

Preventive Maintenance (PM) is an effort that aims at preventing breakdowns caused by time-dependent component failures, typically normal component wear. Examples of actions with which PM is concerned are:

- Replacing belts and chains, seals and gaskets, and bearings.
- Opening and cleaning components such as pumps.
- Changing electric motors.

Machine lubrication and filter replacement are not included in PM because these activities are more relevant to first-level maintenance operations.

Preventive maintenance is the first step from breakdown maintenance on the journey to TM. Breakdown Maintenance is event based, whereas, PM is time based. Intensive PM programs can prevent up to 80 percent of the total number of breakdowns. However, PM alone is not a zero breakdown program, it is rather one of the strategies used to reach the goal of zero breakdowns.

Preventive maintenance has been practiced with success for more than fifty years by leading U.S. and Western companies, and it was adopted in the fifties by Japanese companies. Predictably, world-class companies have all embraced PM as a key element in their strategy to prevent unexpected breakdowns.

Unfortunately, some companies are still wondering whether they can afford PM. Still other companies have timidly attempted to develop PM programs. However, lack of dedication from maintenance and production departments and pressure for maximum machine utilization have often resulted in disorganized PM schedules and increased maintenance costs—without the compensation of significantly reduced breakdown rates. Additionally, lack of support from top management has lead to slowing down these PM programs and in some cases to simply cancelling them.

Figure 9.6 shows the basics steps of a successful PM program.

These steps are discussed in the remainder of this section.

Preventive maintenance schedules are established by the particular maintenance group that maintains a specific production area (equivalent to the maintenance department in traditional structures). For easy access, PM schedules should be stored or posted at the workstations along with the machine log books and technical manuals. A PM schedule is set up for each different machine in the production area and lists the machine identification and a list of PM operations. With each of these PM operations the following pertinent information is included:

- Operation description (for example, change main transmission belt)
- Estimated operation time (for example, 1.5 hours)

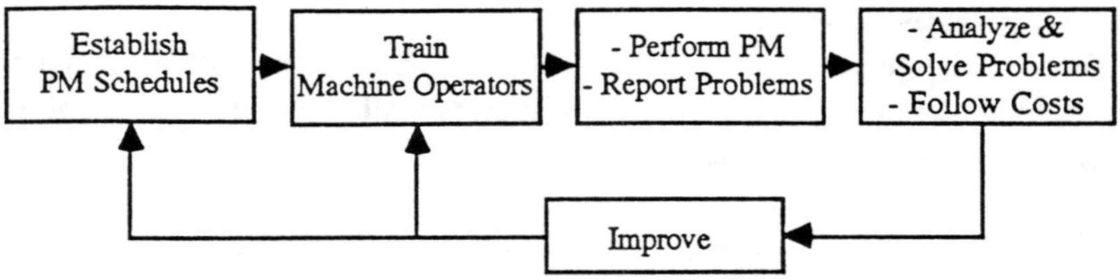

Figure 9.6. Main Steps of PM Program.

- Frequency (for example, every six months or every 2,000 hours)
- Replaced parts identification (for example, belt number 345123)
- Special tools indication (for example, torque wrench 6 DaN-M)
- Who performs the operation (the machine operator or maintenance worker)

When analyzing the maintenance records, input from the industrial engineering group might be helpful, especially when deriving the component's estimated work life from the breakdown history. Because simple formulas from reliability theory are used during the analysis, the maintenance staff should be trained in the proper applications of this theory. Depending on the regularity of equipment utilization (machine constantly used versus machine sporadically used), the PM operation frequency should be expressed in number of weeks, months, or actual operating hours. The latter case requires that hour counters be installed on the machines.

After PM schedules are established by the maintenance group, PM operations are then planned by the production supervisors. These operations can be done by maintenance workers or by the operators, depending on the complexity of the operations and on the current training level of the operators. However, PM operations performed by machine operators can be more flexibly scheduled than when they are performed by the maintenance workers.

Major shifts in the role of maintenance and production personnel can provide an opportunity to break away from traditional work classifications. Machine operators can often realize increased responsibility and rewards, and conflicts between the maintenance and production departments can be eliminated. One of the direct effects of such reorganization is that maintenance workers will not perform any routine work; instead they will play a support role for machine operators and participate only in complex maintenance operations, such as machine check-up, predictive maintenance operations, and emergency repairs. In effect, PM as well as first-level machine maintenance becomes mostly the responsibility of machine operators who are trained and assisted by maintenance workers.

The posted PM report sheets are regularly checked by the production supervisors or by the production team leaders. In addition, production supervisors should periodically check that PM is being performed in a timely manner and help resolve any problems that the operators encountered and marked on the report sheets, such as PM schedules being difficult to understand, not enough time for PM operations, and the operation frequency seeming too low or too high. The cost of PM is also followed by production supervisors.

Maintenance workers need to periodically revise PM schedules and modify (if necessary) information such as lists of operations, operation frequency, and operation assignment (operator versus maintenance worker). Problems encountered when doing PM operations and PM schedule revisions are discussed in periodic meetings (weekly, or biweekly) that include production and maintenance workers and management; thus, PM becomes everyone's problem just as any other production problem. Employee involvement is a prerequisite to successful PM programs.

9.2.4 Employee Involvement

The success of any activity in an organization depends, to a large extent, on the degree of motivation and the involvement of the people in the organization. Motivated people tend to make projects run more smoothly by finding solutions to problems that inevitably arise; while unmotivated people tend to not only let problems remain unsolved, but also create new problems. Motivated people are an asset; unmotivated people are a burden.

Employee involvement is so crucial to the competitiveness of manufacturing companies that the second chapter of this book is entirely focused on this aspect. The reader is invited to refer to that chapter for specifics on EI and work force education. Total Maintenance relies, for the most part, on people-oriented policies that can only be implemented with full employee support and involvement. As detailed in chapter 2, in order to create an environment favorable to EI, three main actions are required:

- Informing the entire work force.
- Educating the work force.
- Implementing a structure dedicated to solving problems.

However, before informing the work force, top managers themselves must be informed; for example, they must understand what TM is, what results can be expected, and how best to attain these results. The best information sources are companies in which TM programs have been in effect for two or three years and are today experiencing dramatic improvement in downtime reduction. Only when top managers believe that similar improvements can be attained in their own company should they inform the work force. Informing the work force serves several purposes:

- Presents the TM program to the employees.
- Explains to the employees what is expected from them.
- Assigns improvement objectives (for example, a monthly reduction in downtime of 6.5 percent).
- Shows management's commitment to the program.
- Fosters employee's interest in the program.

Depending on the number of employees, the entire work force can be informed during a single meeting, or an initial meeting can be held with all the department managers who then conduct individual meetings in their own departments. Experience has shown that workers are more motivated if the project is presented to them by senior management rather than by department heads, because the feeling of sharing common objectives is stronger, and workers can actually witness top management's commitment. Information for TM can also be part of a more general meeting where a global WCM program is presented to the employees.

Proper training should prepare workers to assume new responsibilities. For example, machine operators must be trained for performing first-level and PM operations. The operators should be trained by maintenance workers, first in small group sessions, to give examples and explain concepts, then in individual sessions at the machine to demonstrate processes.

Similarly, maintenance workers should be taught how to train machine operators. Particular efforts must be made in the way the maintenance worker's are informed about the training program: It must be made clear that the changes are not aimed at taking work away from them, but instead that they will perform more varied tasks and will have the opportunity to support production personnel in solving maintenance-related problems. The various information and education steps are illustrated on the diagram in Figure 9.7.

Nevertheless, many problems will arise when TM programs are implemented. Some problems obviously will be because of the new policies and others will be problems that existed before but, because of lack of motivation, have remained unsolved. People in the company should not be discouraged by the emergence of problems, but instead see in the resolution of problems the opportunity for improving operating efficiency and work conditions.

1. Top management education and commitment
2. Employee information and motivation
3. Maintenance personnel training
4. Production workers training (by maintenance workers)

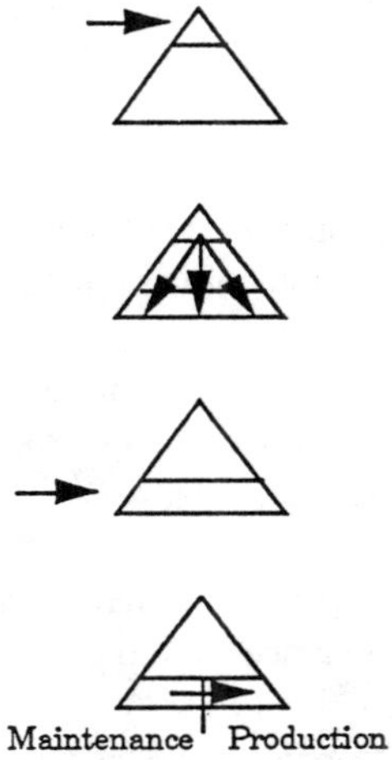

Figure 9.7. Steps for Information and Education.

To ease the problem-solving process, problems should be reported, confronted, and solved at the appropriate level in the company; for example, reducing the number of breakdowns on a specific machine can be attacked by a project team composed of machine operators supported by one or two maintenance workers and coached by a supervisor or technician. Other questions, such as the assistance of machine operators to maintenance workers when repairing machines, should be resolved at an upper level in the company hierarchy. Creating employee teams dedicated to solving these problems plays a major role in the success of TM. These teams should exist at each level of the company's hierarchy; problems that cannot be solved at one level are then transmitted to the next. Permanent teams can meet regularly to address problems reported on PM maintenance sheets, and special project teams can be created to address specific problems. This work in teams serves two major purposes:

1. To ensure that every single problem encountered by workers are addressed and solved.
2. To motivate the employees by giving them the means and time to solve problems.

9.2.5 The New Maintenance Organization

Many companies and plants are organized so that a central maintenance department is in charge of the equipment and buildings in the entire facility and the maintenance workers are commonly organized by craft trades. The central maintenance department answers the requests from the different production departments as breakdowns occur, because usually only the maintenance department is responsible for preventive maintenance. This traditional organization provides advantages associated with economy of scale, yet it also presents many disadvantages:

- Communication delays between production and maintenance.
- Redundancy of information exchanges and deterioration of information quality.
- Lack of concern by maintenance personnel vis-à-vis production priorities.
- Lack of concern by production personnel vis-à-vis machine maintenance requirements.

The redefinition of maintenance activities for TM has a profound impact on the way maintenance should be organized. The significant activity changes are:

- Transferring some responsibilities from maintenance workers to machine operators.

- Developing predictive maintenance.
- Training and supporting operators by maintenance workers.
- Participation of maintenance workers to project teams.

Thus, the new maintenance organization serves two major purposes: (1) To optimize the maintenance response time when machines breakdown, and (2) to increase the level of cooperation and trust between machine operators and maintenance workers. As a result, two key aspects of the new maintenance organization emerge: (1) Maintenance worker specialization by craft trades is reduced or eliminated, and (2) equipment maintenance operations are decentralized.

Craft-trades specialization, often inherited from the past, is not compatible with modern manufacturing technology, which prompts a wide range of technologies to be used in most of today's machines. Companies cannot afford maintenance workers who do not understand comprehensive machine functioning. Because most of the breakdowns are simple to repair and do not require deep knowledge in every craft trade, maintenance workers should be cross-trained; they should be given a solid knowledge base so that they can repair most of the breakdowns (80–90 percent) occurring on the machines in their area. When problems are more difficult, it is still possible to request the assistance of a maintenance worker who is more skilled in a specific technology—even if that worker is assigned to a different production area.

World-Class Manufacturing organizations are characterized by small autonomous operations as opposed to the large functionally specialized departments of traditional companies. Therefore, when striving for WCM concepts, maintenance operations should be decentralized and maintenance workers attached to specific production areas. Because the production organization is based on the concept of a focused factory where each autonomous production entity is in charge of its own support operations (such as receiving, shipping, and detailed scheduling) maintenance becomes one of the functions that is attached to each of the focused production areas. In addition, each of the small dedicated maintenance groups should report directly to their production area manager. Figure 9.8 shows an example of this decentralized manufacturing organization.

The size of a focused maintenance group depends on the need for maintenance services in a particular production area. Actual group size can be based on the number of machines, types of machines, and service history of machines in a production area. Obviously, the modification of maintenance policies will impact the maintenance workers work load by reducing the need for maintenance service by optimizing the maintenance organization and transferring activities from maintenance workers to machine operators. On the other hand, maintenance workers are asked to take on new responsibilities. Considering the change in responsibilities, the size of a focused maintenance group will range from two maintenance

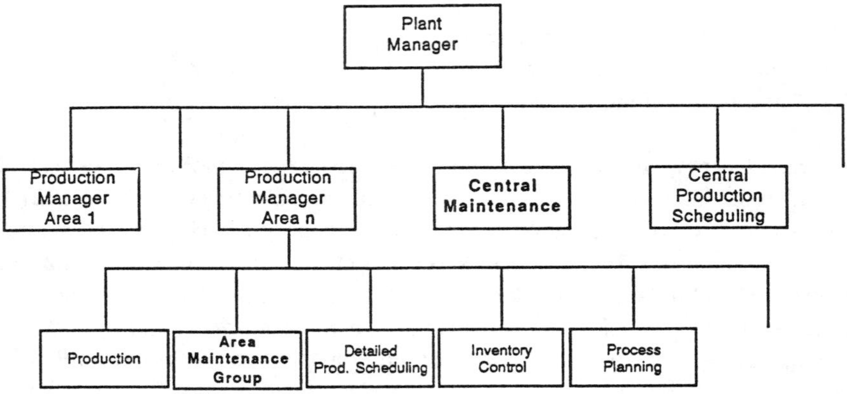

Figure 9.8. Example of Descentralized Organization.

workers to a small group of five or six with one of the workers taking the responsibility of group leader. (The maintenance group should be located on the shop floor next to the machines of which the group is in charge.)

Although most of the maintenance activities are decentralized when implementing TM, there is still a need to have a maintenance structure with a central base. This central base is in charge of maintaining the buildings and the material handling equipment used among the production areas; moreover, it is also in charge of establishing a link between the focused groups. As time passes, each focused group evolves along its own path: Different procedures are implemented in the different production areas, new methods for predictive maintenance are used, and efforts are concentrated on different problems in different production areas. To ensure consistent improvements in maintenance service, it is important that each focused group shares with the other groups the experience it has gained; a new role of the central maintenance structure is to facilitate this exchange of information between the different groups by organizing monthly group leaders' meetings, for instance. Another mission is to organize and follow the in-house training of the maintenance workers in coordination with production managers.

Jeumont-Schneider has experienced an additional benefit of decentralizing maintenance personnel into small production areas, which is an improved recognition of the maintenance function. By having production workers taking over some maintenance tasks and having maintenance workers located closer to the production workers, greater respect is granted to maintenance personnel, which as a result, is better motivated to improve equipment condition.

9.2.6 Predictive Maintenance

Predictive and preventive maintenance aim toward the same goal of preventing unforeseen breakdowns. The role of PM is to regularly change machine components that are subject to wear or likely to fail; whereas, the role of predictive maintenance is to periodically inspect various machine parameters (for example, dimensions, pressures, and intensities) and to schedule servicing operations if necessary (from replacing hydraulic drives to rebuilding machines).

Predictive maintenance consists of scheduled activities: Either time-based activities (inspections), or event-based activities (repairing). For example, if a scheduled electrical inspection shows that the current intensity on an electric motor is outside of the normal operating range, repairing or replacing the motor is scheduled. Predictive maintenance activities are planned by the area's maintenance group in accordance with production schedules.

In some organizations, both preventive and predictive maintenance operations are grouped under the PM label; however, they are fundamentally different. One important distinction is the difference between the type of workers that perform the operations. Preventive maintenance covers routine operations that eventually should be the exclusive responsibility of machine operators. On the other hand, predictive maintenance operations should be performed by maintenance workers who are usually assisted by outside people. In addition, companies—especially smaller companies—usually cannot afford to purchase expensive inspection tools, such as laser measuring devices, and therefore must subcontract this type of operation to specialized companies.

Predictive maintenance programs can be developed based on the equipment manufacturers' recommendations and then improved as more knowledge is gained by the company.

Some examples of techniques that can be applied for detecting component failures and abnormal machine functioning are:

- Analyzing machine vibration, temperature, pressure, electric intensity, and sound.
- Checking the condition of filters and fluid.
- Monitoring dimensions and motion patterns.

Because predictive maintenance is still an emerging field, its development depends mainly on the availability, cost, and ease of use of the analysis tools. The purpose of predictive maintenance is not to replace PM, but to complement its action. In many cases it will remain less expensive to regularly replace machine components rather than to periodically perform inspections that can be highly technical and costly.

9.2.7 Design for Dependability and Maintainability

The next step toward increasing equipment availability for production is to purchase equipment that has been designed with reduced breakdown frequency and repair times in mind. Equipment designers need to analyze problems faced by the users; the client's maintenance personnel should be consulted when machines are designed because their experience is invaluable in identifying mechanisms or parts that are likely to fail and in helping engineers design machines that are easier to maintain. For the same reasons, when equipment is purchased, consulting maintenance personnel is strongly recommended.

Machine dependability can be improved by simplifying design and reducing the number of parts (especially moving parts). Another direction for improvement is to design more robust components and machines; that is, machines that have similar behavior under different operating conditions (for example, temperatures, pressures, and humidity). The task of engineers is to design such machines—but first they must gather information from customers to determine which components fail most often, and under which conditions.

Maintainability can be improved by the users themselves through machine modification, up to a certain limit. Machine manufacturers can further improve maintainability by building it into the design.

Improving the maintainability of existing equipment is a problem that can typically be solved during small group activities. For example, replacing screws with rapid attachments is a simple solution to speed up maintenance operations. An approach similar to the one used for setup-time reduction is applied here with most of the improvements being devised by the maintenance workers themselves. In addition, videotaping a particular maintenance operation might help to better identify solutions by allowing the problem to be viewed by a larger group of people. All preventive maintenance operations can benefit by simplifying actions, such as replacing internal control devices with controls that can be checked directly without having to open a single element (for example, component-wear and fluid-level controls). Other improvements, such as increased component modularity and standardization, are virtually beyond the capabilities of equipment users, but they should be implemented when new machines are designed. Good maintainability must be one the objectives given to the designers. Again, in this endeavor, input from maintenance workers can help in designing machines that are easier to maintain.

Machine design for improved maintenance is a long-term strategy. New contacts have to be established between equipment manufacturers and equipment users so that more information is received from machine users (not only from production managers, but also machine operators and maintenance workers), and more visits are made by design teams to their clients' plants. At the client level, when new equipment is to be purchased, contacts between decision makers and final users have to be reinforced. Help from maintenance workers should be sought to identify equipment that is less likely to fail and easier to maintain.

9.3 CHANGE IMPLEMENTATION

Because the TM program is only one of steps that must be taken by a company in progressing toward WCM, organizing the implementation of TM strategies cannot be totally separated from the organization and planning of other steps,such as focusing activities, implementing workcells, and implementing total quality. However, this section will focus solely on how management should organize the implementation of the TM program and detail the related actions that must be performed at various levels in the company. Although this chapter is dedicated to maintenance, it is important to keep in mind that the described actions take place in the context of an entire reorganization of manufacturing operations.

9.3.1 Planning the TM Project

It is the responsibility of company or plant management to plan the various steps of the TM project, the first priority being to choose a project leader. The project leader then guides the management team through the necessary stages in the planning process:

1. *Define objectives and scope of the project.* The project objectives should be specific and measurable. Examples are, reducing the breakdown rate from forty per week to five per week in a ten month period and reducing the average repair time from fifty-five minutes to twenty minutes in a six-month period. The scope of the project should be the same for TM as it is for the entire operations reorganization whether it be for a department, a plant, a manufacturing division, or the entire company's manufacturing operations.
2. *Define the main orientations for the particular strategies described in section 9.2 of this chapter.*
3. *Form a project structure and train its members.* The size of the project team depends on the scope of the project. Training consists of presenting to the project team the concepts involved in TM and possibly visiting a company that is implementing or has implemented similar concepts.
4. *Define main project deadlines.* These TM deadlines have to be considered in relation to the overall project deadlines. For example, implementing decentralized maintenance groups depends on the reorganization of other operations and can only take place when the focused facilities have been installed.
5. *Inform the work force.* The information meeting allows management to present the project to the entire work force. This is the project's kick-off meeting where it is essential to raise interest and enthusiasm from the work

force and sell the win-win benefits of the program. Typically, reorganizing maintenance activities is one of the subjects of the initial meeting in which the overall reorganization is presented. The TM project leader and the project team should be introduced to the work force during this meeting.

6. *Organize follow-up.* Since top management's involvement and support are crucial to the success of the project, formal follow-up meetings should be scheduled weekly or monthly between top managers and the project leader. In addition to these meetings, informal contacts on the shop floor show that managers are committed to keeping contact with the realities of the project.

9.3.2 Implementation Steps

During the implementation phase of TM, the project team works closely with production and maintenance personnel. The role of the project team is to plan the detailed tasks that need to be performed, to assign responsibility for these tasks to the people in the company structure (production and maintenance managers, supervisors, workers, et cetera.), to facilitate the resolution of conflicts and problems, to make sure that tasks are completed on time, to report progress and problems, to management, and to adjust schedules when delays occur.

A list of typical tasks to be performed when implementing TM is shown in Table 9.1. This list is based on the strategies presented in section 9.2 of this chapter.

The responsibility for the tasks listed in Table 9.1 should be assigned to different people in the existing company structure. For instance, defining performance indicators can be assigned to a group of production managers, while listing the current problems with machine conditions can be assigned to the maintenance workers. It is important that every person in the organization is aware and understands that their involvement is required during the change implementation. Having the people involved in the project as early as possible significantly reduces the risk of having them reject the changes.

When deadlines are assigned, a detailed schedule is built. Deadlines for implementation should be realistic. Most of the changes will be originated during the first year of the project life; however a normal time frame for implementing such programs is about three years. Typically, personnel training requires the most time because workers are needed to continue production even though changes in production are occurring.

Follow-up meetings must be planned regularly between the project team and the people in charge of task completion. Most of the problems encountered during the implementation phase are related to the difficulty of changing people's mentality and the existing company culture. It is important not to lose momentum at this critical point in the project. A natural tendency for people who feel overwhelmed by new problems (because of the reorganization), and existing problems (because of continuing production, for instance), is to let the new problems go unsolved,

TABLE 9.1. Tasks to Be Performed for Implementing TM.

To Measure Performance: - Define performance indicators. - Define procedures for performance measurement, analysis, and communication.
To Improve the Equipment Condition : - List the current problems with machine conditions, and define and prioritize actions. (If machines are to be moved for cell formation, take the opportunity to check the geometries and take the necessary actions while the machines are out of production.) - Define the maintenance plans for every machine for the total machine life. The maintenance plans should list all operations covered by first-level, preventive, and predictive maintenance. - Define the format of machine log books and first-level, preventive, and predictive maintenance reports. - Define procedures for : • reporting on log books and maintenance reports, • checking them, and • taking appropriate actions from the reported information. - Define the scope of the relationships between equipment purchasing groups and maintenance personnel.
To Improve the Organization: - Define the size of the new maintenance structures concerning both: • the production area maintenance groups, and • remaining central maintenance cell. - Plan the moving of maintenance groups to (focused) production areas. - Define the responsibilities of maintenance and production personnel. - Define the new maintenance procedures (e.g. what to do in case of breakdown?).
To Train the Personnel:* - Define the training needs and then create a timetable for production and maintenance personnel.
To Organize Teamwork:* - Define the structure to implement and support teamwork. - Define the scope of the problems to be reviewed. - Define the meeting frequency and its location.

*Training and teamwork could be planned with a larger perspective than simply TM, as covered in Chapter. 2.

thereby progressively sliding back to past habits. The role of the project team is to keep momentum in the project.

At least a part of the project team should remain functioning until the end of the implementation phase to help solve problems. As the number of problems decreases and the new organization starts functioning more smoothly, the permanent structure progressively takes over problem solving and the project structure can eventually be dissolved.

9.3.3 Continuous Improvement

Questioning and improving the performance of maintenance operations should not stop after the dissolution of the project structure. The TM program is a permanent improvement program. The change-implementation phase marks a radical transformation in the way people think and act about equipment maintenance. Thus, a new organization has been created, which allows closer contacts between maintenance and production personnel. New strategies have been put in effect that lead to an increase in equipment dependability and a decrease in repair times.

As people on the shop floor are given more responsibility in maintaining the equipment, they are also given the opportunities to improve the way maintenance operations are performed and to suggest equipment modifications that will facilitate maintenance operations. The purpose of working in teams is to enable the operational personnel (production as well as maintenance personnel) to express problems they encounter, to analyze these problems, to make suggestions, to find solutions, and to present these solutions to management. Management should be prepared to evaluate suggested improvements and to facilitate their implementation.

Regular meetings between the different production areas should also be organized to ensure that improvements in one area are generalized to the other production areas.

Production area managers and plant management should follow the evolution of performance indicators and take actions accordingly. The objectives defined at the beginning of the TM project should be reviewed periodically.

9.4 SUMMARY AND MANAGEMENT GUIDELINES FOR SUCCESS

Total maintenance is a mandatory step on the way to becoming mean, lean, and world-class competitive in the nineties. With production lead times and inventories reduced to their minimal level, companies cannot afford to have machines stopped when they are needed for production. Even relatively short equipment breakdowns can increase production lead time, reduce the average utilization of the machines, and result in operator idle time. A major breakdown can stop other downstream and upstream processes, and lead to missed shipping dates and costly overtime. In addition, poorly maintained equipment can only produce poor quality parts.

To avoid such counterproductive situations, new strategies regarding equipment maintenance that will reduce the machine breakdown rate must be implemented. The bottom line of the TM program is to reorganize maintenance operations so that maintenance problems are eliminated or become easier to solve and can be solved preventively rather than reactively.

A strategy to apply first is to record and utilize the information available from shop-floor processes. The analysis of these records allows management to follow the condition of the equipment, to schedule maintenance operations, to know the effectiveness of new policies, and to appreciate personnel's performance.

An important aspect of the new culture in world-class companies is that operational personnel become involved in activities that have traditionally been the role of so-called support departments. As far as general maintenance is concerned, machine operators become responsible for first-level maintenance and progress to performing most of the preventive maintenance operations. Operators are responsible for the condition of their machine(s) and for reporting problems related to equipment conditions.

Implementing TM leads to the participation of machine operators in the decision-making process. Working in teams with maintenance workers, machine operators can apply their knowledge of the equipment to help prepare maintenance programs or to improve the way the machines can be maintained. Operators trained for equipment maintenance enjoy larger responsibilities, through enlarged tasks. Often machine operators gain a sense of ownership of the equipment they use that, in turn, motivates them to keep it in good working condition. Total maintenance is necessarily linked to EI, with the long-term result that employees gain control of their job tasks and, therefore, can enjoy greater job satisfaction.

A new maintenance structure is created that allows maintenance personnel to be closer to the equipment and to the machine operators. The original centralized maintenance structure is broken down by each particular production area. Each production area maintenance group is under the responsibility of the production manager, thereby reducing the number of potential conflicts between production and maintenance personnel.

This new organization has the advantage of reducing the physical and psychological distance between maintenance worker and machine operators. This advantage is attained as early as during the machine operator's training for first-level and preventive maintenance for which the maintenance personnel plays a major role.

Maintenance workers benefit from TM through a reduction of routine work and of "fire fighting" maintenance. Their work is more planned and at the same time becomes more challenging, because of the increase in technical and interpersonal skills (for predictive maintenance tasks and for machine operators training, for example). Being physically closer to the production areas, and also closer to production needs, maintenance workers gain increased recognition from production personnel. The evolution toward TM undoubtedly adds value to the maintenance function. Being relieved of routine maintenance operations, maintenance personnel can spend more time organizing maintenance operations, performing predictive maintenance, and collaborating on machine design improvement.

The organization of maintenance operations consists of establishing maintenance plans for the entire life of each of the machines (first-level, preventive, and predictive maintenance).

Predictive maintenance consists of periodic inspections of machine characteristics (dimensions, displacements, pressures, intensities, et cetera.) and is performed by maintenance personnel supported by specialized contractors. The role of predictive maintenance is to complement preventive maintenance in reducing the occurrences of unscheduled downtimes.

Machine design improvement, in the short term, is aimed at modifying existing machines in order to (1) reduce the risk of sensitive component breakdowns, and (2) reduce repair times (for instance, by improving accessibility to internal components). A long-term strategy is to involve maintenance workers in designing new machines. Communication is improved between maintenance workers and machine buyers so that more dependable and easier-to-maintain machines are given top priority when purchasing equipment.

On the management side, the win–win benefits, are a reduction in inefficiencies and waste associated with traditional maintenance operations and, as a result, improved productivity. Repetitive problems are identified and eliminated, the productive work force is more flexible, better trained, and has better morale. Better maintained equipment leads to manufacturing better quality products, and allows for higher utilization of the equipment. It is clear that enjoying these benefits can only be the result of important efforts on the management side. These new strategies for world-class maintenance require at least as much involvement and support from top management as designing implementing manufacturing cells, reducing setup times, and the TQ program. The role of top management guided by the project leader is to organize the TM project, select the project structure, inform of the entire work force, closely follow the evolution of the project, and continuously support the people in the project structure. The role of the people in the project structure is to establish a detailed plan of the various implementation tasks, assign these tasks to the most appropriate people in the company's organization, facilitate and follow the completions of the tasks, and report to top management. The major problems during the implementation phase are likely to stem form the difficulties that people experience in changing their way of working and way of considering other people in the organization. Therefore, personnel education and training is allocated sufficient resources.

Evolving toward TM is certainly a major culture shift for many companies. Since the success of this shift relies for the most part on how well the company work force understands and adheres to management's vision, it is crucial that management succeeds in "selling" the TM project to company personnel. In this regard, follow a few tips from the experience of companies successful in implementing change:

- Inform the personnel of projected change as early as possible.
- Explain the reasons for the change, show why the current situation is not acceptable (for example, communicate the number of production hours lost due to machine breakdowns, or the number of overtime hours that were caused by breakdowns).

- Involve the work force and the unions in designing the new system (for example, in preparing the new job descriptions).
- Give a clear picture of the projected change and of its impact on everyone's work (nobody will cooperate with the project team if they do not see where they fit in the new organization).
- Identify key employees that are in favor of the change and that will lead the rest of the work force.
- Show how important the project is to the company (promote the project in the company's newsletter, for example).
- "Debug" the new organization in a pilot area and then expand the change to the rest of the company.
- And finally, make sure that the work force understands that it is long-term efforts, and that it will take several months to several years to fully implement all the steps of a TM program (actually, it is never-ending effort, since the objective is continuous improvement). As other steps toward WCM, implementing TM should be viewed as a Marathon race and not as a 100-meter sprint.

Continuous support from management will help everyone to retain their energy and enthusiasm for the project. The results are worth the efforts. As experienced at Jeumont-Schneider, the implementation of suitable organization, analysis methods, and manufacturing team motivation are necessary to the gradual cultural change essential for the success and efficiency of TM—a key requirement in the search for quality, reduced delivery times, and reduced costs.

10

Supplier Development

Supplier development is now recognized by WCM companies as an essential component for increasing competitiveness. Suppliers are no longer "important" to success, they are *critical* to success. One element of supplier development is to develop and foster cooperative relationships with a selected number of suppliers. When companies—both the customer's and the supplier's—function in a more integrated manner by working together, a synergism results that provides productivity and quality improvements in design, manufacturing, and logistics. These improvements reduce both the supplier's cost to manufacture the product and the customer's price to purchase the product. In essence, when customers and suppliers work together to reduce the waste and inefficiencies in design, manufacturing, and logistics, which typically occur between their organizational boundaries, the resulting partnership provides significant improvements that increase the competitive strength of each member. This chapter discusses the nature and thrust of supplier development in the nineties and presents a vendor survey and certification plan for achieving the partnership-like relationships with quality-certified vendors that are critical for success in WCM.

10.1 DEVELOPING PARTNERSHIP-LIKE VENDOR RELATIONSHIPS

In chapter 2, the concept of working in teams to identify problems and implement effective solutions was discussed, showing the benefits of group versus individual efforts. Chapters 4, 5, and 8 addressed the importance of standardizing and simplifying the processing and flow of materials and information, and presented GT concepts and cellular manufacturing layout strategies for accomplishing these objectives. These same concepts and related benefits are also applicable when developing integrated strategies and common goals between a company and its suppliers. This section addresses the type of relationships that need to be established for greater competitive positions and discusses the basis that exists for common goals and benefits.

10.1.1 Why Suppliers Are Critical for Success

On the average, manufacturing companies spend more than half of a products' manufacturing cost in purchasing raw materials and parts from suppliers. For a manufacturing company to produce and ship a high-quality competitive product on time, it is essential that the purchase materials used to manufacture the product are likewise obtained at competitive levels of quality, delivery, and cost (QDC). Quality and delivery problems caused by vendors induce problems into production that are difficult and costly to accommodate or correct. For example, adjusting to such problems by increasing inspection activities or holding higher levels of inventory adds only cost (not value) to a product. Adding cost is no longer an option in the nineties. Quite simply, a manufacturing company cannot become a world-class competitor unless it has a strong base of suppliers that are likewise able to provide materials and parts at competitive QDC levels.

Suppliers are strategically critical for several reasons:

1. In a world of increasing technological complexity, vendors can offer knowledge and ideas for product innovation beyond the capabilities and resources of any single company. By working as partners in both product development and product improvement, the competitiveness and profitability of both companies can be strengthened and increased.
2. Establishing good supplier relationship provides for operational efficiencies. If the supplier is integrated into the operational requirements of a company's fabrication or assembly line, then a smoother and more reliable flow of both materials and information from the vendor to the company is achieved. When such a relationship is firmly in place, the result is often a substantial competitive advantage reflected in significantly lower inventories, faster overall response times, higher quality, and lower total costs. Most importantly, the new relationship encourages continued improvement in processes and procedures.
3. Establishing good supplier relationships strengthens a company's ability to grow and respond to the increasing competitive demands of the marketplace as more and more companies strive to become world-class. Those who will operate based on the telephone-price-contract types of supplier relationship will have less and less abilities to respond to improving performance standards of manufacturing industries, endure market fluctuations,

reduce costs of inventory and delivery, and prevent quality problems caused by the sources. It's a matter of long-term survival.

Yet many manufacturing companies still consider themselves to be independent from their suppliers. The relationship between them and their suppliers is simply that of buyer and sellers. The primary supplier selection criterion is price, and there is reluctance to change. The realities of the nineties, however, will dictate that manufacturers and suppliers work together to eliminate the wastes and costs that are typically accumulated between their organization boundaries. It will become increasingly more important for manufacturers to select and award to suppliers on the basis of minimizing total cost and maximizing product value, not simply minimizing purchase price. While some companies may claim that their manufacturing operations are already operating satisfactorily without supplier development efforts, the key issue for these companies really comes down to whether it is wise to wait until a competitive opportunity becomes a competitive necessity.

10.1.2 New Criteria for Purchasing: QDC

The ultimate objective of any manufacturing company is to make a profit by providing the customer with goods and services that offer better value than the competition. To achieve this objective, many companies are drastically reducing (or have reduced) the size of their supplier base, thereby, working with a limited number of suppliers to develop partnership-like relationships that are based on the common goals of increased competitive strength and profit. Improving any company's competitive position and profit picture requires a reduction in the levels of waste and inefficiency, which can be achieved, in part, by focusing on improving QDC. This effort involves the following considerations for supplier development.

Quality

Problem prevention at the source is the best strategy for quality improvement. Implementing this strategy involves providing better definitions of the company's quality requirements, working with the vendor's design team, and assisting the supplier in implementing statistical process control and other techniques that focus on defect prevention. In the nineties, considerable focus must also be on enhancing the supplier's quality planning activities. For example, a company may work with a supplier to implement the following activities to improve the quality of each part and/or part family supplied (if they don't already exist):

- Define the process flow
- Identify critical characteristics
- Develop the process FMEA

- Develop a process control plan
- Define annual quality improvement (AQI) objectives

The many other facets of a TQ program for the product and process improvement are addressed in the Vendor Survey Instrument presented in section 10.2.

Delivery

The objective is to have a continuous and reliable flow of quality materials delivered from the supplier directly to the point of requirement (or production line) *just-in-time*. This objective is often more easily accomplished with a dedicated supplier who is in close proximity to the manufacturer. Physical distance usually influences the transportation batch size and the frequency of transportation and delivery. Even though there are ways to reduce the batch size and increase the frequency of delivery, such as by a mixed loading truck arrangement, close suppliers provide advantages over more distant suppliers. For example, reduced storage areas, eliminating double and triple handling, more frequent feedback, quick adjustments to abnormal conditions, and reduced inventories are some of the benefits of frequent small batch deliveries from suppliers. Quite simply, supplier closeness facilitates the delivery of smaller quantities of parts at greater frequencies and with lower transportation costs and less risk.

Proximity also promotes better "face-to-face" contact between the various members of the partnership companies. It is essential that the supplier clearly understands the customer's requirements for delivery, and the consequences of failing to meet those requirements. Especially in the early phase of developing a supplier relationship, frequent cross-visitations are most helpful in tying separate operations together. People from purchasing, production, quality assurance, and engineering need to participate in developing a reliable, cooperative, long-term working relationship. Physical distance makes it more difficult to develop face-to-face contacts. Letters, reports, or phone conversations cannot provide the benefit of "visual control," which is only possible during actual visits to the plant.

Cost

The best way for a company to reduce their material costs (that is, the price of the purchased materials) is to work with their suppliers to reduce each supplier's cost of manufacture. The issue of quality improvement and cost reduction goes beyond the supplier's responsibility for defect prevention and quality planning. Continuous process improvement needs to be practiced on the supplier's side as well as the manufacturer's side. As partners, both need to practice "quality at the source." The earlier problems can be anticipated and prevented, the lower the cost to the supplier and the customer. Such efforts can yield less rework, less scrap, less disruption to production schedules, and fewer resources wasted for both the supplier and the purchaser.

When the number of suppliers is reduced, arm's length relationships turn into closer, mutually beneficial relationships, where cost reduction and cost sharing activities can be practiced much more easily. With a shared understanding, both parties will develop the idea of codestiny and learn to help each other improve their competitive positions through activities such as early involvement of suppliers in product development, cross-visitation to share productivity improvement activities, and joint study teams to develop improvement and cost-reduction programs. However, because of the personal nature of internal costing information, this task involves not only a commitment, but also a view of codestiny that is shared between a company and their suppliers.

The QDC issues are certainly not unrelated. For example, delivery problems can be caused by quality problems at the supplier's base; cost can be affected by both quality and delivery; and so forth. Dr. Deming's chain reaction shown in Figure 10.1, illustrates the nature of the dependencies between QDC and long-term company growth and competitive advantage.

In effect, supplier development involves improving the manufacturing capabilities of one's dedicated suppliers to improve, in turn, the manufacturing capability of one's own company. These are significant common goals. The old purchasing policy of using competitive pricing among vendors is clearly not an effective strategy for achieving quality, delivery, and cost—the joint requirements for purchased materials in WCM.

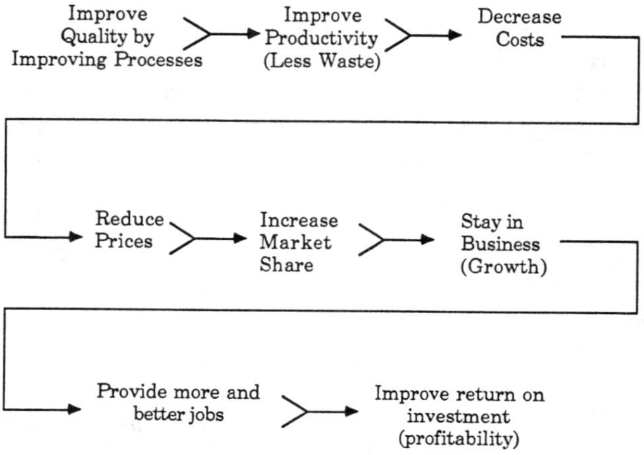

Figure 10.1. Deming's Chain Reaction of Quality-Generated Profitability

10.1.3 Identifying Common Goals

A first impression of supplier development seems to be that it is a time-consuming and risky undertaking. From a company's viewpoint, the time and cost required to seek out, select or classify, qualify, improve, and then certify suppliers is certainly significant. In addition to the vendor-appraisal process, the efforts a company expends in working with a supplier to increase their manufacturing capability is an even longer term task requiring even more significant energies and investment. From the vendor's viewpoint, the time and energy requirements can also be quite significant. Furthermore, the supplier has to absorb the costs associated with implementing many new systems that focus on quality and productivity improvement. For example, the purchaser may require the supplier to implement statistical process control procedures on all applicable manufacturing operations, which typically involves training and adding personnel technically competent to both implement and maintain such programs. A more sensitive issue is the fact that the customer often requests the supplier's information on costing and pricing in order to establish long-term agreements from both standpoints. While all these activities are certainly worthwhile in the long run, the contrast to the old way of doing business, based on competitive pricing, is , at first, quite dramatic. Consequently, it is crucial that clearly defined win–win strategies and arrangements be established in order to counteract and balance the time, money, and risk involved in investing in this new way of doing business. While increasing profit and competitive advantage is certainly an attractive inducement to forming win–win relationships, it is still the element of honesty and openness between the respective companies that determines the success of these new partnership relationships.

It is not always possible to immediately, or even ultimately, establish win-win relationships between any pair of companies. Many small companies, for example, have little leverage because of their minimal impact in buying materials from much larger supplier companies. Consequently, the majority of purchaser–supplier partnerships established to date have been of the type where the larger purchasing company came to the smaller supplier company in order to initiate the possibility of a relationship.

The common goals that exist for both parties in a customer–vendor relationship are to increase quality and productivity, to increase their market share, and to increase their long-term viability and profitability. The purchaser also stands to benefit by having a vendor that provides a steady and reliable stream of quality products on time and with some long-term price stability. In turn, the vendor in the relationship stands to benefit by being selected as the primary vendor (or one of a very few) for a company that provides some assurance of a market for their products over a long-term horizon. Quite simply, both parties have something to offer and something to gain: A codestiny as partners for increasing their competitive advantage in today's international marketplace.

10.2 A SYSTEMATIC APPROACH TO VENDOR SELECTION AND CERTIFICATION

A significant part of supplier development involves the process of vendor selection, qualification, and certification. Although this process varies somewhat in procedure, it has fairly standard elements that typically consist of a multiple-stage process, which involves an initial selection of primary (critical) vendors; performing an initial quality systems survey; a recommendation based on the survey, either to accept or reject the supplier as a "qualified" vendor; a trial performance period; and then awarding the certified-vendor status. Once a vendor is certified, the process of working with vendors continues on an ongoing basis—as with continuous improvement—typically with yearly reviews to evaluate their performance on jointly-agreed-upon annual quality improvement objectives. Figure 10.2 shows the three major steps in this process, namely, selection, qualification, and certification. This section discusses this sequential process in a step-by-step format suitable for presentation to vendors in the early stages of communications.

10.2.1 Vendor Selection

The first stage of vendor selection consists of the following four steps.

Step 1: Categorize existing vendors. In step 1, a company typically categorizes their existing pool of vendors

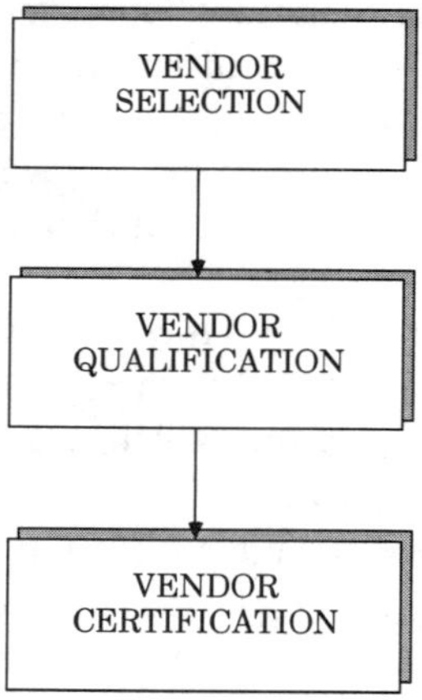

Figure 10.2. Three Stages for Establishing Certified Vendors.

into one of two groups: primary vendors and secondary vendors. The primary vendor category is reserved for a fixed number of suppliers, often the 10–20 percent of all suppliers that provide 80–90 percent of the dollar value of purchased materials. Typically three criteria are used in establishing primary vendor status: The dollar value of purchasing activity, the frequency of purchasing performed (for example, orders per month), and an index describing the purchased material's importance on the quality of the final product and on the company's manufacturing operations. This index can be set up by using a scale numbered from 1 to 10, where 10 represents a category of utmost quality. The secondary-vendor category is for an unlimited number of suppliers, ordered from on a regular basis, that do not qualify as primary vendors.

Step 2: Solicit and seek out possible new vendors.

In starting a vendor qualification-certification program, a company should be open to looking for new vendors with whom partnership-like relationships would provide good win–win paybacks and opportunities for technological and managerial exchanges. For best results, such vendors should be located in relatively close proximity to the point of use, have good quality cost and delivery potential, and be willing to establish codestiny relationships incorporating cost and competitiveness.

Step 3: Submit a letter of introduction and a vendor quality-systems survey to the primary vendors.

Ideally, the survey should be sent to the vendors for the purposes of introduction and information only. At a later date, the customer then uses the vendor survey instrument to conduct a quality audit at the supplier's facilities. In cases of financial constraints, some purchasers submit the vendor survey to the supplier asking them to complete and return it to the company.

Step 4: Visit interested vendors by sending a task team for an initial discussion on the vendor-certification program and processes.

If the vendor approves of the process for certification, then this step, completes the end of the selection process. It is also during this meeting that both parties agree on a timetable for performing the in-house quality systems audit using a vendor survey instrument. (A complete survey instrument is presented in section 10.3.) Alternatively, the vendor may not be interested in or approve of the customer's certification process, whereby no further activity is typically completed with this vendor.

10.2.2 Vendor Qualification

Qualification is the status a vendor receives after passing the initial quality-systems survey. Once qualified, the vendor goes through a trial period to show that they can supply, on time, a quality product that does not require incoming inspection at the customer's facility. The vendor qualification stage consists of steps 5 through 13.

Step 5: The customer performs an in-depth quality-systems evaluation at the vendor's facilities using the survey instrument.

The evaluation at the vendor's facilities typically takes one to two days. Prior to conducting the evaluation, the vendor should supply the customer with a copy of their quality manual and plan; a copy of their process-control flow chart; a copy of their organizational chart, showing accountability for quality, and information covering details for material control. If these items are not available, a plan for their development would be discussed prior to conducting the evaluation at the vendor's facilities.

Step 6: The customer and the vendor define the terms of agreement for qualification and for a sample job.

To prepare the qualification sample bid, the customer provides the vendor with information to use for establishing pricing on the qualifying sample, which encompasses blue prints, workmanship standards, and process requirements. The agreement also includes a general timetable stating the vendor requirements needed to achieve certification.

Step 7: The customer's purchasing department reviews information on pricing and material control.

Step 8: The vendor names a quality contact and a material release representative.

Step 9: The customer and the vendor finalize the process requirements relative to the customer's product specifications.

Step 10: Both parties review the vendor's process and control systems for soundness and integrity.

Step 11: The customer's task team reviews and evaluates the supplier's overall capability.

Step 11 is a review of all information provided by the vendor and represents a major milestone in the process toward certification. If any of the vendor's systems and procedures are felt to be lacking by the customer's task team then the vendor is given an opportunity to take corrective action. If the resulting corrective action also fails to meet the standards of excellence established by the customer's task team, then activity is terminated with this supplier. On the other hand, if the supplier passes this review and evaluation then the process continues.

Step 12: Both parties sign an initial vendor agreement authorizing the vendor to begin working on the sample job.

Once the vendor agreement is reached in terms of quality, cost, and delivery requirements, the vendor begins working on an initial qualification sample of parts. This job, when completed, is then shipped to the customer.

Step 13: Both parties evaluate the qualification sample.

If the product delivered to the customer meets all the requirements, then the vendor is classified as a "qualified vendor." Again, if there are problems with the qualification sample and corrective action fails to alleviate these problems; then, in most cases, further activity in the process toward certification would cease.

10.2.3 Vendor Certification

After the vendor is qualified, the following steps in the process need to be completed before the certification is conferred.

Step 14: The vendor is notified by the customer's purchasing department of conditional approval for the first production lot.

Steps 14 through 18 basically address what is to be done to achieve certification, when it is to be done, and what is required from both the customer and vendor in terms of information, materials, and cooperative efforts.

Step 15: Both parties determine the size and composition of the first production lot.

Step 16: The customer and the vendor determine the final process requirements needed to satisfy product requirements.

Step 17: Both parties approve the vendor's final process capabilities to satisfy the customer's product requirements.

Step 18: Both parties agree on certification requirements.

Step 19: Both parties evaluate and inspect the first production lot.

This evaluation is based on a measurement of QCD for the first production lot. If these criteria are not met by the vendor, then an opportunity for corrective action is typically given. Once again, if corrective action fails to alleviate the problems, further activity is typically terminated at this point.

Step 20: The process control and inspection that was formally provided by the customer is no longer required.

The supplier starts shipping directly to the point of use. When this stage is reached, the customer's quality assurance department notifies purchasing of the direct shipping policy.

Step 21: The customer's quality assurance department monitors quality on an ongoing basis.

If problems are detected with the point-of-use shipment policy, quality assurance immediately initiates corrective action on the part of the vendor. If corrective action fails in this case, all activity with direct shipment will be terminated and the certification process will return to step 19. Furthermore, purchasing will be notified to take appropriate action as was the case prior to the direct-shipment policy.

Step 22: The vendor task team evaluates the supplier's performance for a predetermined time period.

Step 23: The customer's vendor task team performs an in-depth quality-systems audit at the vendor's facility.

During this audit, the task team also evaluates progress the vendor has made on implementing quality-improvement projects that were initially agreed upon as requirements for certification. In essence, the basis for giving certification is a combination of three factors, namely, performance on the quality-system audit, delivery performance over the period of time while the vendor was qualified, and performance in meeting quality improvement activities previously agreed upon.

Step 24: The customer's vendor task team formally certifies the supplier as a certified quality vendor.

Once a vendor is certified, their materials can be shipped directly to the point of use in the customer's facility. Certification, like graduation, is a milestone of accomplishment and a major element in bonding the partnership-like relationship between the customer and the vendor. Since being certified as a quality vendor represents a mark of quality excellence, many companies are able to capitalize in their marketing efforts promoting themselves on the basis of their recognition through certification.

In summary, the process of supplier development, like total quality, is an ongoing process. Once a vendor is certified, the customer's vendor task team continues to work with the vendor on implementing quality-improvement projects for improving the vendor's QCD. Typically on an annual basis, the vendor task team will return to the vendor's facility and perform another in-depth quality-system audit using the survey instrument. Again, in addition to this survey instrument, the same factors that were used for certification are typically used again for recertification.

10.2.4 Multiple Source versus Sole Source

Some companies believe that it is wise for them to buy from multiple sources. By doing this, they believe that they place themselves in a buyer's market where suppliers will fight for their business, and they will have leverage to negotiate on the basis of price. In addition, these companies think that multiple sources may provide a guarantee for availability of parts and material during conditions of shortage. Such a policy is not wise for manufacturing in the nineties. The best strategy for a manufacturer is to sort out and retain the best supplier as the sole source, and establish a long-term relationship based on loyalty and trust, and cooperative efforts for achieving joint competitive advantage.

There are several reasons why a sole-source supplier policy is essential for achieving WCM excellence. For one, a single-source supplier reduces complexity and administration costs. Once the supplier is certified, incoming inspection activities and costs can be significantly reduced (if not eliminated) and often the number of procurement people can be reduced since negotiating with multiple vendors is no longer necessary. Second, significant benefits can result with reduced variability. Even when each one of the multiple suppliers for a single item can provide parts with very small variation, the integrated total variation of the mixtures of parts from the various suppliers may be quite large. Thus three good suppliers in combination can produce a mediocre result. Also when the number of suppliers is large, problems in diagnosis and source tracing become difficult. This principle applies to any suppliers, "over-the-fence" suppliers as well as in-house suppliers. Third, with sole sourcing, supply is often more reliable because larger orders are placed. The manufacturer becomes a more important customer to the selected suppliers. Because a closer relationship has been created and an atmosphere of trust and mutual respect has been formed, the supplier becomes more loyal to make sure that the manufacturer's needs are fulfilled. Frequent deliveries are always more reliable than infrequent deliveries because they become a part of the supplier's daily or weekly routine. Another factor affected by the number of suppliers is the cost of assisting the suppliers to improve their quality. The costs due to conflicting expectation, redundant administration, and confusing coordination in suppliers quality assurance (SQA) are obvious. Finally, component cost is reduced in the long term. In the short term, sometimes the costs will increase, but in the long term the supplier's manufacturing costs are reduced and product cost will go down. Larger orders, improved production techniques, better design, and better quality will significantly help to reduce costs.

To plunge straight into a single-source, noninspection arrangement would be irresponsible if the supplier's capabilities and performance are not known. One needs a tool to assess the supplier's quality systems and procedures, and the supplier's adherence and performance to them.

10.3 QUALITY-SYSTEMS SURVEY

The entire process of achieving certification as a quality vendor is focused on building excellence in one's manufacturing operations. The basis for evaluating performance and one's current market position are the following elements:

- Management systems
- Design, specifications, and change control
- Incoming purchased materials
- In-process operations and practices

- Finished goods
- Measurement and test systems

These basic control elements are viewed as essential building blocks of a TQ control system. They thus serve as the elements for evaluating a supplier's capabilities for certification. This section presents the details of a quality system's survey instrument that has been developed and applied successfully many times, both for the process of qualification and measuring improvement and for ultimately achieving certification. This material is presented in the survey instrument format so that anyone can easily employ it directly in their own operations. The survey contains a number of evaluation factors that are listed under the six major control elements, which are stated above. Each of the evaluation factors are scored in terms of both the company's systems and procedures, and their actual performance and adherence to these systems and procedures, as shown in Table 10.1. Also shown are the subjective levels for evaluation, which range from *excellent, very good,* down to *failure,* and the corresponding points awarded at each level of evaluation. The definition of scoring for each level of evaluation and points is given in Table 10.2

Experience has shown that evaluating and scoring each factor on both systems/procedures and actual performance adherence is very useful in identifying where improvements need to be made. For example, some companies may have defined and implemented very effective system procedures yet the actual adherence to these procedures on the shop floor may be fairly poor. On the other hand, other companies may have very good performance and adherence to their quality systems but these systems themselves are ill-defined and not documented, thereby inciting little assurance of good quality control in the future.

10.3.1 Management Systems

The first section (A) of the survey instrument addresses issues related to the management of quality in the vendor's organization. Each of these ten evaluation factors is listed below, and a brief description of each is also given, identifying both the meaning and the scope of activities to be considered under that particular evaluation point.

1. **Top management's commitment, leadership, and adherence to a policy of continuous quality improvement.**

   Given that continuous improvement is the underlying philosophy of total quality, it is essential that practices and systems supporting this philosophy exist in a company and that management provides the resources and support to foster the ongoing development of and adherence to these

TABLE 10.1. Scoring System for Vendor Quality Systems Survey.

Evaluate and score each of the factors in the Quality Systems Audit Survey as follows:

Evaluation	Score Points*
Excellent	5
Very Good	4
Good	3
Fair	2
Poor	1
Failure	0

Evaluate and score each factor on both systems/procedures and actual performance/adherence

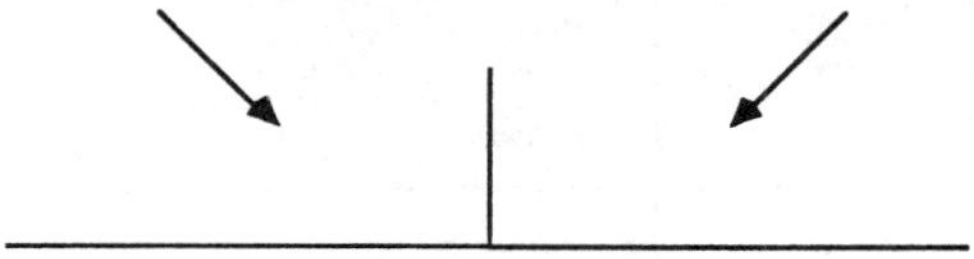

If an item on the survey is considered not to apply, an explanation should be written below the survey factor as to why it can not be rated. If the Vendor Task Team shares in this opinion, the item will be scored at 60% (3 of 5 max.).

*All items should be scored on an absolute basis, that is, evaluate each item in terms of current position compared to a high level of quality excellence.

practices and systems. The company should be able to provide documented examples showing that these practices and systems are understood and used by people at different levels of the organization and that continuous improvement in both product and process quality results from their application. In essence, the development of and adherence to practices and systems based on the philosophy of continuous quality improvement needs to be company policy.

TABLE 10.2. Definition of Scoring for Quality Audit/Survey Questions.

Systems and/or Procedures	Evaluation (Points)	Performance and Adherence
Well defined & complete Concise and well documented Formally reviewed, and updated (Possibly Leading Edge)	EXCELLENT (5)	Well understood and well executed Complete records (history) of adherence and compliance (Execution is a "way of life")
Adequately defined & complete Satisfactorily documented Informally reviewed and updated (No Deficiencies)	VERY GOOD (4)	Generally understood and well executed Satisfactory history of adherence and compliance. (Execution is a "requirement of the job")
Mostly defined and complete Marginally documented (Minor Deficiencies)	GOOD (3)	Mostly understood and followed Satisfactory performance, adherence, and compliance. (Occasional Omissions)
Marginally defined and complete. Little or no documentation (Numerous minor deficiencies	FAIR (2)	Partially understood Performance, adherence and compliance requires some improvement (Frequent Omissions)
Poorly defined & incomplete, but Potential for improvement exists (Major Deficiencies)	POOR (1)	Marginally understood Performance, adherence, and compliance requires significant improvement (Very Frequent Omissions)
Not defined (No concrete plans for implementation)	FAILURE (0)	Not in place

2. A quality manual (and plan?) for processes and procedures.

An up-to-date QA manual, possibly augmented by a quality plan, is a company document detailing how its policy on quality is implemented and executed. This manual should reflect the philosophy of continuous quality improvement through employee involvement and define explicitly the *who, what, when, how, and why* for all quality-control procedures. It is therefore essential that the manual (and plan) be distributed to, understood by, and used by people throughout the organization. Procedures and resources to ensure that the quality manual is kept current must be well defined and apparent.

3. Employee education and training programs to support Total Quality.

Employee education and training programs are the lifeblood of implementing total quality through employee involvement. A well-defined, long-term program for improving the understanding and skills of the human resource is crucial for implementing and sustaining successful programs in quality and productivity improvement. A company should therefore devote the resources necessary to identify, document, and satisfy the organization's educational and training requirements. The results of the effectiveness of both in-house and off-site seminars, classes, and workshops should also be ascertained, documented, and used as a basis for setting new goals and priorities for future education and training. Employees at all levels of the organization should feel that they are being given the knowledge and skills necessary to perform their job responsibilities at a high level of quality and performance.

4. **An emphasis on quality systems for defect prevention.**

Quality systems that focus on the prevention of defects are now recognized as a more effective and less costly approach to achieving conformance quality, that is, the degree to which a product performs to its intended design requirements. Management needs to recognize this concept and, accordingly, give priority and resources for the development of practices and systems that emphasize quality defect prevention. The prevention concept can be applied in both product design and development and in production, often through the use of design of experiments, statistical process control (SPC), and other statistically based methodologies. The factor for evaluation in this case is the degree to which management recognizes and is committed to fostering and promoting the development and use of defect-prevention methods and practices.

5. **A program for annual quality improvement (AQI) for the elimination of process waste.**

A project-oriented program consisting of employee-based quality-improvement teams is a very effective way to identify processes that contribute to chronic levels of waste in manufacturing, and to implement changes for achieving a level of quality performance never previously attained. This type of AQI program needs to be supported and directed by management because the AQI project teams require time and money to be effective. This factor of evaluation thus addresses the question of how well the AQI program is organized, directed, supported, and functioning and what level of success is obtained from these investments.

6. **Statistical methods for problem identification and problem solving and the measurement of quality improvement.**

Statistical methods are now widely recognized and used for identifying and solving quality problems and for measuring the level of quality improvement obtained. For example, the techniques of Pareto analysis—cause-and-effect diagrams, scatter diagrams, flow charting, and brainstorming—are often useful for problem identification and problem solving; tools such as histograms, pie charts, line graphs, and bar charts can provide good displays of statistical information. The degree to which these methods and tools are encouraged by managers and used effectively in the organization is the key issue for evaluation.

7. **Documentation control of process requirements and specifications.**

   Providing people with accurate and timely information defining the requirements and specifications of a manufacturing process is important for ensuring that the job is performed correctly and consistently, both now and in the future. A company, therefore, needs to provide the resources, in terms of both proper procedures and personnel, to ensure that this documentation is available, distributed where needed, and up-to-date.

8. **An organizational structure for fostering participative quality management.**

   A company's ability to treat the labor force as an experienced and knowledge-based asset for problem solving, instead of as a burden, is a strong indicator of their capabilities to achieve quality excellence. Management's role is to recognize the potential value of the human resource and to provide responsibility, recognition, and rewards (the RRR's for quality and productivity improvement) to the employees for their role in quality assurance and quality improvement—often through a team approach. By making quality "everyone's job responsibility," the human resource is utilized more effectively, resulting in a synergism for implementing total quality and gaining a competitive advantage. It is thus appropriate for people from all levels in the organization to evaluate the effectiveness of participative management.

9. **A formal program for cost of quality.**

   A cost-of-quality program provides a financial basis for measuring and monitoring the costs incurred in providing a quality product to the marketplace and the cost-effectiveness of corporate policies, practices, and systems relating to quality improvement. This financial feedback provides a basis for setting new policies, objectives, and project priorities. A cost-of-quality system should recognize and track four different categories of quality costs: appraisal, preventative, internal failures, and external failures. It is essential that the cost-of-quality program be capable of measuring the component costs accurately so that any trends in the data reflect actual changes in the total quality system and not just a change in one's ability to capture and measure "quality cost." To be most effective, results from the cost-of-quality program should be presented to the people who control or influence the corresponding elements of quality, thereby serving as a source of valuable feedback. Hence, this factor evaluates management's commitment to and effectiveness in using cost information to monitor and control the many activities and the future direction of its total quality system.

10. The internal quality-system audit.

A periodic (that is, semiannual or annual) internal audit of the completeness and effectiveness of a company's quality assurance and control policies and procedures is a valuable exercise that provides feedback for identifying weaknesses and, subsequently, implementing improvements. To be effective, however, the audit must be performed by an unbiased and qualified evaluator or evaluation team. The results of the audit should be presented and reviewed by management and then distributed to all affected personnel—ideally with recognition for a job well done and, when appropriate, with a request for a plan of corrective action. The audit is an important element of a TQ program, since it provides the QA department with the assurance of quality.

10.3.2 Design, Specifications, and Change Control

The second section (B) of the survey instrument evaluates the effectiveness of the vendor's systems and procedures for maintaining quality in product design, product specification distribution and control, and engineering change control. Specifically, the survey evaluates the vendor's ability to define the performance and quality requirements of the customer's product. The survey then translates these high-level requirements into the critical final-product control specifications for all levels of product manufacturing and testing. The vendor's procedures for performing process capability studies, which determine the quality effectiveness of existing or new equipment for new products and components, are addressed along with the vendor's procedures for performing design review and making possible design changes based on results from both product performance evaluation and input from the customer. Procedures relating to distribution, control, and revisions of documentation in the form of part drawings and product specifications are also addressed for both existing and new jobs, as are the procedures for tracing products and components to a given design level. In effect, this section of the survey addresses the vendor's ability to maintain and improve the quality of the overall design process by detecting and solving quality problems during the design stages rather than during the manufacturing stages of the product's life.

For each of the following ten factors, the evaluator needs to consider (1) Are existing procedures well defined, well documented, and complete in light of the requirements placed on the product design,product specification distribution and control, and engineering change control systems? (2) Are existing procedures well understood and used consistently and effectively by all appropriate personnel?

1. **Systems for defining and communicating a customer's quality requirements into critical final-product control specifications.**

2. **Procedures to perform process capability studies during the development of new products and components.**

3. **Design review procedures.**

4. **Procedures for making customer and design review changes.**

5. **Failure mode effect analysis (FMEA) performed for new product designs.**

6. **Print and engineering change control system.**

7. **System to distribute and control drawings and related specifications on parts and materials.**

8. **System for drawings and specifications related to process improvement and design revisions.**

9. **System for new-job startup process and documentation control.**

10. **System for product identification and lot traceability to the design level.**

10.3.3 Incoming Purchased Materials

Section C of the vendor's survey instrument addresses the quality-related ramifications and effectiveness of the vendor's policies and procedures for purchasing and controlling materials used in their manufacturing operations. The premise is simply that a vendor will not be able to supply customers with quality value-added products in a highly dependable fashion unless they have effective systems and procedures in place for purchasing and controlling the materials they use in their manufacturing processes. This section thus evaluates the vendor's policies, systems, and practices for ensuring the QDC of materials they purchase from their suppliers (subsuppliers). The vendor's procedures for controlling the quality of materials through all stages of manufacturing are evaluated along with the procedures used to handle nonconforming materials and initiate appropriate corrective action. The survey also addresses the vendor's systems and capabilities to plan their materials requirements and to schedule and release material through their manufacturing facilities in order to satisfy delivery requirements. In total, this section of the survey evaluates the capabilities and effectiveness of the vendor's systems and procedures for purchasing and material control in order to assess their ability to provide customers with quality value-added products in a highly dependable fashion.

For each of the ten evaluation factors that are subsequently listed, the following two questions should be considered (1) Are the existing policies, systems, and procedures consistent, well defined, and documented clearly; and are they sufficiently sophisticated and complete to be effective, yet simple enough to apply and utilize? (2) Are the policies, systems, and procedures well understood and followed, providing effective controls to ensure the vendor's ability to meet quality requirements and contractual obligations?

1. **Assessment of the subsupplier's capability and capacity to meet contractual agreements.**

2. **Qualification of subsuppliers prior to issuing an order.**

3. **Certification program established with periodic surveys for recertification of subsuppliers.**

4. **Receiving inspection instructions and documentation, with feedback for any problems.**

5. **Procedures for comparison of purchase order specifications to part-number design levels.**

6. **Formal program for initiating, documenting, and implementing corrective actions with preventative measures.**

7. Identification, isolation, and disposition of nonconforming materials.

8. Reinspection and traceability of reworked parts and products.

9. Materials planning, scheduling, and job release control system.

10. Material storage control for purchased components and supplies.

10.3.4 In-Process Operations, Control, and Practices

Process control focuses on the procedures and information required to ensure quality performance on the many activities in shop-floor manufacturing. A number of the factors for evaluation in section D of the vendor survey instrument address the completeness, utilization, and effectiveness of quality specifications and instructional information to provide proper procedures for shop-floor personnel to follow in performing equipment changeover, part-processing operations, QC inspections, and corrective actions for internal part quality and system failures. Effective process control is essential in providing the shop with the information required to consistently perform work in a quality manner, supporting the reduction in process variability. This section also addresses the vendor's procedures for maintaining their equipment and facilities to support high-quality processing and the vendor's procedures for measuring the process capability of their equipment.

In scoring the following ten factors under process control, the evaluator should not only consider how well these systems and procedures are defined, documented, and utilized, but also consider the degree to which shop-floor personnel understand, support, and follow these systems and procedures and feel that there is value in their use.

1. Process sheets and work instructions of each operation of each part, incorporating visual information where possible.

2. Operators being knowledgeable and involved regarding adherence to quality requirements and customer needs.

3. Generation and distribution of up-to-date setup instruction sheets and methods for all equipment.

4. Instruction sheets and documentation for first- and last- piece inspection and in-process inspections.

5. Work flow and material identification and control.

6. Inspection, scrap audit records, and feedback control.

7. Customer return and rework procedures and their facilities.

8. Statistical process control and corrective actions.

9. Procedures for performing, documenting, and distributing process-capability analysis.

10. Preventative maintenance on production equipment and facilities.

10.3.5 Finished Goods

Section E addresses QC capability for finished materials in the vendor's facilities and operations by evaluating the following six factors, which cover the systems and procedures for inspecting, storing, preserving, packing, distributing, and tracking finished components and products. In effect, this section focuses on the capabilities and performance of the vendor's QA system for all components and products from the time they leave production, extending even into the tracking of quality and customer service and satisfaction in the field.

1. Material storage control system (FIFO?).

2. Packaging and preservation instructions.

3. Material distribution control (test, verify, record).

4. Storage facilities for quality preservation.

5. **Final inspection and control at shipping (customer shipping instructions).**

   ___|___

6. **Audit history and tracking of product quality.**

   ___|___

10.3.6 Measurement and Test Systems

The sixth section (F) contains the following four factors, which evaluate the systems and procedures for controlling all measuring and test equipment. In order to make reliable decisions based upon the outcome of the measurement and testing of components and products, it is necessary that the equipment utilized also be reliable. Quite simply, controlling quality requires the capability to measure quality. Thus, the vendor's procedures for verifying accuracy and repeatability of all gauges and test equipment warrant evaluation in terms of both completeness and adherence.

1. **Gauge control program (incoming, in-process, final).**

   ___|___

2. **Calibration schedule and records for all measurement and test equipment.**

   ___|___

3. **Traceability to National Standards.**

   ___|___

4. **Deterioration tracking and maintenance program (fixtures, tooling patterns, etc.).**

   ___|___

Table 10.3 shows an example of a survey score sheet. In each of the six sections of the survey, the number of points is recorded and multiplied by the weighting value, giving a total for each of the six categories, which are evaluated in terms of both procedures and performance. As shown in the table, the maximum number of weighted points is currently set at 1,000 and all sections have equal weighting using a factor of 2. Using a weighting value has been useful in the past since some companies wish to apply a greater weighting value to some sections than others, and this provides an easy way of adjusting relative importance in the survey and still maintaining a value of 1,000 (or 100 percent) as the upper limit.

This evaluation form can be used in three different ways.

1. Send it out as a questionnaire to selected vendors, asking them for self-eval-

TABLE 10.3. Survey Score Sheet.

	Section	Survey Rating	Weighting Value	Total	Max. Possible
A.	Management Systems				
	Procedures	______	X 2	______	100
	Performances	______	X 2	______	100
B.	Design, Specifications, and Change Control				
	Procedures	______	X 2	______	100
	Performances	______	X 2	______	100
C.	Incoming Purchased Materials				
	Procedures	______	X 2	______	100
	Performances	______	X 2	______	100
D.	In-Process Operations, Control, and Practices				
	Procedures	______	X 2	______	100
	Performances	______	X 2	______	100
E.	Finished Goods				
	Procedures	______	X 2	______	60
	Performances	______	X 2	______	60
F.	Measurement and Test Systems				
	Procedures	______	X 2	______	40
	Performances	______	X 2	______	40

Procedures	Performance	Overall
Total______________	Total______________	Total______________
Max. possible 500	Max. possible 500	Max. possible 1000
Percentage________	Percentage________	Percentage________

uation. The questionnaire, when returned, is reviewed. Then a follow-up meeting is held with the vendors to discuss the levels of capability and performance needed to receive a qualified vendor status.

2. A vendor-certification team uses it when they visit a vendor for immediate evaluation related to either qualification, certification, or recertification status.

3. Use it in-house as a basis for performing an internal audit of the existing quality systems and procedures to determine any inadequacies and opportunities for continuous improvement.

The survey, in effect, provides a means to objectively evaluate the effectiveness of the quality systems/procedures and performance/adherence present in a manufacturing organization. This evaluation, in turn, provides a basis for decision making, corrective actions, and ongoing quality improvement. It has proven to be a valuable tool for becoming a mean, lean, and world-class competitor in manufacturing in the nineties.

Index